# Environmental External Costs of Transport

Springer-Verlag Berlin Heidelberg GmbH

Rainer Friedrich • Peter Bickel (Eds.)

# Environmental External Costs of Transport

With 76 Figures and 122 Tables

Springer

Professor Dr. Rainer Friedrich
Peter Bickel
University of Stuttgart
Institute of Energy Economics
and the Rational Use of Energy (IER)
Heßbrühlstr. 49 a
70550 Stuttgart
Germany

Library of Congress Cataloging-in-Publication Data Applied For
Die Deutsche Bibliothek - CIP Einheitsaufnahme
Friedrich, Rainer: Environmental external costs of transport / Rainern Friedrich; Peter Bickel. .

DOI 10.1007/978-3-662-04329-5



http://www.springer.de

Originally published by Springer-Verlag Berlin Heidelberg New York in 2001
MyCopy version of the original edition 2001


Cover Design: Erich Kirchner
Typesetting: Camera-ready by the authors

SPIN: 10791849 30/3130xz – 5 4 3 2 1 0 – Printed on acid free paper
www.springer.com/mycopy

# Preface

Transport is very important for the economy and our welfare. However, transport also causes a lot of problems, including air pollution. Such problems should be taken into account, when making decisions. A prerequisite for doing so is, that the impacts are known, quantitatively measured and allocated to the different activities in transport. Furthermore, they should be transformed into monetary units to be used as a basis for cost-benefit analyses or as an aid for setting taxes and charges, that reflect the external costs.

This book describes a methodology for calculating impacts of transport activities and external costs caused by air pollution and presents numerous applications of this methodology for different transport technologies, locations and policy case studies. The approach has been developed and results have been calculated within the research project 'ExternE Core/Transport', financed to a large extent by the European Commission, Directorate General Research. We would like to thank especially Pierre Vallette and Pekka Jarviletho from the EC for their advice and support.

A considerable number of experts with expertise in the different disciplines of this highly interdisciplinary work contributed to this book. The editors would like to thank the authors (see list on p. XV) for their contributions; it is especially remarkable, that the authors helped to make this book an integrated whole instead of a number of independent contributions.

Stuttgart, February 2001 Rainer Friedrich and Peter Bickel

# Contents

# List of Authors

| Responsible authors | Organisation | Contribution |
|---|---|---|
| **Stephan Schmid**<br>**Peter Bickel**<br>**Rainer Friedrich**<br>**Wolfram Krewitt** | IER[a] | Executive Summary, Objectives, The Impact Pathway Methodology, Emission Calculation, Marginal Costs Germany, Results of Marginal Cost Case Studies, Aggregated Costs Germany, Policy Case Study Germany |
| **Paul Watkiss**<br>**Mike Holland** | AEA Technology[b] | The Impact Pathway Methodology, Quantification of Total and Average Externalities (Aggregation), Marginal Costs UK, Aggregated Costs (UK and conclusions), Policy Case Studies (UK and summary) |
| **Kees Dorland**<br>**Richard Tol**<br>**A.Q.A. Omtzight**<br>**A.A. Olsthoorn**<br>**P.J.M. Beukering** | IVM[c] | Emission Calculation (Inland shipping), Global Warming, Marginal Costs Netherlands, Aggregated Costs Netherlands, Policy Case Study Netherlands (in collaboration with M.A. Janssen (Department of Spatial Economics, Vrije Universiteit Amsterdam)) |
| **Dionysis Assimacopoulos**<br>**Giorgos Vossiniotis** | LIEE[d] | Atmospheric Transport Modelling (Gaussian Dispersion Modelling, The ROADPOL Model), Marginal Costs Greece, Policy Case Study Greece |
| **David Simpson** | DNMI[e] | Atmospheric Transport Modelling (local and regional scale ozone) |
| **Jutta Graf** | DLR[f] | Atmospheric Transport Modelling (global scale ozone), Aircraft Transport |
| **Ari Rabl**<br>**Joseph V. Spadaro** | Ecole des Mines de Paris (FR) | Atmospheric Transport Modelling (Validation of Dispersion Modelling), Treatment of Up- and Downstream Processes, Uncertainty, Marginal Costs France, Policy Case Study France |
| **Fintan Hurley**<br>**Brian Miller** | IOM[g] | Health Effects |
| **Johan Tidblad**<br>**Vladimir Kucera** | SCI[h] | Impacts on Building Materials |
| **David Howard**<br>**Mike Hornung** | ITE[i] | Impacts on Terrestrial Ecosystems |

| Responsible authors | Organisation | Contribution |
|---|---|---|
| **Alistair Hunt**<br>**Anil Markandya** | University of Bath (UK) | Economic Valuation, Sustainability Indicators and Alternative Assessment Techniques, Quantification of Total and Average Externalities (Aggregation) (Economic Aspects) |
| **Ståle Navrud** | ENCO[j] | Sustainability Indicators |
| **Leo de Nocker**<br>**Luc Int Panis** | VITO[k] | Marginal Costs Belgium, Aggregated Costs Belgium, Policy Case Studies Belgium |
| **Thomas Otterström**<br>**Sirpa Torkkeli** | EKONO[l] | Marginal Costs Finland, Policy Case Study Finland |
| **Yolanda Lechon**<br>**Rosa Saez**<br>**Helena Cabal** | CIEMAT[m] | Policy Case Study Spain |

[a] Institute of Energy Economics and the Rational Use of Energy, University of Stuttgart, Germany
[b] AEA Technology PLC acting through ETSU, Culham, United Kingdom
[c] Institute for Environmental Studies, Vrije Universiteit, Amsterdam, Netherlands
[d] Laboratory of Industrial and Energy Economics, National Technical University of Athens, Greece
[e] The Norwegian Meteorological Institute, Oslo, Norway
[f] Deutsche Forschungsanstalt für Luft- und Raumfahrt, Weßling, Germany
[g] Institute of Occupational Medicine, Edinburgh, United Kingdom
[h] Swedish Corrosion Institute, Stockholm, Sweden
[i] Institute of Terrestrial Ecology, Grange over Sands, United Kingdom
[j] Environmental Consultants a.s., Sandvika, Norway
[k] Vlaamse Instelling voor Technologisch Onderzoek, Mol, Belgium
[l] Electrowatt-Ekono Oy., Espoo, Finland
[m] Centro de investigaciones energeticas medio ambientales y tecnologicas, Madrid, Spain

# 1 Introduction

Transport is an important source of emissions of many pollutants to the air, e.g. nitrogen oxides, fine particles, carbon monoxide, carbon dioxide, sulphur dioxide, various volatile organic compounds and others. The emissions of these substances cause considerable damage affecting a wide range of receptors including human health, flora, fauna and materials. These damages are in general not or not adequately reflected in the transport prices; as long as third parties and not the user of transport himself is effected, they represent external effects. If such effects exist but are not adequately considered, when making decisions - for example about the number and destinations of trips, the choice of the transport mode, or investments in infrastructure - a non-optimal allocation of resources leading to welfare losses may occur. So, external effects should be taken into account when making decisions. In other words,

- to identify optimal targets for the levels of environmental quality to be pursued by environmental policy,
- to compare costs and benefits of measures for improving environmental quality,
- to provide decision support on the form and content of instruments, that use market forces to pursue environmental aims (e.g. taxes, charges, permits, etc.),

the identification, quantification and monetisation of the external effects of the use of transport technologies is necessary. This concept is widely accepted; the white paper of the European Commission on "Fair Payment for Infrastructure Use" in the transport sector (European Commission, 1995b) and the reports of the *High Level Group on Transport Infrastructure Charging*, that had the task of investigating the issue of fair pricing in more detail, strongly support the aim that prices in transport should reflect the full costs including the external costs and that decisions in transport should be based on a full social cost benefit analysis, covering both the sum of internal and external costs and benefits.

Thus, the question has to be answered, how external effects can be quantified and transferred into monetary values. In most earlier and also some recent studies a top-down approach was used, leading to the estimation of average costs. The basis for the calculation is a whole geographical unit, a country for example. For such a unit the total cost due to an externality (e.g. air pollution) is calculated. This cost is then divided by the total amount of activity leading to the externality (e.g. the emissions of the polluting substances). The top-down approach can be implemented and used with reasonable effort. However, the approach has limits; it can be used to address general issues like: 'is train transport in general and average more environmentally friendly than road transport'. If concrete decision aid for alternatives or the optimal form of ecopolitical instruments has to be provided, there are a number of problems associated with its use. The main argument is, that marginal external costs are to a large extent dependent on the location respectively the conditions around the site of the additional transport to be assessed. The damage caused for instance by the emission of pollutants from a certain vehicle depends on the driving pattern, the meteorological conditions, the land use and population density around the site of the emissions and the background concentrations of pollutants, that react with the emitted substances to form secondary pollutants. The damage stemming from a vehicle driving in stop and go traffic in a large city is quite different from the damage of the same car driving on a motorway in a rural area.

As the aim is to estimate the marginal costs of concrete alternatives as accurate as possible, a bottom up approach should be used, which is able to take site and technology dependence into account. The starting point of this approach is the micro level, i.e. an additional transport activity on a particular route segment. The marginal external costs of one additional vehicle are then calculated for a single trip on this route segment by modelling the pathway from emission to impact and costs.

Such a bottom-up approach, the so called *impact pathway-approach* has been developed within a series of research projects financed by the European Commission, Directorate-General (DG) Research; these projects have been termed *ExternE*. The impact pathway methodology was first developed and used for energy conversion processes, for example electricity generation. It has now been further developed and extended for application to the transport sector. This improved methodology and its application is described here. So, the aim of this book is,

- to describe the impact pathway methodology for estimating marginal environmental external costs of transport (i.e. costs caused by air pollution),
- to present results of calculations of marginal external costs for different transport technologies being operated at different locations in Europe,
- to show results of aggregations of marginal costs (e.g. the total costs from air pollution caused by a transport mode of a country in Europe),
- to demonstrate the use and usefulness of the results within a number of case studies.

The following Chap. 2 contains a general description of the methodology, Chaps. 3 to 12 describe the methodology in more detail, emphasising the new developments and improvements made. Following the general structure of the impact pathway approach, Chap. 3 starts with explaining, how emissions are calculated. The emitted substances are then transported in the troposphere and chemically converted into other substances; these processes are simulated with the help of atmospheric transport and chemistry models, that are described in Chap. 4. The result of these calculations are changes in concentration or deposition of pollutants. Using exposure-response relationships, these changes are then converted into additional damage, especially health effects, impacts on building materials and impacts on terrestrial ecosystems. The exposure-response-functions used for these impact categories are presented in chapters 5, 6 and 7 respectively. Chapter 8 deals with the methods and parameters for assigning monetary values to the impact categories. The straight forward way to do this - as long as market prices for external effects are not available - is to measure the preferences of the population respectively their 'willingness to pay' to avoid a certain damage or their 'willingness to accept' compensation to bear a damage. An own chapter (Chap. 9) is devoted to estimating the damage costs of global warming. An attempt is made to estimate the future damage of releasing greenhouse gases to the atmosphere based on current best knowledge.

Transport emissions do not only stem from the use of the transport technology, but also from up- and downstream processes, e.g. the provision of fuel, the production of vehicles and transport infrastructure. An example is the production of electricity used for rail transport. As long as external costs exist or are not internalised in these sectors, they have to be taken into account. A method for doing this is described in Chap. 10. Uncertainties are still considerable when estimating external costs. These uncertainties are discussed in Chap. 11. For most applications, aggregated or average external costs are needed. For example a tax per km might be fixed not for each car separately but for a class of cars (e.g. passenger cars with EURO III standard). Methods for calculating aggregated external costs are described in Chap. 12.

In Chap. 13 the developed methods are applied to calculate marginal environmental external costs. Results are given for a large number of transport technologies for road, rail, air and ship transport and for a large number of locations in Europe. Aggregated external costs – for whole countries or regions – are presented in Chap. 14. Chapter 15 contains examples on how the methodology can be used to support and aid policy decisions. The policy case studies described include studies that analyse the benefits of using new fuels and technologies, for example the use of compressed natural gas in passenger cars and busses and the use of electric vehicles, and studies, that look at taxation of fuels and goods vehicles.

# 2 The Impact Pathway Methodology

The ExternE Project has adopted the *Impact Pathway Approach* for the assessment of external impacts and associated costs resulting from the supply and use of energy. The work described in this book builds on the work done in the ExternE Project, which is described in detail in (European Commission, 1995a) and (European Commission, 1999). The term *impact pathway* simply relates to the sequence of events linking a *burden* to an *impact* and subsequent valuation, as illustrated in Fig. 2.1. Emissions and other types of burden such as risk of accident can be quantified and followed through to impact assessment and valuation, whereby for the transport sector in this project the focus lay on impacts due to airborne pollutants and greenhouse gases. The approach thus provides a logical and transparent way of quantifying externalities. However, only recently, through developments in environmental science and economics, and improvements in computing power has it become a realistic proposition.

## 2.1 Guiding Principles in the Development of the ExternE Methodology

The underlying principles on which the methodology for the ExternE Project has been developed are:

- **Transparency**, to show precisely how results are calculated, the uncertainty associated with the results and the extent to which the external costs of any fuel chain have been fully quantified. Specific cases are selected in order to provide a clear reference for the date used.
- **Consistency**, of methodology, models and assumptions (e.g. system boundaries, exposure-response functions and valuation of risks to life) to allow valid comparison to be made between different fuel chains and different types of impact within a fuel chain. With the extension of the analysis to the transport sector, a comparison becomes possible even between mobile and stationary energy conversion processes.
- That analysis should be **comprehensive**, it should be sought to at least identify all of the effects that may give rise to significant externalities.

In order to comply with these principles, much of the analysis is based in the first instance around the assessment of the effects of individual transport technologies which are closely specified with respect to:

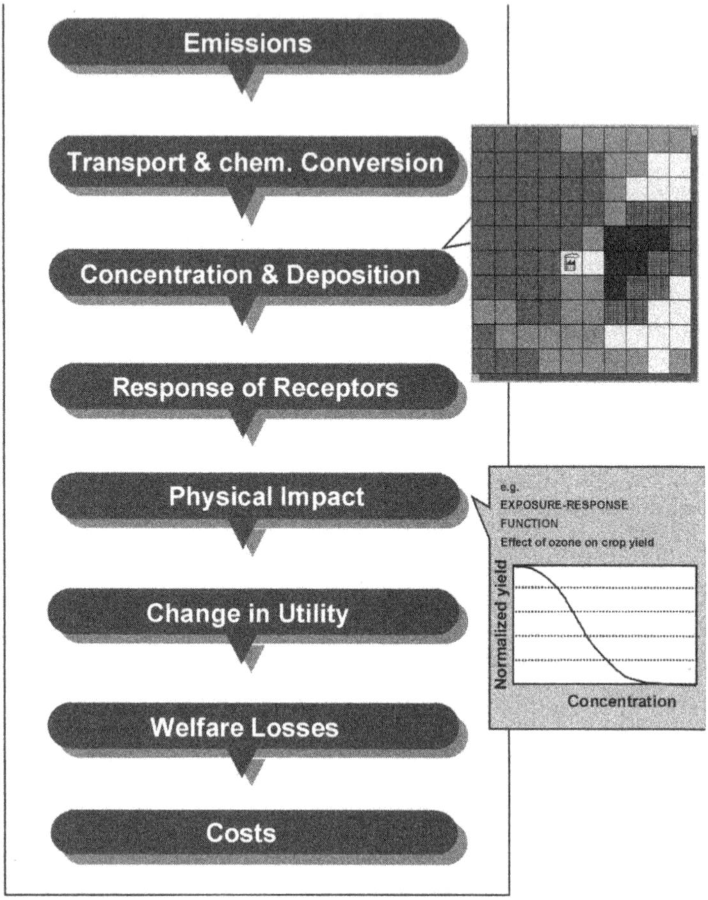


**Fig. 2.1.** The Impact Pathway Approach

- the technologies used;
- the location of the transport activity and its emissions;
- the type of fuel used;
- the emissions due to production, maintenance and disposal of vehicles and infrastructure.

Each of these factors is important in determining the magnitude of impacts and hence associated externalities.

## 2.2 Quantification of Impacts

The specific aspects of quantifying impacts due to transport activities are given in Chaps. 3 to 10 (emission calculation, dispersion modelling, health impacts, impacts on building materials, impacts on terrestrial ecosystems, economic valuation, global warming, up- and downstream processes). The complexity of the analysis varies greatly between impacts. In some cases externalities can be calculated by multiplying as few as 3 or 4 parameters. In others it is necessary to use a series of sophisticated models linked to extensive databases.

Common to all of the analyses conducted on the impacts of pollutants emitted from fuel chains is the need for modelling the dispersion of pollutants and the use of a dose-response function of some kind. Again, there is much variation in the complexity of the models used.

The most important pollutant transport models used within ExternE relate to the atmospheric dispersion of pollutants. They need to account not only for the physical transport of pollutants by wind but also for chemical transformation. The dispersion of pollutants that are in effect chemically stable in the region of the emission can be predicted using Gaussian plume models. These models assume source emissions are carried in a straight line by the wind, mixing with the surrounding air both horizontally and vertically to produce pollutant concentrations with a normal (or Gaussian) spatial distribution. For low level emission sources the use of these models is typically constrained to a distance of 30 km from the source.

However, pollutant transport usually extends over much greater distances. A different approach is needed for assessing regional transport as chemical reactions in the atmosphere become increasingly important. This is particularly so for acidifying pollutants. For this analysis receptor-orientated Lagrangian trajectory models have been used. The outputs from the trajectory models include atmospheric concentrations and deposition of both the emitted species and secondary pollutants formed in the atmosphere.

A major problem of previous project phases, the lack of a regional model of ozone formation and transport that was applicable to the European situation, could be solved. The source-receptor ozone model (SROM) allows to calculate ozone statistics. The SROM is based on source-receptor relationships from the EMEP MSC-W oxidant model for five years of meteorology. Input for SROM are national annual $NO_x$ and anthropogenic NMVOC emission data from 37 European countries, while output is calculated for individual EMEP 150x150 $km^2$ grid squares by employing country-to-grid matrices. Available ozone statistics include AOT40 for crops, AOT40 for forests, AOT60 and 6-monthly mean of daily maximum 6-hour average ozone concentration.

The term 'dose-response' is used somewhat loosely in much of this work, as what is really meant is the response to a given exposure of a pollutant in terms of atmospheric concentration, rather than an ingested dose. Hence the terms 'dose-response' and 'exposure-response' should be considered comparable. A major issue with the application of dose-response functions concerns the assumption that they are transferable from one context to another. For example, some of the functions for health effects of air pollutants are still derived from studies in the United States. Is it valid to assume that these can be used in Europe? The answer to this question is to a certain degree unknown - there is good reason to suspect that there will be some variation, resulting from the affluence of the affected population, the exact composition of the pollutant mix that the study group was exposed to, etc. Indeed, such variation has been noted in the results of different epidemiological studies. However, in most cases the view of our experts has been that transference of functions is to be preferred to ignoring particular types of impact altogether - neither option is free from uncertainty.

Dose-response functions come in a variety of functional forms, some of which are illustrated in Fig. 2.2. They may be linear or non-linear and contain thresholds (e.g. critical loads) or not. Those describing effects of various air pollutants on agriculture have proved to be particularly complex, incorporating both positive and negative effects, because of the potential for certain pollutants, e.g. those containing sulphur and nitrogen, to act as fertilisers.

Ideally these functions and other models are derived from studies that are epidemiological - assessing the effects of pollutants on real populations of people, crops, etc. This type of work has the advantage of studying response under realistic conditions. However, results are much more difficult to interpret than when working under laboratory conditions, where the environment can be closely controlled. Although laboratory studies provide invaluable data on response mechanisms, they often suffer from the need to expose study populations to extremely high levels of pollutants, often significantly greater than they would be

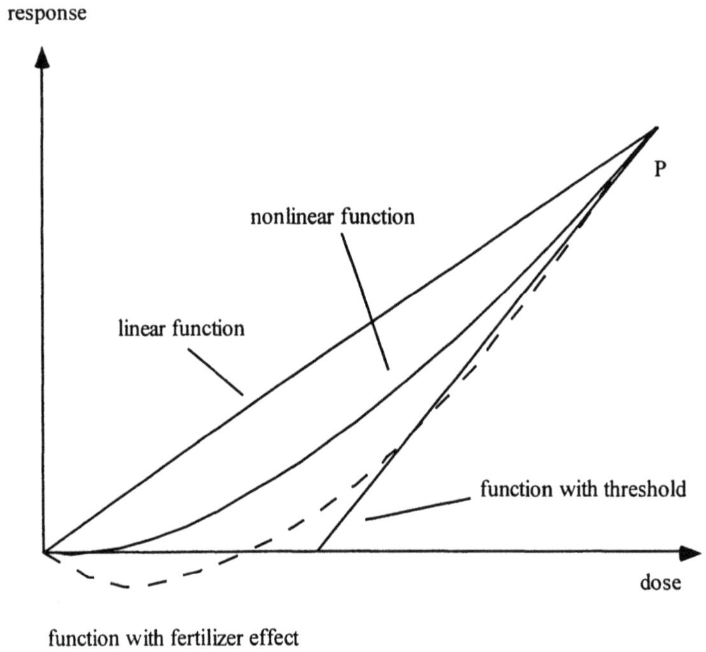


Fig. 2.2. A variety of possible forms for dose-response functions

exposed to in the field. Extrapolation to lower, more realistic levels may introduce significant uncertainties, particularly in cases where there is reason to suspect that a threshold may exist.

## 2.3 Economic Valuation

The rationale and procedures underlying the economic valuation applied within the ExternE Project are discussed in detail in European Commission (1995a and 1999). The approach followed here is to quantify individual 'willingness to pay' (WTP) for environmental benefit, or 'willingness to accept' (WTA) payment in lieu of environmental harm.

A limited number of goods of interest to this study - crops, building materials, etc. - are directly marketed, and for these valuation data are easy to obtain. However, many of the more important goods of concern are not directly marketed, including human health, ecological systems and non-timber benefits of forests. Alternative techniques have been developed for valuation of such goods, the main ones being hedonic pricing, travel cost methods and contingent valuation.

Hedonic pricing seeks to quantify the worth of one good through variation in the price of another. The best example is probably the economic estimation of people's aversion to noise by reference to variation in house prices between noisy and less noisy environments. Travel cost approaches quantify the worth of something that attracts visitors (ecosystems, landscapes, historic buildings, etc.) by consideration of the money and time spent on visits. Contingent valuation involves setting up a hypothetical market, for example by questionnaire, to elicit the preferences of those interviewed. All of these techniques involve uncertainties, though they have been considerably refined over the years.

The base year for the valuation described in this book is 2000, and all values are referenced to that year. The unit of currency used is the Euro and the (Euro-) Cent.

## 2.4 Assessment of Uncertainty

Uncertainty in externality estimates arises from a number of sources, including:

- the variability inherent in any set of data;
- extrapolation of data from the laboratory to the field;
- extrapolation of exposure-response data from one geographical location to another;
- assumptions regarding threshold conditions;
- lack of detailed information with respect to human behaviour and preferences;
- political and ethical issues, such as the selection of discount rate;
- the need to assume some scenario of the future for any long term impacts;
- the fact that some types of damage cannot be quantified at all.

It is important to note that some of the most important uncertainties listed here are not associated with technical or scientific issues, instead they relate to political and ethical issues, and questions relating to the development of world society. It is also worth noting that, in general, the largest uncertainties are those associated with impact assessment and valuation, rather than quantification of emissions and other burdens. Chapter 11 will address the issue of uncertainty in depth.

## 2.5 Priority Impact Pathways

A considerable number of priority impact pathways was identified and addressed (see Table 2.1). Within the current project the focus lay on the refinement and application of the methodology for quantifying the external costs due to energy conversion processes. Therefore, only impacts of airborne pollutants and greenhouse gases were assessed in detail. For a comprehensive assessment of transport externalities, of course, impacts due to accidents, noise, etc. should be taken into account.

**Table 2.1.** Priority Impact Pathways considered

| EMISSION/ BURDEN | POLLUTANT | IMPACT |
|---|---|---|
| Benzene | Benzene | Human health |
| 1,3-butadiene | 1,3-butadiene | Human health |
| Carbon monoxide | Carbon monoxide | Human health |
| Combustion particulates | Particulates | Human health |
| Formaldehyde | Formaldehyde | Human health |
| Ethene | Ethene | Human health |
| Lead and compounds | Lead | Human health |
| $NO_x$ | $NO_2$ | Human health |
| $SO_2$ | $SO_2$ | Human health |
| $SO_2/NO_x$ | Aerosols | Human health |
| Greenhouse gases | Global warming | Various |
| $VOC/NO_x$ | Ozone | Human health |
| $VOC/NO_x$ | Ozone | Crops |
| $SO_2$ | $SO_2$ | Crops |
| $SO_2/NO_x$ | Acidity/nitrogen | Crops |
| $SO_2/NO_x$ | Acidity/nitrogen | Ecosystems |
| $SO_2/NO_x$ | Acidity | Fisheries |
| $SO_2$ | $SO_2$ | Materials |
| $SO_2/NO_x$ | Acidity | Materials |
| Combustion particulates | Particulates | Materials soiling |
| $SO_2/NO_x$ | Aerosols | Materials |
| Land use for infrastructure | | Loss of habitat |
| Infrastructure networks | | Habitat fragmentation |

# 3 Emission Calculation

Comparability of case studies of externalities is an important issue in the context of European policy decision making. At this, the use of incompatible approaches for emission calculation is an often criticised matter. Therefore it was attempted to avoid the use of national emission factor methodologies as far as possible, and to present a list of reference technologies for road transport, were the same basis was used.

A major source of information was provided by the several outcomes of the MEET project (Methodologies for Estimating Air Pollutant Emissions from Transport) funded by the European Commission, DG TREN. Here in particular, the aim was followed to "develop a set of methodologies for the calculation of emissions which have been accepted by most of the European experts" (Joumard, 1999).

However, some subjects remained, where additional effort had to be undertaken in order to cover all priority pollutants and all modes. In the sections below, the methodologies for emission calculations are described for road, rail, inland waterway and air transport.

## 3.1 Road Transport

Emission factors based on the 'average-speed' model (MEET, see Samaras et al., 1998), the 'instantaneous emission approach' (UBA handbook, see INFRAS, 1999), drive-cycle averages or reconstruction (ESTEEM project – DG Research, JOULE programme: see European Commission, 1998a) are currently in use. Parameters like average speed and traffic situation as well as road gradient on distinct road links, load for heavy goods vehicles and vehicle mileage for catalyst cars are considered. Cold start emissions can be taken into account in cities.

In view of a Europe-wide application of the impact pathway approach, the data given by the MEET project is the most recent source of information which incorporates measurements from 4 different laboratories throughout Europe. MEET's results will also be fully incorporated in updates of the CORINAIR ('Co-ordination d'Information Environmental -Air')/ COPERT ('Computer Programme to Calculate Emissions from Road Traffic') activities carried out on behalf of the European Environment Agency's European Topic Centre on Air Emissions in the future (Kyriakis et al., 1997).

Emission factors derived from measurements on representative samples of vehicles are limited to technologies complying to the EURO I standard (91/441/EEC). Predictions of the UBA handbook, CBS emission factors/scenarios of the Auto/Oil project (Dings, 1996), and MEET for more recent and future standards have been subject to comparison (see e.g. Schmid, 1998). They turned out to differ considerably between methodologies. Due to the questions targeted and the site-dependency of impacts, large-scale emission prediction tools like FOREMOVE ('Forecast of Emissions from Motor Vehicles')/CASPER ('Calculation Scheme to Predict Emissions into the Air') have not been used in the impact pathway approach so far.

For the reasons given above, it was recommended to use the MEET emission functions in ExternE. In coherence with the computer tool development, MEET emission functions for road transport were integrated in the EcoSense Transport model. Conventional as well as alternative fuelled vehicles and predictions for future emissions reductions were included. Examples for emission functions for typical urban and highway drives in Germany are given in Table 3.4.

Some of the pollutants under assessment within ExternE are however not or not directly covered by the MEET methodology. These are:

- $SO_2$ emissions,
- particle emissions of gasoline vehicles,
- VOC split.

$SO_2$ emissions were calculated based on the sulphur content of the fuel, which can be defined by the user in the EcoSense Transport Emission Module. Speed-dependent particle emission functions for gasoline vehicles were added to the model, referring to information from Helsinki Metropolitan Area Council (provided by Torkkeli, 1999). Comparisons by Int Panis (1999) proved their reliability compared to other sources (e.g. Klein, 1993).

**Table 3.1.** VOC components relevant for direct health impacts as fraction of total VOC. Source: Obermeier (1999) adapted from Hassel (1998).

| Vehicle | Pollutant | speed range km/h | | | |
|---|---|---|---|---|---|
| | | 0-55 | 55-100 | 100-120 | 120-200 |
| Gasoline car, conventional | Benzene | 4.07% | 4.94% | 5.85% | 7.20% |
| | 1,3 Butadiene | 0.55% | 0.44% | 0.45% | 0.48% |
| | Ethene | 6.02% | 8.90% | 11.39% | 13.64% |
| | Formaldehyde | 3.02% | 2.97% | 3.20% | 1.37% |
| Gasoline car, TWC | Benzene | 2.07% | 6.32% | 6.78% | 17.99% |
| | 1,3 Butadiene | 0.01% | 0.07% | 0.12% | 0.35% |
| | Ethene | 0.41% | 2.70% | 3.31% | 11.25% |
| | Formaldehyde | 6.56% | 1.55% | 1.61% | 0.09% |
| Diesel car | Benzene | 4.70% | 4.66% | 4.44% | 7.56% |
| | 1,3 Butadiene | 1.03% | 0.41% | 0.13% | 0.24% |
| | Ethene | 24.51% | 17.88% | 12.92% | 23.39% |
| | Formaldehyde | 17.80% | 14.72% | 10.80% | 10.80% |

**Table 3.2.** VOC components relevant for ozone modelling. Source: Krüger (1999)

| Vehicle | Pollutant | speed range km/h 0-50 | 50-90 | 90-200 |
|---|---|---|---|---|
| Gasoline car, conventional | Methane | 4.84% | 4.72% | 6.75% |
| | NMVOC | 95.16% | 95.28% | 93.25% |
| Gasoline car, TWC | Methane | 45.19% | 22.36% | 15.38% |
| | NMVOC | 54.81% | 77.64% | 84.62% |
| Diesel car | Methane | 7.83% | 4.80% | 3.88% |
| | NMVOC | 92.17% | 95.20% | 96.12% |

Emission factors for volatile organic compounds (VOC) are given as total, containing up to several hundred of single compounds, as well as methane. Therefore split factors were applied to compute emissions of those components, which are relevant for direct health impacts, mainly cancer, and global warming. Pollutants like benzene, formaldehyde, 1,3 butadiene and ethene are calculated with data from Obermeier (1999) which were adapted from corresponding measurements undertaken by TÜV Rheinland (Hassel et al., 1998). Furthermore, VOC is divided into NMVOC and methane using data summarised by Krüger (1999). NMVOC is used for ozone modelling and methane is considered for global warming. All split factors are based on total VOC.

This data was included in the EcoSense Transport model for the automated emission calculation and recommended for the case studies. A summary of the numbers is given in Table 3.1 and Table 3.2.

To ensure the comparability of case studies, a number of reference technologies were defined (Table 3.3). Therefore, differing impacts of vehicles under assessment in locations throughout Europe are clearly restricted to real site-depending parameters, like e.g. vehicle speed, population, meteorology etc.

The year 2000 was recommended as the reference year for the case studies. $SO_2$ emissions were calculated based on the fuel consumption of the vehicle, assuming a sulphur content of 0.015 mass-% for gasoline and 0.035 mass-% for diesel (after 98/70/EG came into force 1.1.2000). Where $CO_2$ emission functions were not used

**Table 3.3.** Reference technologies for application in case studies.

| Vehicle category | Fuel | Size class | Vehicle class / emission standard |
|---|---|---|---|
| Passenger cars | Gasoline | < 1.4 l | ECE 15/04, EURO I, EURO II, EURO III |
| Passenger cars | Gasoline | 1.4 – 2 l | ECE 15/04, EURO I, EURO II, EURO III |
| Passenger cars | Diesel | < 2 l | Uncontrolled, EURO I, EURO II, EURO III |
| Light duty vehicles | Diesel | < 3.5 t | Uncontrolled, EURO II |
| Heavy duty vehicles | Diesel | 32-40 t | Uncontrolled, EURO II |
| Urban bus | Diesel | | Uncontrolled, EURO II |
| Coach | Diesel | | Uncontrolled, EURO II |

**Table 3.4.** Emission formulas and emission factors for typical urban and highway drives in Germany for 2 different passenger cars: Gasoline EURO I cc<1.4 l and diesel EURO I (<2.5t). All formulas are taken from (MEET, 1999), except particle emission functions for gasoline passenger cars taken from (Torkkeli, 1999).

| Vehicle | Pollutant | Emission formula | Emission factor [g/km] |
|---|---|---|---|
| PC, gasoline, EURO I, < 1.4l, urban drive (39 km/h) | CO | 9.846 - 0.2867*V + 0.0022*V^2 | 2.0109 |
| | $CO_2$ | - | 163.0700 |
| | FC | 93.672- 1.5100*V + 0.01090*V^2 | 51.3610 |
| | $NO_X$ | 0.5595- 0.01047*V +0.000108*V^2 | 0.3154 |
| | PM | 0.059964 - 0.00418601*V + 0.00014063*V^2 - 0.00000217*V^3 + 0.000000013*V^4 | 0.0120 |
| | VOC | 0.628- 0.01377*V + 0.0000852*V^2 | 0.2206 |
| PC, diesel, EURO I, (< 2.5t), urban drive (39 km/h) | CO | 1.4497- 0.03385*V + 0.00021*V^2 | 0.4490 |
| | $CO_2$ | 286-4.07*V+0.0271*V^2 | 168.4900 |
| | FC | 90.079 - 1.282*V + 0.0085350*V^2 | 53.0630 |
| | $NO_X$ | 1.4335- 0.026*V + 0.0001785*V^2 | 0.6910 |
| | PM | 0.1804 - 0.004415*V + 0.0000333*V^2 | 0.0589 |
| | VOC | 0.1978- 0.003925*V + 0.0000224*V^2 | 0.0788 |
| PC, gasoline, EURO I, < 1.4l, highway drive (115 km/h) | CO | 9.846 - 0.2867*V + 0.0022*V^2 | 5.9705 |
| | $CO_2$ | - | 203.7600 |
| | FC | 93.672- 1.5100*V + 0.01090*V^2 | 64.1750 |
| | $NO_X$ | 0.5595- 0.01047*V +0.000108*V^2 | 0.7838 |
| | PM | 0.022232 - 0.00036696*V + 0.00000241*V^2 | 0.0119 |
| | VOC | 0.628- 0.01377*V + 0.0000852*V^2 | 0.1712 |
| PC, diesel, EURO I, (< 2.5t), highway drive (115 km/h) | CO | 1.4497- 0.03385*V + 0.00021*V^2 | 0.3342 |
| | $CO_2$ | 286-4.07*V+0.0271*V^2 | 176.3500 |
| | FC | 90.079 - 1.282*V + 0.0085350*V^2 | 55.5240 |
| | $NO_x$ | 1.4335- 0.026*V + 0.0001785*V^2 | 0.8042 |
| | PM | 0.1804 - 0.004415*V + 0.0000333*V^2 | 0.1131 |
| | VOC | 0.1978- 0.003925*V + 0.0000224*V^2 | 0.0427 |

explicitly, 3.175 kg $CO_2$ per kg fuel was used, to estimate $CO_2$ emissions from fuel consumption (Umweltbundesamt and INFRAS, 1997).

## 3.2 Rail Transport

In European railway transport two traction types are very common: electric traction and diesel traction. For electric traction the emissions are determined by the power plants generating the electricity used. They are not related to the location where a train is running, but to the locations of the power plants. Therefore, externalities due to a train kilometre are calculated by combining the externalities associated with the production of one kWh of electricity and the electricity use per train kilometre.

**Table 3.5.** Typical emissions and fuel consumption factors for diesel railway locomotives (taken from (Jørgensen and Sorensson, 1997), Table 10.14)

| Emission | Power Specific: g/kWh | | Fuel Specific: g/kg | |
|---|---|---|---|---|
| | Range | Average | Range | Average |
| CO | 1 - 10 | 3.9 | 5 - 40 | 22 |
| HC | 0.5 - 4.0 | 2.0 | 3 - 25 | 11 |
| $NO_x$ | 6 - 16 | 10.7 | 30 - 70 | 53 |
| Particles | 0.2 - 1.2 | 0.6 | 1 - 6 | 3 |
| $SO_2$ | 0.2 - 2 | 0.8 | 1 - 10 | 4 |
| $CO_2$ | 604 - 700 | 640 | 3180 | 3180 |
| Fuel consumption | 190 - 220 | 200 | - | - |

For diesel trains the emission calculation is similar to the emission calculation of road vehicles with internal combustion engine. The MEET project provided European average emission factors for diesel locomotives. Depending on the type and composition of a train, i.e. the number and types of coaches/waggons and its load, the mass, capacity, speed, etc. may vary considerably. Therefore, emissions are not given per kilometre but per kWh of power used or per kg of fuel used. Table 3.5 shows the emission factors given in (Jørgensen and Sorensson, 1997). The ranges given vary by an order of magnitude (except $CO_2$ and fuel consumption). For this reason, country specific emission data was used, where available. Together with data on the power or fuel use of a train the resulting emissions can be calculated.

The data available on locomotive emissions is much less detailed than for road vehicles. As only data for generic trip types is available, it was not possible to differentiate by speed or gradient.

## 3.3 Inland Shipping Transport

In the literature on emission factors little information on inland shipping emission factors is available. For task 5.3: emission factors, the available literature was reviewed and emission factors for the different stages of the transport cycles were presented.

Below, a summary is given on how to calculate current and future emission factors for different stages of inland shipping transport cycles and different ship classes based on Dorland (2000), details may be found in Schmid (1998).

### 3.3.1 Steady State Operation

Data on the average load capacity, load factor, engine capacity and fuel consumption is presented in Table 3.9. Class A to G represent barge data. With the base data presented in Table 3.6, Table 3.7 and Table 3.9 the emission factors for

**Table 3.6.** Fuel use and fuel emission factors (g/kg gas oil).

| Technology base year | 1990 | 1995 | 2000 | 2005 | 2010 |
|---|---|---|---|---|---|
| fuel use (%) | 100 | 98 | 94 | 93 | 90 |
| *Pollutant* | | | | | |
| $CO_2$ | 3,130 | 3,130 | 3,130 | 3,130 | 3,130 |
| $NO_x$ | 55 | 55 | 50 | 45 | 40 |
| $PM_{10}$ | 3.0 | 2.0 | 1.0 | 1.0 | 1 |
| $SO_2$ [a] | 3.4 | 3.4 | 3.4 | 3.4 | 3.4 |
| HC | 3.0 | 2.0 | 1.0 | 1.0 | 1 |
| NMHC [b] | 2.88 | 1.92 | 0.96 | 0.96 | 0.96 |
| CO | 3.0 | 3.0 | 2.7 | 2.2 | 1.6 |

[a] Assumed Sulphur content of the gas oil is 0.17% m/m.
[b] Decrease assumed linear with decrease in HC emission factors.

steady state operation can now be calculated using Equation 3.1. It is assumed that until 2010 no innovation in ship transport technology, leading to a sharp decrease in emission factors, will take place (van Wee, 1998, Dijkstra, 1998). The scaling factors for the different steady state transport stages (95% MCR, 50% MCR, 25% MCR) to be used in the equation are given in Table 3.7. The scaling index can probably not be used for inland ships as the index was obtained for a broad range of medium and large marine ships. However, no data for inland shipping is currently available. Furthermore, applying this index neglects that the 'maximum engine output capacity'/'engine design capacity' ratio is lower for marine ships than for inland ships (0.85 vs. 0.95). This probably does not lead to a large error (<10%).

$$EF_{x,y,z} = \frac{P_x \cdot \%MCR \cdot F_b \cdot FC_y}{C_x \cdot L \cdot 100 \cdot V_{x,95\%MCR} \cdot \sqrt[3]{\dfrac{F}{F_{95\%MCR}}}} \cdot EF_{b,y,z} \cdot EFC_z \qquad (3.1)$$

With:

| | | |
|---|---|---|
| $C_x$ | = | Ship load capacity (tonnes) |
| $EFC_{z,\%MCR}$ | = | Pollutant specific and engine load specific emission factor correction factor (%) |
| $EF_{x,y,z}$ | = | Pollutant and year specific base fuel emission factor (g/kg) |
| $EF_{b,y,z}$ | = | Base fuel emission factor for each pollutant in a given year (g/kg) |
| $F_{95\%MCR}$ | = | Base fuel use in 1990 (g/kWh) |
| $FC_y$ | = | Year specific fuel use correction factor (%) |
| $P_x$ | = | Ship engine power at 95%MCR (= 0.95*design engine power) (kW) |
| $V_{x,95\%MCR}$ | = | Ship class specific speed at 95%MCR (km/h) |
| L | = | Capacity load factor (%) |

x = ship class
y = year of base emission factor
z = pollutant ($CO_2$, $NO_x$, etc.)

**Table 3.7.** Emission factor scaling index for marine ships.

| Pollutant | Steady state operation | | | Transient manoeuvring | 'Hotel'[a] |
|---|---|---|---|---|---|
| | 85% MCR | 50% MCR | 25% MCR | | |
| $CO_2$ | 100 | 100 | 100 | 100 | 100 |
| $NO_x$[a] | 100 | 117 | 130 | 90 | 100 |
| $PM_{10}$ | 100 | 105 | 110 | 120 | 100 |
| $SO_2$ | 100 | 100 | 100 | 100 | 100 |
| HC | 100 | 121 | 148 | 150 | 100 |
| NMHC | 100 | 121 | 148 | 150 | 100 |
| CO | 100 | 145 | 218 | 380 | 100 |
| Fuel use (%) | 100 | 50 | 25 | 35 | |

[a] The hotel state is a state in which the ship is in rest and used for living only.
Source: Lloyd's Register (1995)

## 3.3.2 Non Steady State Operation

For the 'non steady state operation' transport stages, except 'approach' and 'depart', it is assumed no distance is travelled. This is a reasonable assumption as the actual distance travelled is negligible compared to the distance of a normal transport trajectory. The emission factors can now be estimated with Equation 3.2. The time and the engine load factors of each stage are important parameters in this equation. For typical inland shipping transport stages they are presented in Table 3.8.

$$EF_{x,y,z} = T \cdot F \cdot SC_x \cdot \frac{\%MCR}{L} \cdot EF_{b,y,z} \qquad (3.2)$$

With:

$EF_{x,y,z}$ = Emission factor for each transport stage (g/ktonne)
$EF_{b,y,z}$ = Base emission factor (g/kg fuel)
$F_{x,y}$ = Fuel use (g/kWh)
$SC_x$ = Ship class specific capacity at 100% load factor (kW/tonne)
T = Transport stage time (hour)
%MCR = Engine capacity use at transport stage (%)
L = Load factor (%)
x = ship class
y = year of base emission factor
z = pollutant ($CO_2$, $NO_x$, etc.)

**Table 3.8.** The time spent on different 'non steady state' transport stages and engine load factors.

| Transport stage | | Engine load (% MCR) | Time (seconds) |
|---|---|---|---|
| Docked | 'Hotel' | 10 kW[a] | 40 hours[a] |
| | Transhipment | n.q. | 5400[a] |
| Docking | approach | 50 | 1080 |
| | Manoeuvring | 35 | 3300 |
| Undocking | Manoeuvring | 35 | 900 |
| | depart | 50 | 360 |
| Passing a lock | approach | 50 | 900 |
| | Manoeuvring in | 35 | 1560 |
| | idle in lock[b] | 25 | 540 |
| | Manoeuvring out | 35 | 240 |
| | depart out | 50 | 270 |

[a] For inland shipping a factor 10 lower power need is assumed as for marine ships. Marine ships on average use 10 kW power for 'Hotel' operation (Dijkstra and Dings, 1996).
[b] The manoeuvring emissions are multiplied by 3.
n.q. = not quantified
Source: Lloyd's Register (1995)

Approach and depart transport stages can be seen as 'steady state operation' stages. The emission factors in these stages should thus be estimated using Equation 3.1.

## 3.4 Aircraft Transport

Emission factors for aircraft engines are based on engine certification data and are available from the ICAO Engine Exhaust Emission Database (ICAO/CAEP, 1999). Emission index (EI) given in g/kg fuel by ICAO are limited to the 'classical' pollutants CO, $NO_x$ and hydrocarbons (HC). Furthermore, fuel flow in kg/s is given per engine tested. $SO_2$ emissions are calculated in the case studies using the EI 1.0 g/kg given in European Commission (1998b).

Very few experiments have been reported where the exhaust gas from aircraft turbines has been analysed in detail to identify specific species profiles. The VOC profile varies with the trust setting of the aircraft and therefore on the actual operation stage of the aircraft, like idle/taxi, take-off etc. For the case studies, average VOC-split factors (weight) based on the LTO cycle were taken from (Shareef et al., 1988, in EMEP/CORINAIR, 1996-1999)(see Table 3.10).

Particle emissions of aircraft engines are not given in the ICAO database. However, the smoke number (SN) is measured, which is an indicator for the mass of emitted particles. The smoke number is calculated from the reflectance of a filter paper measured before and after the passage of a known volume of a smoke-bearing sample. Hagen et al. (1992) state that modern jet engines emit 0.01 – 0.1 g of soot per kg fuel, with a peak diameter of 30 – 100 nm. On the basis of available

**Table 3.9.** Inland shipping technical parameters (Steady State Operation 95% MCR).

| Class (t) | Average load capacity [a] (t) | Design engine capacity [b] (kW) | Engine output 95% MCR (kW) | Specific capacity at 95% MCR (kW/t) | Maximum speed 95% MCR, 100% load (km/h) | Fuel use 100% load factor (kg/kt.km) | Load factor[c] % (%) | Fuel use corrected for actual load factor (kg/kt.km) |
|---|---|---|---|---|---|---|---|---|
| A 251-450 | 350 | 350 | 333 | 0.95 | 10.0 | 17.0 | 74.0 | 23.0 |
| B 451-650 | 550 | 248 | 235 | 0.43 | 10.0 | 7.7 | 74.0 | 10.4 |
| C 651-850 | 750 | 338 | 321 | 0.43 | 12.0 | 6.4 | 91.0 | 7.0 |
| D 851-1050 | 950 | 428 | 406 | 0.43 | 12.0 | 6.4 | 91.0 | 7.0 |
| E 1051-1250 | 1150 | 575 | 546 | 0.48 | 14.0 | 6.1 | 74.0 | 8.2 |
| F 1251-1800 | 1550 | 775 | 736 | 0.48 | 14.0 | 6.1 | 74.0 | 8.2 |
| G >1800 | 2250 | 1125 | 1069 | 0.48 | 14.0 | 6.1 | 74.0 | 8.2 |
| H push tug (2) | 5400 | 2160 | 2052 | 0.38 | 13.9[e] | 4.9 | 74.0 | 6.6 |
| I push tug (4) | 10800 | 3312[a] | 3146 | 0.29 | 12.6[e] | 4.1 | 74.0 | 5.6 |

| | | |
|---|---|---|
| energy content fuel [d] | 42.7 | MJ/kg |
| engine efficiency[d] | 0.47 | |
| base fuel use[d] | 179 | g/kWh |

[a] NEA (1997)
[b] Dings et al. (1997)
[c] Brink and Wee (1997)
[d] Dijkstra and Dings (Dijkstra and Dings, 1996)
[e] Brink and Wee (1997)

data, Fried developed a model to estimate a soot emission index from the smoke number and the fuel flow, which was used for the case studies (Fried, 1999). They are approximately on the upper limit compared to data given by Hagen, except for take-off, where they are a factor of 10 larger.

**Table 3.10.** VOC split for aircraft jet engines emissions, based on an average LTO cycle, percentage of total VOC

| | Benzene | Formaldehyde | 1,3 Butadiene | Ethylene |
|---|---|---|---|---|
| LTO-average | 1.9 | 15 | 1.8 | 17.4 |

Source: Shareef et al. (1988)

# 4 Atmospheric Transport Modelling

## 4.1 Local Scale

The local scale analysis focuses on distances up to 20 km from the emission source. Gaussian plume models are commonly used for the estimation of local scale atmospheric dispersion of primary pollutants as they combine relatively accurate results and low computing time. The effects of secondary pollutants, within this distance, were found to be rather insignificant. Sulphate aerosols generally have low formation rates, so the error from ignoring their impacts on local scale was considered to be small. Nitrate aerosols were considered a potentially greater problem due to the high $NO_x$ emission levels from vehicles in comparison to other sources. However, an analysis for the investigation of nitrate impacts revealed that damages arising within the local scale area are less than 10% of the total, even in large urban areas (Lee and Watkiss, 1998). On the basis of these results, it was decided that the secondary particulate formation would not be quantified within the local scale area. However this issue should be addressed in more detail in subsequent work.

### 4.1.1 Gaussian Dispersion Modelling

Plume diffusion equations using the assumption of turbulence uniformity result in concentration distribution having Gaussian profiles. The concentration distribution of the pollutants emitted by a point source of effective height $h_e$, at a constant rate Q, is calculated by:

$$C(x,y,z,h_e) = \frac{Q}{2\pi\sigma_y\sigma_z U}\exp\left[-0.5\left(\frac{y}{\sigma_y}\right)^2\right]\cdot V \qquad (4.1)$$

With

| | | |
|---|---|---|
| C | = | concentration of the pollutant at the receptor point |
| $h_e$ | = | effective height of the emission source |
| $\sigma_y$, $\sigma_z$ | = | dispersion coefficients |
| Q | = | emission's rate |
| U | = | average wind speed |

V = vertical term accounting for the effects of the restriction on vertical plume growth at the top of the mixing layer, assuming multiple reflections

The determination of $\sigma_y$ and $\sigma_z$ is crucial for the proper use of the Gaussian models. Several methods have been developed for the calculation of the vertical dispersion coefficients as functions of distance and meteorological conditions (U.S. Environmental Protection Agency, 1995, Griffiths, 1994, Zannetti, 1990, Benson, 1982).

The calculation of long term concentrations is based on the Gaussian sector average equation (U.S. Environmental Protection Agency, 1995). The area surrounding the emission sources of pollutants is divided into sectors of equal angular width corresponding to the sectors of the seasonal or annual distribution of wind directions. Seasonal or annual emissions from the sources are partitioned among the sectors according to the frequencies of occurrence of wind blowing toward the sectors. The concentration fields calculated for each source are summed up to obtain the total concentration. Assuming that the meteorological data consist of $N_i$ wind speed classes, $N_j$ wind directions classes and $N_k$ atmospheric stability categories the average concentration is given by:

$$C = \frac{Q}{\sqrt{2\pi} R \Delta\theta} \sum_{i=1}^{N_i} \sum_{j=1}^{N_j} \sum_{k=1}^{N_k} \frac{f_{i,j,k} S_j V_{i,k}}{u_i \sigma_{z_k}} \tag{4.2}$$

where:

C = long term average concentration
Q = pollutant's emission rate
R = radial distance between the receptor and the emission source
$\Delta\theta$ = sector width in radians
$f_{i,j,k}$ = frequency of occurrence of the ith wind speed category, the jth wind direction category and the kth atmospheric stability category
$u_i$ = mean wind speed for the ith wind speed category
$\sigma_{zk}$ = standard deviation of the vertical concentration distribution for the kth stability category
$S_j$ = smoothing function
$V_{i,k}$ = vertical term for the ith wind speed category and the kth atmospheric stability category

Modelling pollutant concentrations from linear sources is achieved by dividing the source into finite sections each one of which is assumed to be a point source. Complex road networks and curved roads are simulated by linear segments that are supposed to be independent line sources. The pollutant concentration at a specific receptor point is the sum of the contributions of all the sections of the linear source located downwind the receptor. The contribution of each sector to the total concentration is calculated through Equation 4.1 or 4.2. In order to estimate the total pollutant concentration Equation 4.1 or 4.2 is integrated over the line source. The total concentration of the pollutant at the receptor point is calculated by:

$$C(x,y,z,H) = \int_{r1}^{r2} C_r dr \tag{4.3}$$

where:

| | | |
|---|---|---|
| C | = | concentration of the pollutant at the receptor point (x, y, z) |
| Cr | = | concentration of the pollutant at the receptor point due to a point source S which lies on the line segment at a distance r from the starting point r1 |

## 4.1.2 The ROADPOL Model

ROADPOL is a Gaussian line source dispersion model predicting the annual or seasonal average concentrations of pollutants and the population exposure in a range of up to 20 km away from the emission source. The basis of the model is Equation 4.2 which is numerically integrated along the line source. ROADPOL operates under a Geographic Information System (G.I.S.) environment exploiting its powerful features in handling complex road geometry, and location-specific parameters (receptors' density, meteorological characteristics). The input data needed are the road geometry, the pollutant emissions, the receptors co-ordinates, the population distribution around the emission source, as well as meteorological data such as frequencies of occurrence of each wind speed, direction, atmospheric stability class and mixing layer height. Geographical data such as the road segment co-ordinates and the population allocation are retrieved from G.I.S. maps.

The first step for the calculation of pollutant concentrations is the generation of a grid for the calculation of pollutant concentrations. The grid is adapted to the road geometry and has variable cell density (the grid is denser close to the source, where the concentration gradient is steeper) in order to get more accurate results from dispersion modelling without excessively increasing the computing time. Buffer zones around the road examined define the areas having grid cell sizes ranging between 125m and 2000m. The pollutant concentrations are calculated by integrating Equation 4.2 along the road examined. Four different methods, depending on the surface roughness of the area examined and on vehicle induced turbulence, for the calculation of the vertical dispersion coefficients are implemented in the model. The methods available are (a) the Pasquill-Gifford curves (Zannetti, 1990), (b) the Briggs formulas for urban areas (Zannetti, 1990, Griffiths, 1994), (c) the Briggs formulas for rural areas (Zannetti, 1990), and (d) the Benson equations (Benson, 1982). Finally the population exposure to the pollutants is estimated using a map representing the population allocation in the area. The exposure estimation is performed by geographic operations that provide the relative geographical location of the population and the pollutant concentrations.

## 4.1.3 Ozone

The assessment of the effect of emission reductions on ozone concentrations at the local scale probably represents the most difficult problem within atmospheric transport modelling for ExternE. The difficulties are caused by the strong non-linearities associated with ozone formation, particularly concerning $NO_x$ emissions (e.g. Seinfeld and Pandis, 1998). A release of $NO_x$ within an urban area may well cause an immediate decrease in ozone levels near the source (because of the titration reaction, $NO+O_3 \rightarrow NO_2$). Slightly downwind (for example in suburban areas), this $NO_x$ release may still have caused an ozone decrease in most cases, but in some photochemically very active cities (typically Mediterranean cities such as Milan or Athens) chemical reaction processes may enhance ozone levels. Far downwind, in rural areas, this $NO_x$ release would cause increased ozone in most parts of Europe, with the exception of those rural areas still heavily influenced by anthropogenic $NO_x$ sources.

The effects of VOC emissions are much more straightforward, although still rather variable in magnitude. Increased VOC emissions almost always lead to increased ozone levels. In high-$NO_x$ areas, ozone levels are strongly affected by VOC emissions. In low-$NO_x$ areas, ozone levels are less sensitive to VOC emissions.

Whether ozone increases or decreases as a result of $NO_x$ and VOC emission changes depends heavily on a range of factors, including the distribution of emissions near the source, the ambient $NO_x$ level (high or low-$NO_x$), and the meteorological conditions. The same release may cause an increase in ozone one day and a decrease in ozone on the next day.

Evaluation of the effects of control measures then requires either detailed modelling for each case of interest, or extrapolation from a sufficiently large number of studies to have some confidence in the reliability of the predictions. Unfortunately, there is no systematic assessment of the effect of local emission-control measures on ozone concentrations in and around urban areas. Many studies have been conducted of course, but the over-riding problem has always been that studies have covered only very short time periods and specific conditions.

**Table 4.1.** Effect of 50% emission reductions on ozone concentrations in city-plume modelling. Note that negative changes represent ozone reductions. Calculations for year 2010 – see (Simpson and Andersson-Sköld, 1997) for details.

| | Base-ozone (year 2010, ppb) | Ozone change (ppb), for scenario: -50% $NO_x$ | -50%VOC | -50% ($NO_x$+VOC) |
|---|---|---|---|---|
| Athens | 87 | 44.9 | -3.2 | 8.0 |
| Cologne | 122 | 22.6 | -11.6 | -8.4 |
| London | 97 | | | -6.8 |
| Lyons | 164 | 32 | -5.7 | -35.4 |
| Madrid | 120 | -10.5 | -20.1 | -17.6 |
| Milan | 184 | -27.0 | -29.9 | -42.4 |

Some examples from the studies may illustrate the problem further. Within the EU Auto/Oil I project, Simpson and Andersson-Sköld (1997) used a Lagrangian model to look at ozone formation downwind of six urban areas. By running simulations for a number of different wind-directions and meteorological conditions, some trajectories giving high ozone concentrations were established and studied in more detail. Table 4.1 shows the results of 50% emission control, applied to the plumes with maximum ozone in the year 2010 scenarios. It is seen that $NO_x$ control may increase or decrease ozone, VOC control always decreases ozone, and combined $NO_x$+VOC control generally reduces ozone (with Athens as the exception, where a small ozone increase is still predicted). More detailed calculations, using the 3-D European Zooming Model (Moussiopoulos et al., 1996) were also conducted within this project, covering 2 days for Athens and one day for Cologne (ibid). For Athens, $NO_x$ control was found to decrease ozone by about 20 ppb on one day (a sea-breeze day, having peak values of ca. 200 ppb), but to increase ozone by about 60 ppb on the other day (a stagnant situation with 140 ppb in the base-case). For the Cologne case, $NO_x$ emissions control again increased ozone. It should be noted though that this study did not account for changes in the regional background ozone, which probably has a significant effect on at least the Cologne simulations.

More recent studies (e.g. Ziomass et al., 1998, Svensson, 1998, Volta and Finzi, 1999) have generally confirmed these findings concerning $NO_x$ and VOC control. In an interesting study, Giovannoni et al. (1995) used two different methods for deriving the meteorology used in simulations of an Athens air pollution episode. The two methodologies gave very different wind-fields, which in turn lead to very different responses in the effectiveness of emission control in the region. Such studies illustrate well that ozone formation and control is a complex phenomenon, and that there are significant uncertainties in attempting to predict local ozone changes, even with advanced tools.

All such studies cover far too limited a time-series to give any general picture of the effects of emission control. An interesting attempt to model longer term ozone around cities has been made by Sahm and Moussiopoulos (1999). The model used was very simplified and thus it is hard to know how reliable the results are, but the conclusion is the same as from other studies: Ozone sometimes goes up, sometimes down!

Given these considerations, currently a simple reliable method of including the effects of emission control on local ozone changes cannot be recommended in this study's framework. Detailed studies are still required, either with full photochemical models, applied over long-periods, or with appropriate detailed dispersion modelling of $NO_x$ concentrations in order to simulate the titration effect.

# 4.2 Regional Scale

## 4.2.1 Reactive and Non-Reactive Pollutants

With increasing distance from the emission source the plume spreads vertically and horizontally due to atmospheric turbulence. Outside the area of the local analysis (i.e. at distances beyond 25 km from the release location), it can be assumed for most purposes that the pollutants have vertically been mixed throughout the height of the mixing layer of the atmosphere. On the other hand, chemical transformations can no longer be neglected on a regional scale. The most economic way to assess annual, regional scale pollution is a model with a simple representation of transport and a detailed enough representation of chemical reactions.

The Windrose Trajectory Model (WTM) used in EcoSense to estimate the concentration and deposition of acid species on a regional scale was originally developed at Harwell Laboratory by Derwent and Nodop (1986) for atmospheric nitrogen species, and extended to include sulphur species by Derwent, Dollard and Metcalfe (1988). The model is a receptor-orientated Lagrangian plume model employing an air parcel with a constant mixing height of 800 m moving with a representative wind speed. The results are obtained at each receptor point by considering the arrival of 24 trajectories weighted by the frequency of the wind in each 15° sector. The trajectory paths are assumed to be along straight lines and are started at 96 hours from the receptor point. The chemical scheme of the model is shown in Fig. 4.1.

In EcoSense, the model is implemented by means of

- a set of parameters and chemical equations in the EcoSense database which defines the model,
- a model interpreter,
- a set of meteorological input data (gridded wind roses and precipitation fields) in the reference environment database,
- emission inventories for $NO_x$, $SO_2$ and ammonia, which are also provided in the reference environment database,
- additional emissions of the power plant from the reference technology database, or
- additional emissions of the transport activity.

The 1990 meteorological data were provided by the Meteorological Synthesising Centre-West of EMEP at The Norwegian Meteorological Institute (Nordeng, 1986, Hollingsworth, 1987). 6-hourly data in the EMEP 150 km grid of precipitation and wind (at the 925 hPa level) were transformed to the EMEP 50 km grid and averaged to obtain, receptor specific, the mean annual wind rose (frequency distribution of the wind per sector), the mean annual windspeed,

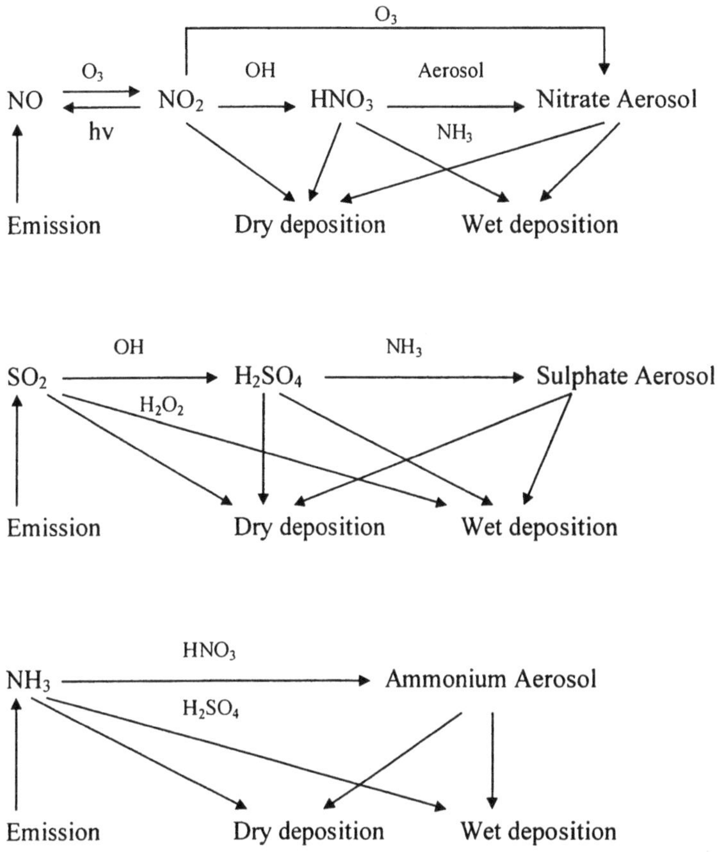


**Fig. 4.1.** Chemical scheme in WTM, adopted from Derwent (1988).

and total annual precipitation. The background emission scenario used for this study was taken from CORINAIR 1994, which was geographically distributed like the CORINAIR 1990 emission data.

### 4.2.2 Ozone

A large amount of information is available on the effects of $NO_x$ and VOC emission abatement on regional ozone levels in Europe. The EMEP Lagrangian model (Simpson, 1992, Simpson, 1993, Simpson, 1995) has been used to calculate the effects of reducing $NO_x$ and VOC emissions from each country in turn, generating so-called country-to-grid matrices (EMEP, 1999). These matrices have been incorporated within the EcoSense model, and are available at the EMEP website http://www.emep.int. The EMEP Lagrangian and Eulerian oxidant models have also been used to evaluate the effects of different emission sectors in turn (Reis et al., 1999, EMEP, 1999). The main emission sectors (SNAP 1-10) have been investigated, with sector 7 (road-traffic) studied in much more detail. These scenario runs were conducted for the year-2010 scenario developed under the INFOS project (Friedrich and Reis, 2000). Some summary results from these studies are given in Table 4.2 and Table 4.3.

**Table 4.2.** Emission projection (year 2010 base-case) and mean reductions in European AOT40 (forests) and AOT60 obtained when EU sector-emissions are set to zero

| SNAP code | Emissions sector | Emissions [ktonnes] $NO_x$ | NMVOC | Mean AOT reduction [a] $\Delta AOT40_f$ [ppb.h] | $\Delta AOT60$ [ppb.h] |
|---|---|---|---|---|---|
| 1 | Public power, cogeneration and district heating | 1041 | 41 | 134 | 9 |
| 2 | Commercial,institutional and residential heating | 499 | 730 | 148 | 18 |
| 3 | Industrial combustion | 1315 | 79 | 68 | 4 |
| 4 | Production processes | 340 | 1450 | 681 | 72 |
| 5 | Extraction and distribution of fossil fuels | 110 | 946 | 357 | 38 |
| 6 | Solvent use | 0.9 | 2756 | 1000 | 110 |
| 7 | Road transport | 3990 | 1820 | 1138 | 100 |
| 8 | Other mobile sources and machinery | 1518 | 510 | 312 | 28 |
| 9 | Waste treatment and disposal | 121 | 244 | 83 | 14 |
| 10 | Agriculture | 33 | 0[b] | 16 | 1 |

[a] the mean ΔAOT values are calculated over all of Europe. For comparison the European mean base-case 2010 $AOT40_f$ is ca.7200 ppb.h, and mean AOT60 is 230 ppb.h.
[b] agricultural VOC emissions are set to zero for all countries, to avoid double counting forest VOC emissions from managed forest (often reported as agricultural VOC) with the biogenic emissions calculated within the model.

**Table 4.3.** Emission projection (year 2010 base-case) from road transport sectors and calculated reductions in mean EU-15 AOT40 and AOT60 from each sector. Results given from 1-D (Lagrangian) and 3-D (Eulerian) EMEP MSC-W oxidant models

| SNAP | Sector | Emissions [ktonne] $NO_x$ | NMVOC | ΔAOT40 [ppb.h] 1-D | 3-D | ΔAOT60 [ppb.h] 1-D | 3-D |
|---|---|---|---|---|---|---|---|
| 7.0 | Road Transport (Total) | 3989 | 1819 | 1720 | 1138 | 242 | 100 |
| 7.1 | Passenger cars | 616 | 479 | 315 | 151 | 53 | 20 |
| 7.2 | LDV | 175 | 90 | 133 | 66 | 17 | 6 |
| 7.3 | HDV | 3180 | 383 | 800 | 437 | 157 | 36 |
| 7.4 | Mopeds | 0.9 | 137 | 42 | 76 | 7 | 11 |
| 7.5 | Motorcycles | 16 | 118 | 77 | 53 | 9 | 6 |
| 7.6 | Evaporation | 0 | 610 | 144 | 355 | 26 | 43 |

Note: numbers given in parentheses correspond to base-case European-mean $AOT40_f$ and AOT60. Other numbers refer to differences from these base-case values. Thus the base-case mean AOT40 for the 1-D model is 7500 ppb.h, whereas the mean AOT40 on removal of all road-traffic emissions is 7500-1720 ppb.h, ca. 5800 ppb.h.

These studies clearly suggest that road-traffic emissions have a significant role to play in ozone formation over Europe, even by the year 2010. Indeed, road-traffic as a whole is found to be the most important source-sector, and within this sector the biggest contributors are found to be heavy-duty vehicles and evaporative emissions. Thus, both $NO_x$ and VOC are found to be important. Negative $NO_x$ effects in some areas are offset by positive effects elsewhere.

## 4.3 Global Scale (Ozone)

The global three-dimensional dynamic-chemical climate model ECHAM3/CHEM was used to estimate the effects of changes in $NO_x$ emissions over Europe on the global ozone concentration. ECHAM3/CHEM can be used for studies of global climate and chemistry changes due to anthropogenic emissions.

On the basis of a version of the numerical weather prediction model of the European Centre for Medium-Range Weather Forecasts (ECMWF) the spectral atmospheric general circulation model ECHAM3 has been developed and completed with a chemistry model (Steil et al., 1998). Prognostic variables are vorticity, divergence, temperature, surface pressure, water vapour, cloud water, and up to 16 different tracers. ECHAM3 has 19 vertical layers between the surface and the model top at 10 hPa. A horizontal resolution of spectral T21 was chosen, corresponding to a resolution for dynamic processes of approximately 1000 km. Based on the Mainz 2-D chemistry model (Brühl and Crutzen, 1993) a comprehensive 3-D chemistry module has been developed. The model handles a comprehensive chemical system, involving 107 photochemical reactions and 37 species, for long-term integration of the 3-D circulation model.

Ozone concentration changes were assessed for a 50% reduction of European $NO_x$ surface emissions and 50% reduction of European aircraft emissions. First a basic model run (with entire global emissions) was carried out. Then a model run was performed with 50% reduction of European surface emissions. The total surface emissions for the base case scenario were 31.1 Tg N/year, the reduced surface emission 28.4 TgN/year. This means, if the European surface emission are reduced by 50%, the global surface emission are reduced by 2.7 TgN/year. The efficiency *eff* of reducing European surface emissions is about 5.60 $TgO_3$/TgN/year following the equation

$$eff = \Delta O_{3,tot} / e_{cur} \quad (4.4)$$

$\Delta O_{3,tot}$ is the change of the ozone concentration due to total surface emissions and $e_{cur} = e_{tot} - e_{red}$ .

In addition, the effect of reducing aircraft emissions was quantified. Due to the fact, that atmospheric chemistry is non-linear it is not possible to estimate marginal costs by simply simulating the effect of emissions from one airplane. For this reason a control run with ECHAM3/CHEM has been performed, with emissions of the entire world fleet of aircrafts and emissions from other sources, like surface-emissions or lightning. Global Nitrogen aircraft emissions sum up to 0.56 TgN/year. In comparison surface Nitrogen emissions for this scenario are 31.2 TgN/year. A second model-run was performed, assuming 50% reduction of European aircraft emissions, which leads to global emissions of 0.51 TgN/year. The marginal changes are calculated according to the formula:

$$\Delta X_{m\arg} = \frac{\Delta X_{tot} - \Delta X_{red}}{e_{tot} - e_{red}} \tag{4.5}$$

Where X stands for a chemical species like $NO_x$, $CH_4$ or $O_3$, e tot is the total aircraft emission, ered the reduced aircraft emission, $\Delta X_{tot}$ the resulting change of specie X due to total aircraft emissions and $\Delta X_{red}$ the resulting change of specie X due to reduced aircraft emissions. Total $NO_x$ aircraft emissions result in a global ozone increase of 26.7 Tg, if the European aircraft emissions are reduced, then the change of ozone is 25 Tg. The efficiency

$$eff = \frac{\Delta O3_{tot} - \Delta O3_{red}}{e_{tot} - r_{red}} \tag{4.6}$$

of reducing European aircraft emissions is 34.9 Tg $O_3$/Tg N/year, and thus 7 times higher than the efficiency of reducing European surface emissions.

In addition, temperature changes due to the total fleet emissions were calculated with the climate model version ECHAM4/MLO. Assuming a linear approach the marginal temperature changes were derived from this results and external costs calculated using the model FUND (see Chap. 9).

## 4.4 Validation of Dispersion Modelling

Validation is crucial for the credibility of the ExternE Project, but it is difficult. In fact, the damages from individual pollution sources are so small that they cannot possibly be measured directly. The only feasible validation can come from verification of elements of the calculations. In practice the most computationally intensive and difficult part of an externality calculation is the dispersion of the pollutant.

One approach to validating the EcoSense dispersion calculations is to compare them with other dispersion models that have been validated. Therefore the total damage for sources in France were calculated, using two different models for regional dispersion: EMEP (Barrett et al., 1995) and the Windrose Trajectory Model (as integrated into the EcoSense software (Krewitt et al., 1995)). The results agree within about 20%. The Windrose Trajectory Model is a modified version of the Harwell Trajectory Model (Derwent and Nodop, 1986), updated and adapted for the calculations of ExternE.

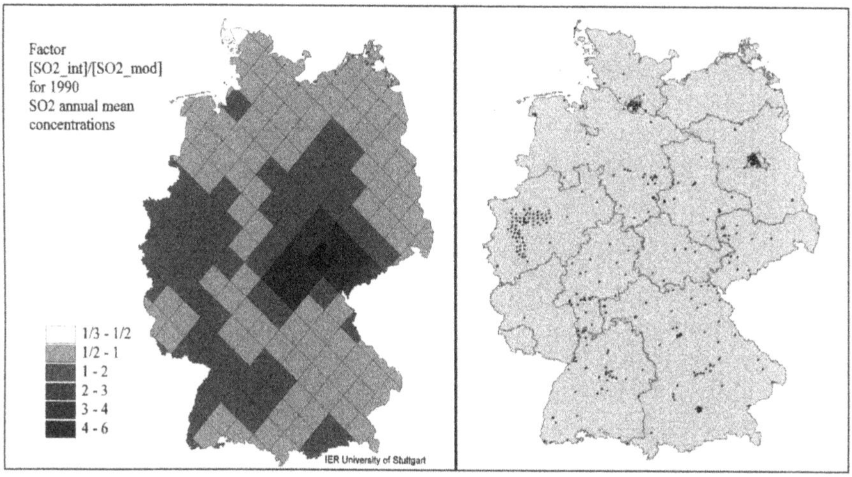


**Fig. 4.2.** Comparison of modelled ("mod") and measured ("int") annual mean concentrations of $SO_2$ and respective measurement stations for which observed concentration data were reported by the German Federal Environmental Agency for 1990.

A related comparison of interest has been reported by Derwent et al. (1989) who have compared four models with each other and with measured data. In addition to the Harwell Trajectory Model, they have considered EMEP (Eliassen and Saltbones, 1983), IE (Asman and Janssen, 1987) and TREND (Egmond et al., 1986). While there are substantial differences between the observed concentrations and the model calculations as well as between the models, there is agreement within an order of magnitude, and for most species the agreement is better than an order of magnitude. Agreement is generally better for $SO_2$ and aerosols and poorest for NO, $NO_2$ and $NH_3$. Here, too, the comparisons involved episodic values rather than long term averages. Comparisons for long term averages can be found in the reports of EMEP (see for instance Barrett et al. (1995)). They generally indicate agreement within a factor of two or better.

Ultimately dispersion should of course be verified by comparing the concentrations calculated by EcoSense with measured concentration data. For this purpose one needs to aggregate over all pollution sources within the geographic range (on the order of a thousand km or more for the classical air pollutants) to check whether the incremental concentrations calculated by EcoSense are consistent with the observed ambient concentrations. This has been done for the Windrose Trajectory Model (WTM), for a primary pollutant ($SO_2$) and for a secondary pollutant (sulphate aerosols).

Measured $SO_2$ concentration data in Germany for 1990 were interpolated to the model grid and compared to the concentration derived with the model (see Droste-Franke et al., 1998). Figure 4.2 shows the ratio of interpolated to modelled $SO_2$ concentrations. The measured data seem to be in some regions over- (ratio <1) and in some underestimated (ratio > 1) by the model results. In most areas the agreement is better than a factor of two. The only exception is seen in some

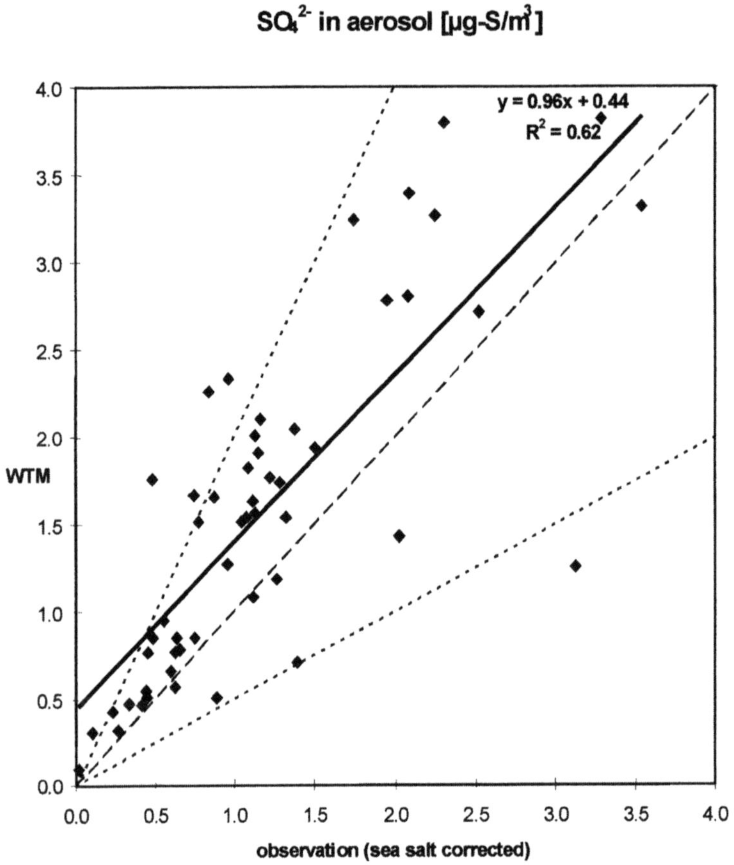


**Fig. 4.3.** Comparison of measured and modelled sulfate concentrations.

regions of East Germany where the underestimation reaches a factor of six. However, these are regions with very high $SO_2$ emissions from power plants, and with the resolution of the 50 × 50 km² EcoSense grid cells the spatial relation between these sources and the measurement stations is not modelled with sufficient detail for a primary pollutant near a point source. Therefore disagreement in this region is not significant.

In Fig. 4.3 the modelled sulphate concentrations are compared to observed concentration data for measurement stations all over Europe. Extra WTM model runs were performed to calculate the concentration levels at the locations of the measurement stations. The values taken into account were restricted to data measured within the mixing layer assumed for the WTM calculations. The two dashed lines represent an over- and underestimation of a factor of two. More than 85% of the data points lay between these two lines so that the most concentration data calculated by the model show smaller deviations to the observed data than a factor of two, with a slight overestimation.

This analysis shows that the EcoSense dispersion calculations agree with measured $SO_2$ and sulphate data within a factor of two. Since overestimation in some regions can compensate underestimation in others, as far as the calculation of external costs is concerned, the overall agreement between EcoSense results and the true values can only be better, as far as atmospheric modelling is concerned. This point is examined further in Sect. 11.2.

# 5 Health Effects: Exposure-Response Functions

## 5.1 Introduction

In this chapter, both for the 'classical' pollutants (particulate matter, PM; ozone, $SO_2$, $NO_2$, CO) and for carcinogenic substances, a set of exposure-response (E-R) functions is provided for acute and for chronic effects of air pollution. This set of E-R functions has been developed over various stages of ExternE and related projects. Chapter 4 of European Commission (1995a) gives a comprehensive description of background literature and E-R functions at that time. Summaries of the current set of E-R functions may be found in Friedrich et al. (1998) and Markandya et al. (1999). The choice of studies and E-R functions is described most fully in the revised ExternE Methodology report (European Commission, 1999; Chap. 8). The present relatively brief description draws variously on these reports, while highlighting those areas (e.g. with regard to chronic mortality) where implementation for the present study differs from that reported elsewhere.

## 5.2 Summary of Methods

The extensive epidemiological literature on the health effects of ambient air pollutants, and other evidence (e.g. toxicology, human chamber studies), was surveyed to form a view on:

- What ambient air pollutants are associated with adverse health effects (acute or chronic), and for what specific endpoints;
- Which of these associations may reasonably be interpreted as causal, because quantifying the effects of incremental pollution implies causality;
- Where causality is accepted for a pollutant-impact pathway, what individual study or meta-analysis gives a representative value for quantification of that effect;
- How if at all should the E-R functions from individual studies or meta-analyses be adapted for use in ExternE; and
- What combination or set of quantifiable E-R functions provides a suitable basis for quantifying the public health effects of incremental air pollution from transport.

The evaluation has focused on strategic issues. Examples of key strategic questions include the following:

- Do different sorts of particles (primary emissions from transport; sulphates; nitrates) have the same effects on health, per $\mu g/m^3$ concentration in ambient air?
- Are the health effects associated with ambient particles and ozone additive?
- Do the associations, found in some epidemiological studies, between health and (variously) $SO_2$, $NO_x$ and CO imply causal relationships? If so, are the estimated effects additive to those for particles and ozone?
- Does longer-term exposure to ambient particles give rise to effects ('chronic effects') different from and additional to the more-or-less immediate ('acute') effects associated with particle concentrations on a given day? If so, can the effects be estimated reliably (enough)?
- Can the same set of E-R functions be applied reliably throughout Europe?
- Do results from studies conducted elsewhere, and notably in North America, give reliable estimates of effects in Europe also?
- Is there a threshold for the effects on health of ambient particles? of ozone?

Judgements at all of these stages are the focus of debate currently among scientists and policy makers concerned with the health effects of air pollution. In these brief notes it was tried to identify, though not discussed in detail, the most important aspects of that debate. Furthermore, judgements on any of these issues are likely to have a greater impact on final answers than the fine-tuning of individual E-R functions. Thus, the reliability of the E-R set as a whole, rather than any individual component of it, has been the principal focus of our attention. For example, some E-R functions (e.g. linking PM or $O_3$ with restricted activity days (RADs), or PM with chronic bronchitis) where the direct evidence is weak, but a relationship is to be expected in the light of strong evidence for related endpoints, were included. This is an application of the coherence argument of Bates (1992).

For any endpoint, an E-R function is chosen which is considered to be credible and adequate, in the sense that its application will lead to estimated impacts which are of about the right magnitude. It is not claimed that the functions used are in some sense the best possible. Specifically, attention was not focussed on fine-tuning E-R functions (e.g. in the light of emerging new studies, or by carrying out formal meta-analyses rather than by choosing a single 'representative' function) when the results were likely to lead at most to small changes in estimated impacts.

In proposing individual E-R functions, the focus was however on studies of appropriate design, with good data, and suitable methods of statistical analysis, including adjustment for confounding factors, checks on model robustness, and where practicable the simultaneous investigation of the effects of multiple pollutants from the general air pollution mixture. Where necessary the chosen E-R functions were re-scaled, if the time-period of daily measurements in the initial study (e.g. 1-hour peak, 8-hour average) did not conform to the units in which incremental pollution is modelled in this study.

## 5.3 Representation of Particles

### 5.3.1 Should E-R functions for Particles from Transport be Expressed in Terms of $PM_{10}$ or $PM_{2.5}$ or Some Other Index?

Ambient particles are a complex mixture, varying in size and in composition. This leads to E-R functions for particles in terms of total suspended particles (TSP); $PM_{10}$ (inhalable particles); black smoke (BS); finer fractions such as $PM_{2.5}$ or sulphates; and other indices. From these possibilities, E-R functions for ambient particles are generally expressed in terms of $PM_{10}$ (see e.g. European Commission (1995a); COMEAP, 1998). This is because $PM_{10}$ has been or is the basis of standards (and so of routine monitoring) in many countries and many epidemiological studies are now available in that metric.

However, the particles of interest in the present project are not typical of ambient $PM_{10}$ generally. Rather, they arise as primary emissions from transport; or as secondary particles generated subsequently (i.e. as sulphates from $SO_2$, or as nitrates from $NO_x$). In terms of particle size, particles from all three sources lie principally, if not exclusively, within the finer, $PM_{2.5}$, fraction of $PM_{10}$. There is some evidence that, per μg/m³, the relatively fine fractions (e.g. $PM_{2.5}$) are associated with greater risks than $PM_{10}$ generally. Subject to availability of E-R functions, there is therefore a case in terms of particle size for quantifying the effects from all three sources using E-R functions expressed in terms of $PM_{2.5}$ rather than in terms of $PM_{10}$.

However, other aspects of particles may also affect their toxicity. Insoluble particles may be more toxic (other things being equal) than soluble ones, and toxicity may be increased by, or even largely due to, transition metals adsorbed on to the particle surface.

Evidence to date on these issues is far from conclusive. However, they suggest that, per μg/m³ increment in ambient concentrations, the most severe health effects are associated with primary particulate emissions, and the least severe with nitrates. (It may be that health effects associated with nitrates are negligible; there is little or no direct epidemiological evidence about this.) Many epidemiological studies have shown associations of sulphates with adverse health; but they do not show if the effects are a consequence of sulphates as such, or whether sulphates in these studies are a surrogate for particles or other pollutants.

Against this background, in the present study the following conventions were adopted:

- each incremental μg/m³ of primary particles emitted during transport is treated as if had the toxicity of a μg/m³ of $PM_{2.5}$;
- each incremental μg/m³ of sulphates is also treated as if it had the toxicity of $PM_{2.5}$; and
- each incremental μg/m³ of nitrates is treated as is it had the (lower) toxicity of $PM_{10}$.

### 5.3.2
### Developing E-R Functions Expressed as $PM_{2.5}$ rather than $PM_{10}$

At present, epidemiological studies examining the relationships between particles expressed as $PM_{2.5}$ and health come almost entirely from studies in North America. This does not as yet constitute a comprehensive body of evidence for quantification. Also there may be a problem (see below) in transferring E-R functions for acute effects from North America to Europe.

Consequently, the North American functions for $PM_{2.5}$ were not used. Instead, it was directly converted from the exposure-response functions for $PM_{10}$ by multiplying by 1.67. This conversion factor is based on Dockery and Pope (1994) where it was derived principally from the US experience. There is some indirect European evidence that this is a reasonable factor to use in Europe also.

The use of black smoke (BS) as a surrogate for $PM_{2.5}$ in Europe was considered, in order to give access to the APHEA BS results for quantifying $PM_{2.5}$ effects. However BS and $PM_{2.5}$ represent different sub-fractions of $PM_{10}$ and the BS to $PM_{2.5}$ relationship may vary considerably by site, climate and other circumstance. It was decided therefore that this indirect approach to estimating a $PM_{2.5}$ effect via BS studies was too unsound to be used.

## 5.4
## The Gaseous Pollutants

### 5.4.1
### Ozone

The APHEA results have confirmed a relationship of ambient ozone with acute mortality and hospital admissions in Europe. The overall evidence strongly supports the view that the acute health effects of ozone can and should be quantified; and that the estimated health impacts from ozone should be added to those of particles.

Epidemiological studies represent the effects of daily ozone characterised in various ways; e.g. as 1-hr daily max.; as 5-hr daily average; as 8-hr. daily average; as 24-hr daily average. In European Commission (1995a), E-R functions expressed in terms of 1-hr daily max (early evidence of ozone effects came from human chamber studies, with short-term exposures; and the early epidemiology was also mostly in terms of daily 1-hr max) had been used. More recent epidemiology has examined longer time-frames, because these may be more relevant biologically.

Now, in order to link with the modelling of incremental pollution, E-R functions expressed in terms of 6-hr daily average are given. This involved applying conversion factors to functions derived originally using different indices. Data from some of the European APHEA-study cities suggest, as a conversion factor, that 6-hr daily average ozone is approximately 0.8 times 1-hr daily maximum. Consequently any E-R function, expressed as percent change in health effect per ppb of 1-hr max ozone, was scaled upwards by a factor of 1.25 to give the equivalent percent change per ppb of 6-hr daily average.

### 5.4.2
### $SO_2$

In Chap. 4 of European Commission (1995a) no health effects associated with $SO_2$ as a gas were quantified; the evidence at that time was ambiguous. The APHEA results have however established an association of $SO_2$ with acute mortality, and probably with hospital admissions also, from which E-R functions can be derived. The evidence from epidemiological studies carried out in the USA is much less convincing.

It is however still unclear if these associations with $SO_2$ are causal: it may be that $SO_2$ is acting as a surrogate for other pollutants, especially fine particles (e.g. sulphates) not well quantified in the particle measurements available for study. However, in the APHEA studies the size of the apparent $SO_2$ effect does not depend on the background concentrations of ambient particles, suggesting that it is not a surrogate for PM. Therefore E-R functions for $SO_2$ were used as additive to those for particles and of ozone. This is the viewpoint also taken by COMEAP (1998).

### 5.4.3
### CO

There is relatively little epidemiological evidence concerning CO, so that it is difficult to place in context the results from a few (well-conducted) studies which report positive associations. Those studies do provide the basis for E-R functions, but they do not give strong guidance on how representative or transferable these functions are. Specifically, whereas in many studies CO is not examined as a possibly causative pollutant, there are also well-conducted studies which do consider CO and yet do not find a CO-related effect.

On present evidence, it is recommended that the functions for CO and acute hospital admissions for congestive heart failure be used; but that functions for CO and acute hospital admissions for ischaemic heart disease and acute mortality are only used for sensitivity analyses. In epidemiological studies the effects of CO are represented as 1-hr daily maximum and 8-hr daily average. Consequently, E-R functions expressed in terms of 1-hr max CO and 8-hr daily average have been scaled up by a factor of 2 and 1.5 respectively.

### 5.4.4
### $NO_2$

In European Commission (1995a), the epidemiological evidence regarding $NO_2$ was assessed. Some studies reported $NO_2$ effects. However, the broad thrust of the evidence then was that apparent $NO_2$ effects were best understood not as causal, but as $NO_2$ being a surrogate for some mixture of (traffic-related) pollution. It had been concluded that a direct effect of $NO_2$ should not be quantified (indirectly, $NO_x$ did contribute, as a precursor to nitrates and to ozone).

The APHEA results reported positive associations between $NO_2$ and daily mortality or respiratory hospital admissions in several European cities. Consequently, it is possible to propose E-R functions. APHEA however also supports the view that the apparent $NO_2$ effects may be due to particles; or at least, are highly dependent on background particle levels. Against this background, it is recommend that the E-R relationships for $NO_2$ are not used, except for sensitivity analyses. In the epidemiological studies from APHEA the impacts of $NO_2$ are described in terms of 1-hr maximum. Consequently, the E-R functions in terms of 1-hr maximum have been scaled up by a factor of 1.67 to express results in the metric of 24-hr average.

### 5.4.5
### Conversion Factors From ppb to Gravimetric Units

In order to express all functions per unit pollutant, in gravimetric units, the exposure-response functions are expressed in μg/m³. The conversion factors used were:

1 ppb $O_3$ = 1.997 μg/m³ of $O_3$;
1 ppb $NO_2$ = 1.913 μg/m³ of $NO_2$;
1 ppm CO = 1.165 mg/m³ of CO.

## 5.5
## E-R Functions for a European Implementation

### 5.5.1
### Can the Same Functions be Used throughout Europe, without Serious Error?

For particles (black smoke) and for $SO_2$, the APHEA results for acute mortality show important differences between estimated effects in 'Western' and 'Eastern' Europe, with percent change per unit exposure lower in the East than in the West. It is not sure why these differences occur; or how specific they may be to acute mortality. (There are not sufficient data for other endpoints with which to compare.) Some analyses of the effects of transport in Germany were carried out in an earlier phase of ExternE Transport, to check if results were sensitive to whether 'Western' or 'Eastern' Europe functions were applied to incremental pollution in the East. Results showed that it did not matter greatly whether E-R functions from, because most of the population at risk and incremental pollution is Western Europe. Consequently, 'West' European functions were used here. Studies do not show if the effects of ozone and (if real) of CO and of $NO_2$ vary by East-West Europe. The E-R functions proposed come from Western European cities.

### 5.5.2
### Transferring E-R Functions from North America

Epidemiological studies in the later 1990s pointed to differences in risks of acute effects (expressed as % change per μg/m³ pollutant) between Europe and North America. For acute mortality, the estimated effect of particles (expressed as or converted to $PM_{10}$, from APHEA studies conducted principally in terms of Black

Smoke) was generally lower in Europe than in the USA. (Very recent US results, by Dominici et al., in press, suggest that US risk estimates may after all be similar to those estimated from APHEA). The situation regarding ozone is less consistent. The estimated effect of ozone on acute mortality is much higher in Europe than the US, where evidence is from high-ozone regions only. The relative difference for respiratory hospital admissions is not nearly so great as for mortality.

Consequently, where practicable European E-R functions were used in the present study. Where it has been necessary to use US or Canadian E-R functions, in many instances the risk estimates were scaled downwards for particles, or upwards for ozone, for use in Europe: see Holland et al. (1999) for details. The magnitude of appropriate scaling is subjective, but we think it is better than no scaling at all, i.e. an implied scaling factor of one.

## 5.6 Thresholds

For many of these pollutants, there clearly is a threshold at the individual level; e.g. daily pollution will not lead to sudden death in healthy individuals. There is not however good evidence of a threshold at the population level; i.e. it appears that, for a large population even at low background concentrations, some vulnerable people are exposed some of the time to concentrations which do have an adverse effect. This understanding first grew in the context of ambient particles, where the 'no threshold' concept is now well established.

For European Commission (1995a) the limited quantification of ozone effects did include a threshold. This, however, was principally because of difficulties in ozone modelling, rather than on the basis of epidemiology. Overall, the APHEA results do not point to a threshold for the acute effects of ozone; and in the present study quantification of all health effects assumes no threshold.

## 5.7 'Chronic' Mortality: Effects as Estimated from Cohort Studies

There is controversy about whether longer-term exposure to ambient particles causes premature mortality in ways that are not captured by the time series studies of acute mortality and if so, whether and how the impacts can be estimated using results from a small number of cohort studies carried out in the USA. However the effects, if real, are the dominant ones among all health endpoints considered in this study, and so the issues are considered in relatively more detail than for other effects.

### 5.7.1
### Chronic Mortality and Particles: Is there a Causal Relationship?

Mortality effects of long-term exposure to air pollution have frequently been studied by examining relationships cross-sectionally between death rates and air pollution in cities. These studies generally reported positive associations for particles. Their major limitation lies in their ecological design; that is, aggregated measures are used for groups of individuals. This limits adjustment for confounding factors and in other ways also complicates interpretation. Nevertheless, the ecological cross-sectional studies showed consistently positive associations between ambient particles and mortality at different times and places (Lipfert, 1994).

More recently, the relationship of mortality to air pollution has also been studied in prospective cohort studies (Abbey et al., 1991; Dockery et al., 1993; Pope et al., 1995). These studies allow adjustment for possible confounding factors (e.g. smoking habit, gender, race, socio-economic status) on an individual basis. Both Dockery et al. (1993) and Pope et al. (1995) found associations between average particle concentrations and mortality from cardiopulmonary disease, and possibly lung cancer, but not deaths from other causes. More recent results from Abbey et al. also suggest some particulate effects. Together with other evidence, these studies do suggest that longer-term exposure to ambient particles does influence premature mortality in ways unaccounted for by time series studies. This was also the view of COMEAP (1995), though it remains disputed (Gamble, 1998).

### 5.7.2
### Choice of Study and E-R Function for Quantification

Pope et al. (1995) involves about 500,000 individuals in 151 US cities, compared with approximately 8,000 people in six cities studied by Dockery et al. (1993). This difference in scale reduces chances of confounding city differences in air pollution with other between-city differences. Also, Pope et al. give results that are intermediate between the higher estimates of Dockery et al. and the lower risks identified by Abbey and co-workers; and so risk estimates were based on Pope et al., despite the better pollution measurements in Dockery et al.

Pope et al. (1995) found that, after taking account of confounders, the adjusted mortality rate was 15%-17% higher in the most polluted areas compared to the least polluted. Risks are expressed as the % change in hazard (age-specific death rates) per $\mu g/m^3$ and are given in terms of $PM_{2.5}$ and of sulphates. These can in turn be re-expressed in terms of $PM_{10}$ using the usual conversion factors (e.g. Dockery and Pope, 1994).

When converted to $PM_{10}$ equivalent, i.e. expressed per $\mu g/m^3$ $PM_{10}$, the coefficient derived from $PM_{2.5}$ is about twice as high as that derived from sulphates. It appears that $PM_{2.5}$ is more likely than sulphates to capture the fractions of ambient PM which are most detrimental to health; and so risk estimates derived in terms of $PM_{2.5}$ are more likely to transfer well than estimates in terms of sulphates. Consequently, quantification was based on Pope et al.'s E-R function in terms of $PM_{2.5}$, and subsequently converted to $PM_{10}$ equivalent.

However that function was adapted and scaled downwards by a factor of three, to take account both of possible over-estimation of risk estimates in the original study through using recent rather than historical estimates of exposure and the apparently higher particle effects in time series studies in the USA compared with Europe.

### 5.7.3 Estimating Years of Life Lost (YOLL)

The years of life lost (YOLL) attributable to air pollution were estimated by linking the regression estimates from Pope et al. (1995) with the populations-at-risk and age-specific death rates in four European countries (Germany, Italy, Netherlands, UK), using life table methods. Effects were estimated separately for women and for men, for the currently-alive population in each of the four countries, using recent (1995) demographic data (numbers alive, annual death rates) in 1-year age-groups, up to age 94 inclusive, the latest years being extrapolated where necessary. The aim was to evaluate the effect on mortality of a 1-yr. increment or reduction of 10 $\mu g/m^3$ $PM_{10}$; i.e. with concentrations reverting to their original values after one year. This reversion is an unrealistic scenario in practice; but it permits comparability with other effects evaluated in the present study.

The life table applications involved following the 1995 population right through until everybody now alive had reached 94 years, under two sorts of scenario:

1. a baseline scenario assuming current hazard rates remain unchanged in future;
2. a changed scenario whereby hazards were modified to take account of a pollution change.

Differences in results between scenarios 1. and 2. were considered to be the pollution effect. This was summarised as life-years lost, scaled per 100,000 current live population, per 10 $\mu g/m^3$ $PM_{10}$ reduction. Because the pollution change was for one year only, there are no effects on as yet unborn cohorts.

Results varied according to the degree of latency assumed between the pollution change and the resultant change in age-specific death rates. The cohort studies are, strictly, uninformative about this latency. Various assumptions were examined, including immediate effects, and delays variously of 10, 20 and 30 years. (The estimated impacts reduce with degree of latency). For implementation, results close to an average between zero and 20-year latency were chosen.

Results from all four countries were similar. Consequently, a single set of values was used and applied in all countries studied. Without any scaling-down of the estimated $PM_{2.5}$ effect, this amounted to an estimated 470 YOLLs per 100,000 current population (all ages) per 1-yr. increase of 10 $\mu g/m^3$ $PM_{10}$. After scaling, these results were reduced by a factor of three.

### 5.7.4
### Estimating Costs

The corresponding costs were calculated before the results were summarised; i.e. economic valuation is linked with the detailed life table data, not with the summarised data. This enables discounting for how far in the future the life-year is lost. The same VLYL value was applied at all ages.

After scaling the estimated pollution effect downwards by a factor of three, the lost life years equal costs of 23.5 Million Euro at a discount of 0%, of 14 Million Euro at a discount of 3%, and of 6.3 Million Euro at a discount rate of 10%. These results are standardised for a population increment of 100,000 people (all ages), for a pollution increment of 10 $\mu g/m^3$ $PM_{10}$, and for a VLYL of 150,000 Euro. They were then scaled further in the course of implementation to take account of actual pollution increments, actual population sizes, and appropriate VLYL, the scaling in each dimension being a linear one.

## 5.8
## Other E-R Functions for the Classical Pollutants

### 5.8.1 Implementation

E-R functions for the classical pollutants were implemented also for many other endpoints, notably acute mortality, respiratory hospital admissions, cerebrovascular hospital admissions, cardiovascular hospital admissions, emergency room visits (ERVs), restricted activity days (RADs), acute effects in asthmatics, respiratory symptoms in the general population, and chronic respiratory disease. In terms of actual implementation, the various functions were used more-or-less directly, by appropriate linkage with incremental pollution and relevant population-at-risk, using the EcoSense program. Only with regard to chronic mortality were the more complex intermediate calculations, based on life-table methods, needed.

### 5.8.2 Acute Mortality

The E-R function for $PM_{10}$ and acute mortality was calculated as the mean of coefficients from Amsterdam (Verhoeff et al., 1996) and Köln (Spix and Wichmann, 1996) which were similar to high and low estimates respectively from the APHEA study. Estimates for ozone were derived from Barcelona (Sunyer et al., 1996); for $SO_2$ as the mean of Athens (Touloumi et al., 1996) and London (Anderson et al., 1996).

For sensitivity analyses the E-R function for $NO_2$ was obtained from the mean of Barcelona (Sunyer et al., 1996) and London (Anderson et al., 1996); for CO from Touloumi et al. (1994) in Athens.

### 5.8.3 Respiratory Hospital Admissions

The E-R relationship between $PM_{10}$ and respiratory hospital admissions was derived from data from Paris, Dab et al. (1996) within the APHEA study. Data from London (Ponce de Leon et al., 1996) from APHEA provided E-R for ozone, $SO_2$ and $NO_2$, the latter only included in sensitivity analyses.

### 5.8.4 Cardiovascular Hospital Admissions

Direct evidence of PM effects on cardiovascular hospital admissions is consistent with the longer-established mortality effects. The E-R function for $PM_{10}$ and cerebrovascular hospital admissions comes from Wordley et al. (1997) in Birmingham, UK. Schwartz and Morris (1995) reported results for Detroit on the relationship between cardiovascular admissions for congestive heart failure and ischaemic heart disease in those aged 65 or more. Exposure-response functions for $PM_{10}$ and CO in relation to admissions for congestive heart failure were derived from this study. Those functions relating to ischaemic heart disease are for sensitivity analysis only.

### 5.8.5 Emergency Room Visits (ERVs)

Because of differences in health care systems, ERVs apply much more in North America than to Europe, and so impacts are estimated for sensitivity analyses only. The E-R function for COPD and $PM_{10}$ was however derived from a European study: Sunyer et al. (1993) in Barcelona. A function for $PM_{10}$ and ERVs for asthma was derived from Schwartz et al. (1993) and Bates (1990), while that for $PM_{10}$ and ERVs for childhood croup was also European (Schwartz et al., 1991). Relationships for ERVs and ozone are from Cody et al. (1992), with baseline values from Bates et al. (1992)..

### 5.8.6 Restricted Activity Days (RADs)

As noted in European Commission (1995a), the evidence linking RADs quantitatively with air pollution is among the weakest of all endpoints, because the underlying studies are cross-sectional rather than longitudinal in nature. However no better studies had been published in the meantime. The E-R function for $PM_{10}$ and RADs was therefore based as before on Ostro (1987). Because particle effects estimated in the US may be higher than in Europe, the Ostro estimates were reduced to one half of their original value. Minor restricted activity days in relation to ozone were derived from Ostro and Rothschild (1989).

### 5.8.7 Acute Effects in Asthmatics

The health impacts on asthmatics of increased use of medication (bronchodilator usage), increase in respiratory symptoms, and cough were considered in two European studies (Dusseldorp et al. 1995, Roemer et al. 1993) and one US study, Pope and Dockery (1992). The estimate derived from the US has been adjusted downwards to a half of its original value. The exposure-response function for $O_3$ and asthma attacks is from Whittemore and Korn (1980).

### 5.8.8 Respiratory Symptoms in the General Population

The results from a large-scale study in California, USA (Krupnick et al., 1990) were used.

### 5.8.9 Chronic Morbidity in Adults

A PM effect on chronic bronchitis in adults is to be expected, given the evidence on chronic mortality; and there is some direct evidence. In European Commission (1995a) results from a cross-sectional study by Schwartz (1993) which was unsatisfactory in its linkage with valuation were used. More recently, and following Ostro, an E-R function from Abbey et al. (1995) was used. This gives new cases per year of chronic bronchitis per $\mu g/m^3$ for $PM_{10}$, and so links more easily with valuation, though there are issues of discounting which have not yet been thought through fully in the present implementation. Also, Abbey et al. is a study in a community of Seventh Day Adventists, with a distinctive lifestyle, and so results may not transfer well. Because of evidence suggesting that the estimated risks for particles (per $\mu g/m^3$ $PM_{10}$) are higher in the USA than in Europe, the risks as estimated by Abbey et al. were scaled down by a factor of two, for use in the present study.

### 5.8.10 Chronic Morbidity in Children

The exposure-response functions for chronic morbidity in children have been derived from Dockery et al. (1989). Note however that these are best understood as additional episodes of illness, and not the development of a chronic condition; and have been valued accordingly.

## 5.9 Additivity across Health Endpoints

Within endpoint, effects are considered as being additive across pollutants (particles, ozone, $SO_2$) unless otherwise stated; see earlier sections. Between endpoints for particles, assume that all respiratory hospital admissions (RHA), congestive heart failure admissions (CHF), and cerebrovascular admissions (CVA) also involve restricted activity days (RAD). In adjusting RADs to take account of hospital admissions, each admission was converted into equivalent hospital days, giving:

$$\text{Net RAD} = \text{RAD} - (\text{RHA} * 10) - (\text{CHF} * 7) - (\text{CVA} * 45). \quad (5.1)$$

This adjustment makes very little difference to the final RAD results. Between endpoints for ozone, all asthma attack days (AA) are also MRADs. The function for MRADs refers to adults (80%) only, while that of AA refers to all asthmatics (3.5%). Hence

$$\text{Net MRAD} = \text{MRAD} - (\text{AA} * 0.8 * 0.035). \quad (5.2)$$

As noted above, estimated mortality impacts from cohort studies also include acute mortality effects. Avoidance of double-counting is best achieved by not including time series acute mortality estimates when cohort study estimates are included. This has a much greater impact on final estimates than any adjustment to RADs.

## 5.10 Other Pollutants, especially Carcinogens: Methodology

### 5.10.1 Introductory Remarks

As well as the 'classical' pollutants considered above, other substances released into the atmosphere from transport sources were considered. These include aldehydes (including formaldehyde and acrolein); ethene and ethylene oxide; MTBE (methyl tertiary-butyl ether); VOCs (volatile organic compounds) in general, and benzene and 1,3 butadiene in particular; polycyclic aromatic hydrocarbons (PAHs); diesel exhaust particulates; and lead.

Both the effects of, and evidence for, these other pollutants are markedly different from those concerned with particles, $SO_2$, $NO_x$, CO and ozone. Firstly, we are dealing principally with chronic rather than with acute health effects; though, rarely, there may also be identifiable acute effects at ambient outdoor concentrations. (Acute effects at very high concentrations occur for several of the substances.) Secondly, the basis of epidemiological evidence is generally much weaker than for the classical pollutants, and usually involves extrapolation.

### 5.10.2 Objectives of the Evaluation

Health effects with the following specific objectives were evaluated:

- For each of the pollutants listed, to identify what health endpoints may be related;

- To indicate the strength of evidence for each health effect, at concentrations likely to be present in ambient air; and
- In the case of potential carcinogens, or other established health effects, to provide quantitative estimates of unit risks.

Results, guided by existing recent reviews, have been written up in a series of ExternE internal notes (Pilkington and Hurley, 1996, 1997). These include summaries of the results of a few key (occupational) studies to indicate the kind of information underlying these evaluations. The present text summarises the most important findings for ExternE Transport.

### 5.10.3 Unit Risk Factors for the Occurrence of Cancer

A unit risk factor (URF) is the estimated probability that a person of 'standard' weight of 70 kg will develop cancer due to exposure (by inhalation) to a concentration of 1 μg/m$^3$ of a pollutant over a 70-year lifetime. In the present study the effect of a 1-year increment in exposure is evaluated. It was assumed that a 1-year increment increases the lifetime risk of cancer by 1/70 of the URF, i.e. by 0.014 of the URF.

### 5.10.4 Reliability of Scientific Evidence Underlying URFs

As noted above, the scientific evidence usually consists of animal studies and some epidemiological studies of workers exposed to high concentrations. There are major methodological issues when using either occupational and/or animal studies for quantitative human risk assessment; see, e.g., US EPA (1996); or HEI (1995) with regard to diesels specifically. The overview by Nauss et al. in HEI (1995) is particularly useful. Issues to be considered include that:

- the reliability of risk estimates in occupational studies depends crucially on the reliability of estimated long-term exposures of the study subjects, and this can vary considerably;
- use for public health risk estimation requires also extrapolation both to low concentrations and to possibly more susceptible individuals;
- quantitative use of risk estimates from animal studies may also involve low-dose extrapolation, and quantitative animal-to-human scaling.

It follows that unit risk estimates depend not only on the availability and reliability of human and animal evidence on the carcinogenicity of the pollutant, but also on the models and assumptions used in extrapolating from that evidence, and more generally on the judgement of the group of experts making the classification. These difficulties have led to substantial diversity in the acceptance of quantified risk estimates for development of cancer. As a broad generalisation, quantified risk estimation for carcinogens is much more strongly established in the USA than in Europe. For example, in the UK the Department of Health's Expert

Committee on Cancer has chosen not to use quantified risk assessment in its evaluations.

The relative importance of human data compared to animal data varies considerably in the estimation of unit risk for the pollutants which are relevant here. Very briefly, the most reliable human data available are for benzene. There are reasonable human data available for diesel particulates, PAHs and 1,3 butadiene, though quantification of diesel risks based on the human studies is seriously limited by limitations in the exposure data of the principal studies.

### 5.10.5 The Approach of the US Environmental Protection Agency (EPA)

Among the URFs proposed by various expert groups, those proposed by the US Environmental Protection Agency (EPA) are commonly quoted. These will be used here as well for all substances under consideration, to ensure consistency. In extrapolating from high to low exposures, the EPA has traditionally used the linearised multistage model, which assumes no threshold, and (as its name suggests) a linear dose-response relationship in the low-dose region. It may therefore over-estimate risks at low exposures. However, the risks for a baseline exposure other than 1 $\mu g/m^3$ should simply be a scaling of the risks for 1 $\mu g/m^3$; and it does not matter whether the pollution to be evaluated is background (as in the definition of URF) or an increment (as in ExternE Transport). This is a great simplification.

## 5.11 Summary Evaluations of various Transport-Related Pollutants

### 5.11.1 Benzene

As an acute poison benzene produces narcotic effects. Regarding carcinogenicity, benzene has been shown to be clearly genotoxic (able to damage DNA) in both in vivo and in vitro experiments. Benzene is classified by IARC as Category 1, a known human carcinogen. However, risk quantification is complicated by lack of quantitative data, short follow up at low exposure concentrations, co-exposures to other potential carcinogens, and the fact that the body breaks down benzene to metabolites which seem to be more toxic than the parent substance. Individual variation in susceptibility or metabolism may therefore influence the risk at any given exposure.

There are many occupational studies investigating exposure to benzene and development of cancer, especially leukaemia. The US Environmental Protection Agency (EPA) risk assessment for benzene gave a unit risk factor of $8 \cdot 10^{-6}$ per $\mu g/m^3$ (US EPA, 1990). Many different risk estimates have been derived, using different assumptions about the pattern of exposures, the shape of the E-R function, and so the extrapolation to low concentrations. These are similar to the estimates of Crump (1994), giving a range of lifetime risk of leukaemia, at an airborne concentration of 1 $\mu g/m^3$, of $4.4 \cdot 10^{-6}$-$7.5 \cdot 10^{-6}$.

There is no convincing evidence of chronic non-cancer effects at ambient concentrations.

### 5.11.2
### 1,3 Butadiene

1,3 butadiene is potentially carcinogenic to both the white and red cell systems. Animal studies have shown that it is carcinogenic both in rodents and in mice. There is however wide discrepancy in metabolism between different species, complicating extrapolation to humans. Although the available animal evidence for 1,3-butadiene, and comparison with substances of similar chemical structure would support the classification of butadiene as a human carcinogen, the available human data is limited; and 1,3-butadiene is classified by IARC as Category 2a - Probable human carcinogen (IARC Monographs Volume 54, 1992). Irritant effects also occur, but only at concentrations much higher than those relevant to ExternE Transport.

1,3-butadiene is a major ingredient of synthetic rubber, and being volatile, the route of absorption is primarily inhalation. The epidemiological evidence consists mostly of mortality studies which use qualitative estimates or exposure categories rather than estimates of actual lifetime exposures, and with limited consideration of other workplace exposures. There is no evidence available on cancer risks to the general population from ambient exposures. The human studies cannot be used directly in quantified risk assessment because sufficiently reliable estimates of past exposures are not available. Thus, the US EPA (1990) unit risk factor (URF) of 3 (or 2.8)$\cdot 10^{-4}$ per $\mu g/m^3$ lifetime exposure is based on multi-stage modelling of animal (mice) experimental data. An updated estimate by RIVM (1994) of $0.7$-$1.7 \cdot 10^{-5}$ is much lower. However, early sensitivity analyses within ExternE indicated that even a 20-fold difference in estimated URF for 1,3 butadiene has little impact on final estimated costs.

### 5.11.3
### Polycyclic Aromatic Hydrocarbons (PAHs)

These are ring compounds resulting from the incomplete combustion of organic material and which jointly share carbon atoms. They include a wide range of substances including benzo[a]pyrene (BaP). The relationship between BaP and other PAHs differs for various types of emission, but has been shown to be relatively similar in the ambient air of several towns and cities.

There is strong evidence, including from epidemiological studies (e.g. Redmond et al., 1972; Hurley et al., 1983; Armstrong et al., 1994) to suggest that certain components of PAHs, and specifically benzo[a]pyrene, are carcinogenic in humans; and that nitroaromatics as a group pose a hazard to health. In 1986 IARC and the US National Cancer Institute concluded that PAHs were a risk factor for lung cancer in humans. Benzo[a]pyrene specifically, rather than PAHs as a group, is labelled as a probable human carcinogen.

As these compounds form complex mixtures and are also absorbed onto particulates, it is difficult to quantify levels of human exposure and so is difficult to estimate risks reliably. Benzo[a]pyrene is the only PAH for which a suitable database is available, allowing quantitative risk assessment. The EPA unit risk factor of lung cancer for BaP is $1 \cdot 10^{-4}$ per ng/m$^3$ (US EPA, 1990). Limitations in the use of benzo[a]pyrene as an indicator of PAH toxicity in air pollution are that some PAH is bound to particulates, and that some of the gaseous components are not included. WHO (1987) estimated a URF of $8.7 \cdot 10^{-5}$ per ng/m$^3$; i.e. almost identical to that used by US EPA.

### 5.11.4 Diesel Exhaust Particulates

Diesel exhaust contains a complex mix of gases and particles (including CO, $NO_x$, $SO_2$, and PAHs). The composition varies considerably and has changed dramatically over the last 30 years. The carbon core of diesel particles serves as a nucleus for the condensation of organic compounds, for example PAHs, which are by-products of combustion and which total 10-15% of the particle by mass. About 20-30% of particles with an aerodynamic diameter less than 0.25µm will be deposited in the lungs.

Rats develop an increased incidence of lung tumours when exposed chronically to high concentrations of diesel exhaust that overload particle clearance defences. There is growing evidence that the tumours may be because of the particle core rather than the organic constituents. Diesel soot has been shown to be a direct acting mutagen, i.e. able to cause changes in genetic material without being activated by the body's metabolism. The available human (occupational) studies are limited both by poor exposure data and limited consideration of confounders, for example environmental tobacco smoke, diet and socio-economic factors. However, the results of studies published over the past 15 years consistently show a weak association between exposure to diesel exhaust and lung cancer. There is some evidence of excess bladder cancer also.

Diesel emissions are currently classified by IARC as Category IIa; probably carcinogenic to humans. For diesel particulates, the EPA (1994) unit risk estimate derived using a linearised multistage model is $3.4 \cdot 10^{-5}$ based on a lifetime exposure to 1 µg/m$^3$ diesel particulate matter. As discussed by Nauss et al. (1995), the US EPA in 1990 based its risk calculations on the risk of benzo[a]pyrene (BaP), because scientific opinion at that time considered that organic constituents of diesel exhaust were responsible for observed carcinogenicity in rats. The EPA has since decided to use the non-organic carbon core of particles. A range of risk estimates is summarised by Nauss et al. (1995), showing that estimates based on animal data vary by as much as two orders of magnitude, depending on the assumptions made about exposure and the extrapolation models selected.

Respiratory irritation and airway hyper-reactivity can also be caused by several of the components of diesel exhaust. There are however no usable E-R functions, due to the variable and complex mixture of substances. There is very limited evidence of an association with asthma, and little evidence for other (non-pulmonary) health effects.

### 5.11.5 Formaldehyde

Formaldehyde is highly water soluble and most of inhaled formaldehyde is deposited in the lining of the nose. It is a potent irritant, and no clear threshold has been defined for these effects. Available evidence suggests that ambient levels of formaldehyde could produce irritant symptoms to the eyes and respiratory tract in a sub-group of the general population. It is unlikely to cause asthmatic symptoms in healthy subjects at exposures encountered in environmental settings, but could potentially exacerbate symptoms. Thus, an occasional mild effect (e.g. symptom day) among sensitive people cannot be ruled out where incremental formaldehyde from transport adds to existing relatively high background levels. However, effects are likely to be small and are difficult to quantify; and no E-R functions are proposed here.

Formaldehyde is classified as IARC Category 2A, probable human carcinogen (IARC Monographs, Volume 62, 1995). There is however no convincing evidence of an effect at low ambient exposures; and possible mechanisms suggest that in the absence of damage to the respiratory tract tissue, any cancer risks at low ambient concentrations are negligible (WHO, 1997, in press). The Environmental Protection Agency (EPA) risk assessment for formaldehyde gave a unit risk factor of $1 \cdot 10^{-5}$ per $\mu g/m^3$ (US EPA, 1990). This URF may substantially over-estimate the true risks from ambient pollution; it is recommend to be used for sensitivity analyses only.

There is limited evidence that formaldehyde may contribute to the development of asthma, especially where there is co-exposure to environmental tobacco smoke (Krzyzanowski et al., 1990). However, effects are not well-established and are not quantifiable reliably.

### 5.11.6 Acetaldehyde and Other Aldehydes

Irritant effects are reported as for formaldehyde, although the threshold for onset and incremental effects is less well documented than for formaldehyde. Acetaldehyde is classified as IARC category 2B (possible human carcinogen), and acrolein as Category 3, due to inadequate evidence for carcinogenicity in humans and animals at the present time. The Environmental Protection Agency (EPA) risk assessment for acetaldehyde gives a unit risk factor of $2 \cdot 10^{-6}$ per $\mu g/m^3$ (US EPA, 1990). Given the limited available evidence, it is proposed that this not be used.

### 5.11.7 Ethylene Oxide/ Ethene

Ethene is metabolised in both animals and humans to ethylene oxide. However this metabolic pathway becomes saturated, and it is estimated that around 5% of inhaled ethene is metabolised to ethylene oxide. Short term high exposures to ethylene oxide cause irritant effects to the eyes, nose, throat, and lungs, with associated headache, nausea, and in-coordination. These are not relevant to ExternE Transport.

Ethene is classified as IARC Category 3 (not carcinogenic to humans). Ethylene oxide is classified as IARC category 2A (probable human carcinogen). The Environmental Protection Agency (EPA) risk assessment for ethylene oxide gave a unit risk factor of $1 \cdot 10^{-4}$ per $\mu g/m^3$ (US EPA, 1990). Given the weakness of evidence, it may be doubted that the cancer effect is real; and it is recommended that this estimate be used for sensitivity analysis only.

### 5.11.8
### MTBE (Methyl Tertiary-Butyl Ether)

MTBE is an oxygenated fuel additive whose use is likely to increase over the next few decades. Information on the health effects of MTBE is derived almost exclusively from studies in rats; in humans, there is limited information even on typical exposures. There is no evidence that MTBE may be carcinogenic.

### 5.11.9
### Volatile Organic Compounds (VOCs), as a Mixture

This is the name given to a wide range of volatile hydrocarbons, for example alkylbenzene, toluene, alkenes, formaldehyde and trichloroethane. VOCs are categorised by the World Health Organisation into four categories on the basis of their boiling point range, with no sharp cut off between the categories. Several of these have been considered individually above; now VOCs are briefly considered as a mixture.

At present there is no standardised way to summarise the combined effects of the many different compounds in the atmosphere. A tentative dose-response relationship has been suggested by Mølhave (1991), based on studies of indoor environments. However, nasal and eye irritation have been reported during episodes of photochemical pollution, and it is possible that certain VOCs contribute. It is unlikely that the acute health effects reported to be associated with VOCs would occur due to sole exposure to ambient levels of VOCs within the normal range. Their presence could have some effect if additive to the effects of other irritants. The consequences are likely to be small, and are not quantifiable.

Some VOCs are categorised as probable human carcinogens by IARC; see the separate evaluations, above. Due to the varying composition of VOCs and current limitations on monitoring, there is insufficient data to provide a reliable estimate of the risk of cancer from exposure to a mixed airborne level of VOCs. Consequently, risk estimates have not been calculated for VOCs as a group.

### 5.11.10
### Lead

There is no consistent evidence that lead is a carcinogen. Inorganic lead is currently classified by IARC as a possible human carcinogen (Category 2B). For organic lead there is as yet inadequate evidence for categorisation as a human carcinogen (Category 3). It was not attempted to quantify a carcinogenic effect.

However, several non-cancer health effects have been proposed or examined. The main evidence concerns lead in air and childhood IQ in children. The pathway is in two parts, which need to be considered together:

1. the relationship between lead in blood and childhood IQ; and
2. the relationship between lead in air and lead in blood.

IQ as an endpoint is a very non-specific measure, and there is controversy about what it really represents. However, relationships with blood lead are better established with IQ, and are less variable across studies, than with more specific endpoints.

- Lead in blood, and childhood IQ: For this part of the E-R function, consider an estimated reduction based on Schwartz (1994) of:

  2.57 IQ points per 10 μg/dl blood lead; in school-age children.

  For definiteness, assume the relationship refers to children at age 10. (In practice, it is based on studies of school-children over a range of ages). Following the results of various meta-analyses, no threshold is assumed.

- Lead in air and lead in blood, in children: This is not straightforward. Especially in children, ingestion is the principal route of exposure. A recent UK review (EPAQS, 1998) uses a conversion factor of:

  1 μg/m3 in air ~ 5 μg/dl in blood and so ~ 1.25 IQ points.

- Discounting/ time of occurrence: The adverse effects of exposure to lead will be greatest when the child's nervous system is developing most rapidly. Assume that the biologically relevant exposure occurs over three years, aged 0-2 inclusive. (There will be pre-natal exposure also; and presumably exposure after age three may have some effect. So clearly, this is a simplification). This implies a time-lag of about nine years, between exposure (at 18 months) and effect (at 10.5 years), which is relevant for discounting.

Several studies have considered the possible relationship between blood lead levels and blood pressure. A meta-analysis of 19 studies by Staessen et al. (1993) reported that a two-fold increase in blood lead was associated with 1 mmHg increase in systolic and a 0.7 mmHg increase in diastolic blood pressure. This suggests a significant weak positive association. Even if real, this is not likely to have important public health implications; and in any case the causative role of lead is considered unlikely. There is still insufficient evidence to implicate low level lead exposure as a cause of adult hypertension, as poor exposure data and inadequate consideration of confounding factors limit findings both in occupational and population based studies. So again, quantification is not attempted.

### 5.11.11
### Platinum and Related Compounds

Platinum is used in vehicle exhaust catalytic converters to control air pollution. There is no body of evidence that platinum is a carcinogen. Variable results have been shown on skin irritancy tests in animals dependent on the type of platinum salt used. Allergic effects in humans as a result of inhalation of platinum salts were first documented in 1804. The main evidence is from occupational studies (e.g. Hunter et al., 1945; Roberts, 1951; Levene, 1971). It is unclear what if any of these effects might occur at ambient concentrations.

## 5.12 Some Remarks on YOLL and Latency

The above substances are considered as initiators of cancer. For those who die from cancer as principal cause of death, the average YOLL are estimated as the difference between the expected age of death in the absence of cancer, conditional on having survived to age of diagnosis of cancer and the average age of death among those who die from the cancer. For each type of cancer endpoint identified in the review of transport pollutants, average YOLL are given accounting for age of peak incidence of the cancer in the general population, average latency and average survival rate. These are by no means definitive scientific assessments. Rather, they are indicative figures only, based on general understanding, to be used as a guide when considering economic assessment, including the effect of discounting.

## 5.13 Concluding Remarks

In conclusion, it is considered that this quantification framework gives a suitable set of E-R functions for the estimation of health impacts. In particular, compared to previous studies now a set of E-R functions is available which draws heavily on European studies; and the use of life-table methods in estimating mortality impacts from cohort studies is consolidated.

Results have shown that many of the individual pollutant-outcome E-R functions have little impact on final results. This has guided the extent to which it was attempted to fine-tune the values provided and identified priorities for further research in improving the present estimates.

The main omission is in not taking account of the acute effects of high, very short-term concentrations of CO. The functions for this impact pathway are non-linear and depend on background levels; and so it is not straightforward to derive an implementable set.

There continues to be a great amount of new epidemiological and other studies illuminating not only exposure-response relationships between ambient air pollution and adverse health, but also possible mechanisms of action. This continuing information brings with it a need to keep under review not only what specific functions might be considered best, but also the strategic judgements which underlie the present quantification of impacts.

**Table 5.1.** Quantification of human health impacts. The exposure-response slope, $f_{er}$, has units of cases events per year per person per μg/m³, except for mortality which is expressed as percentage increase per μg/m³

| Receptor | Impact Category | Reference | Pollutant | $f_{er}$[a] |
|---|---|---|---|---|
| ASTHMATICS (3.5% of population) | | | | |
| Adults | Bronchodilator usage | Dusseldorp et al., 1995 | $PM_{10}$ | 0.163 |
| | | | Nitrates | 0.163 |
| | | | $PM_{2.5}$ | 0.272 |
| | | | Sulphates | 0.272 |
| | Cough | Dusseldorp et al., 1995 | $PM_{10}$ | 0.168 |
| | | | Nitrates | 0.168 |
| | | | $PM_{2.5}$ | 0.280 |
| | | | Sulphates | 0.280 |
| | Lower respiratory symptoms (wheeze) | Dusseldorp et al., 1995 | $PM_{10}$ | 0.061 |
| | | | Nitrates | 0.061 |
| | | | $PM_{2.5}$ | 0.101 |
| | | | Sulphates | 0.101 |
| Children | Bronchodilator usage | Roemer et al., 1993 | $PM_{10}$ | 0.078 |
| | | | Nitrates | 0.078 |
| | | | $PM_{2.5}$ | 0.129 |
| | | | Sulphates | 0.129 |
| | Cough | Pope and Dockery, 1992 | $PM_{10}$ | 0.133 |
| | | | Nitrates | 0.133 |
| | | | $PM_{2.5}$ | 0.223 |
| | | | Sulphates | 0.223 |
| | Lower respiratory symptoms (wheeze) | Roemer et al., 1993 | $PM_{10}$ | 0.103 |
| | | | Nitrates | 0.103 |
| | | | $PM_{2.5}$ | 0.172 |
| | | | Sulphates | 0.172 |
| All | Asthma attacks (AA) | Whittemore and Korn, 1980 | $O_3$ | 4.29E-3 |
| ELDERLY 65+ (14% of population) | | | | |
| | Congestive heart failure | Schwartz and Morris, 1995 | $PM_{10}$ | 1.85E-5 |
| | | | Nitrates | 1.85E-5 |
| | | | $PM_{2.5}$ | 3.09E-5 |
| | | | Sulphates | 3.09E-5 |
| | | | CO | 5.55E-7 |
| CHILDREN (20% of population) | | | | |
| | Chronic cough | Dockery et al., 1989 | $PM_{10}$ | 2.07E-3 |
| | | | Nitrates | 2.07E-3 |
| | | | $PM_{2.5}$ | 3.46E-3 |
| | | | Sulphates | 3.46E-3 |

| Receptor | Impact Category | Reference | Pollutant | $f_{er}$[a] |
|---|---|---|---|---|
| ADULTS (80% of population) | | | | |
| | Restricted activity days (RAD)[b] | Ostro, 1987 | $PM_{10}$ | 0.025 |
| | | | Nitrates | 0.025 |
| | | | $PM_{2.5}$ | 0.042 |
| | | | Sulphates | 0.042 |
| | Minor restricted activity days (MRAD)[c] | Ostro and Rothschild, 1989 | $O_3$ | 9.76E-3 |
| | Chronic bronchitis | Abbey et al., 1995 (after scaling down: see text) | $PM_{10}$ | 2.45E-5 |
| | | | Nitrates | 2.45E-5 |
| | | | $PM_{2.5}$ | 3.9E-5 |
| | | | Sulphates | 3.9E-5 |
| ENTIRE POPULATION | | | | |
| | Chronic Mortality (CM) | Pope et al., 1995 (after scaling down: see text) | $PM_{10}$ | 0.129% |
| | | | Nitrates | 0.129% |
| | | | $PM_{2.5}$ | 0.214% |
| | | | Sulphates | 0.214% |
| | Respiratory hospital admissions (RHA) | Dab et al., 1996 | $PM_{10}$ | 2.07E-6 |
| | | | Nitrates | 2.07E-6 |
| | | | $PM_{2.5}$ | 3.46E-6 |
| | | | Sulphates | 3.46E-6 |
| | | Ponce de Leon, 1996 | $SO_2$ | 2.04E-6 |
| | | | $O_3$ | 3.54E-6 |
| | Cerebrovascular hospital admissions | Wordley et al., 1997 | $PM_{10}$ | 5.04E-6 |
| | | | Nitrates | 5.04E-6 |
| | | | $PM_{2.5}$ | 8.42E-6 |
| | | | Sulphates | 8.42E-6 |
| | Symptom days | Krupnick et al., 1990 | $O_3$ | 0.033 |
| | Cancer risk estimates | Pilkington et al., 1997; based on US EPA evaluations | Benzene | 1.14E-7 |
| | | | Benzo-[a]-Pyrene | 1.43E-3 |
| | | | 1,3-butadiene | 4.29E-6 |
| | | | Diesel particles | 4.86E-7 |
| | Acute Mortality (AM) | Spix et al. / Verhoeff et al., 1996 | $PM_{10}$ | 0.040% |
| | | | Nitrates | 0.040% |
| | | | $PM_{2.5}$ | 0.068% |
| | | | Sulphates | 0.068% |
| | | Anderson et al. / Touloumi et al., 1996 | $SO_2$ | 0.072% |
| | | Sunyer et al., 1996 | $O_3$ | 0.059% |

[a] Sources: (European Commission, 1995a) and (Hurley et al., 1997).

[b] Assume that all days in hospital for respiratory admissions (RHA), congestive heart failure (CHF) and cerebrovascular conditions (CVA) are also restricted activity days (RAD). Also assume that the average stay for each is 10, 7 and 45 days respectively.
Thus, net RAD = RAD - (RHA*10) - (CHF*7) - (CVA*45).

[c] Assume asthma attacks (AA) are also minor restricted activity days (MRAD), and that 3.5% of the adult population (80% of the total population) are asthmatic. Thus, net MRAD = MRAD - (AA*0.8*0.035).

**Table 5.2.** Human health E-R functions for sensitivity analysis only. The slope, $f_{er}$, has units of events per year per person per $\mu g/m^3$, except for mortality expressed as percentage increase per $\mu g/m^3$.

| Receptor | Impact Category | Reference | Pollutant | $f_{er}$[a] |
|---|---|---|---|---|
| ELDERLY, 65+ (14% of population) | | | | |
| | Ischaemic heart disease | Schwartz and Morris, 1995 | $PM_{10}$ | 1.75E-5 |
| | | | Nitrates | 1.75E-5 |
| | | | $PM_{2.5}$ | 2.92E-5 |
| | | | Sulphates | 2.92E-5 |
| | | | CO | 4.17E-7 |
| ENTIRE POPULATION | | | | |
| | Respiratory hospital Admissions (RHA) | Ponce de Leon et al., 1996 | $NO_2$ | 1.40E-6 |
| | ERV for COPD | Sunyer et al., 1993 | Nitrates | 7.20E-6 |
| | | | $PM_{10}$ | 7.20E-6 |
| | | | Sulphates | 1.20E-5 |
| | | | $PM_{2.5}$ | 1.20E-5 |
| | ERV for asthma | Schwartz, 1993 and Bates, 1990 | Nitrates | 6.45E-6 |
| | | | $PM_{10}$ | 6.45E-6 |
| | | | Sulphates | 1.08E-5 |
| | | | $PM_{2.5}$ | 1.08E-5 |
| | | Cody, 1992 and Bates, 1990 | $O_3$ | 1.32E-5 |
| | ERV for croup in pre school Children | Schwartz et al., 1991 | Nitrates | 2.91E-5 |
| | | | $PM_{10}$ | 2.91E-5 |
| | | | Sulphates | 4.86E-5 |
| | | | $PM_{2.5}$ | 4.86E-5 |
| | Cancer risk estimates | Pilkington et al., 1997 | Formaldehyde | 1.43E-7 |
| | | | Ethylene oxide | 1.43E-6 |
| | Acute Mortality (AM) | Touloumi et al., 1994 | CO | 0.00145% |
| | | Sunyer et al./Anderson et al., 1996 | $NO_2$ | 0.034% |

[a] Sources: (European Commission, 1995a) and (Hurley et al., 1997).

# 6 Impacts on Building Materials

It has been known for several centuries that air pollutants emitted by burning of fossil fuels have a serious impact on buildings. The effects include loss of mechanical strength, leakage and failure of protective coatings due to degradation of materials. Also, the disagreeable appearance in all larger towns of soiled but otherwise beautiful buildings is caused by deposition of particulate matter arising from atmospheric pollution.

For several materials that are frequently used in buildings dose-response functions have been obtained. A dose-response function links the dose of pollution, measured in ambient concentration and/or deposition, to the rate of material corrosion. They are of outmost importance for development of systems for classification of corrosivity of environments, for mapping of areas with increased risk of corrosion, and for calculation of cost of damage caused by deterioration of materials. In order to be able to calculate costs a damage function needs to be obtained. A physical damage function links the rate of material corrosion (due to the pollution exposure given by the dose-response function) to the time of replacement of maintenance of the material. Performance requirements determine the point at which replacement or maintenance is considered to become necessary. If the performance requirements can be described in terms of a critical degradation level, it is possible to transform a dose-response function into a damage function. The mentioned definitions of dose-response and damage functions are the official terminology of UN ECE Workshop (1997). In addition, the term exposure-response function has been used within the ExternE project. It refers to a function expressing the maintenance frequency in terms of the critical damage and pollution parameters. It may be regarded as a special damage function where frequency instead of time is the explained variable.

In the following sections the state of the art of impacts of atmospheric pollutants on building materials is briefly summarised. The description is focused on dose-response functions, their transformation into exposure-response functions and interpretation for calculating costs. The text is divided in two sections dealing with degradation and soiling separately and one concluding section describing the combined effects. Soiling is only one of the effects of particulate matter, which play an important role in the atmospheric corrosion process. However, these other effects are at present stage not sufficiently quantifiable.

## 6.1
## Degradation of Building Materials

There are many parameters that can influence the damage to materials, it is an interplay between chemical, physical and biological parameters. The present study will focus on the aspects specific to the urban situation, i.e., man-made pollutants and their interplay with natural climatic factors. It is, however, important to recognise that corrosion is a process that occurs even in the absence of pollutants and it is important to quantify to which extent urban conditions affects and accelerates the "natural" or background corrosion of materials.

Sulphur and nitrogen compounds including secondary pollutants and particulates are the most important pollutants acting as corrosive agents. It has clearly been demonstrated that pollutants enhance the natural corrosion process for several both metallic and non-metallic materials. Systematic laboratory exposures in the 1930's demonstrated the corrosive effect of $SO_2$ on metals. This was later also proved by field exposures and $SO_2$ was for a long time considered to be the main corrosive pollutant. Today, $SO_2$ is no longer regarded as the only important corrosion stimulator. Instead, its effect in combination with other gaseous pollutants such as $NO_2$, $O_3$, and their reaction products needs to be considered.

### 6.1.1
### Damage Mechanisms

Two types of deposition processes are recognised in atmospheric corrosion, dry deposition and wet deposition, depending on the way pollutants are transported from the atmosphere to the corroding surface. Wet deposition refers to precipitation whereas dry deposition refers to the remaining processes, including gas-phase deposition and particle deposition. For sheltered conditions, wet deposition and run-off are excluded and the corrosion, i.e., the destruction of the base material, is identical to the formation of the corrosion products. Different climatic parameters influence the deposition rates and are also important in the degradation process.

#### *6.1.1.1*
#### *Climatic Effects*

Wind and water can cause erosion and temperature fluctuations result in freezing-thawing cycles. The volume expansion of water turning to ice results in significant stresses to any porous material. Sun radiation can also be a direct factor, especially for the degradation of polymers. Climate has a significant role in degradation of materials, hence the term "weathering".

The dry deposition of pollutants is greatly influenced by temperature and relative humidity since they are the main factors that determine the presence of moisture in absence of rainfall. The time of wetness (TOW) is a commonly used concept for inorganic materials, in particular for metals, that refers to the time when corrosion occurs, i.e., when a moisture layer is present. The degradation process may be considered as discontinuous, only occurring when the surface is sufficiently wet. This concept is useful when describing the degradation process and when classifying different climatic regions from a corrosion point of view but is difficult to calculate from readily available meteorological data. Therefore, recently developed dose-response functions include annual averages of the more easily available parameters temperature and relative humidity.

### 6.1.1.2 Dry Deposition Effects

Dry deposition refers to the process by which particles and gases are transferred from the atmosphere to the material surface. The deposition velocity, i.e., the ratio of surface flux to concentration for a particular gas or particle, depends not only on the conditions in the atmosphere but also on the thickness of the moisture layer, the reactivity of the material and the properties of the corrosion products.

Sulphur dioxide is one of the main contributors to the degradation of materials. $SO_2$ is dissolved in the moisture layer forming sulphite and, after oxidation, sulphate. This process results in an acidification of the moisture layer, which enhances the corrosion process. $SO_2$ is partly oxidised in the atmosphere and contributes therefore to the wet deposition acidity. Sulphate is also frequently present in corrosion products. The $SO_2$ deposition rate depends mostly on the material, it is often higher for materials sensitive to degradation, and varies between 0.01 and 2 cm/s.

The role of nitrogen oxides in the degradation of materials has not yet been fully elucidated. Nitrogen is mainly emitted as NO formed in the combustion process and then further oxidised to $NO_2$ and $HNO_3$ by photochemical reactions and possibly neutralised by $NH_3$ forming particulate nitrates ($NH_4NO_3$). Nitrates are not frequently found in the corrosion products. For unsheltered surfaces this can be explained by the high solubility of many nitrate precipitates. Even if $NO_2$ in itself is much less detrimental to most materials it can contribute to the acidity of precipitation in a similar way as $SO_2$.

Ozone ($O_3$) is a principal pollutant in the past mainly associated with the degradation of natural rubber but most organic materials containing carbon double bonds, such as painted surfaces, polymers and textiles are sensitive to its effect. It is, however, a general oxidant and thus, for inorganic materials $O_3$ has a synergistic effect in combination with $SO_2$ as has $NO_2$.

Particles containing $NH_4NO_3$, $(NH_4)_2SO_4$ and $NH_4HSO_4$ play an important role in atmospheric corrosion and this is related to their ability to increase the time of wetness due to their hygroscopic properties. In addition to the prolonging of the time of wetness, ionic particles enhance the corrosion by providing corrosion stimulators. However, the present dose-response functions do not include the effects of particles as a separable part.

### 6.1.1.3 Wet Deposition Effects

The effect of wet deposition can be either detrimental or beneficial, depending on the conditions. The wet deposition has two effects on the corrosion process. On one hand, it transports chemically active compounds present in the rain to the surface, thereby increasing the corrosivity of the moisture layer. On the other hand, it washes away chemically active compounds previously deposited on the surface, with the opposite effect. Thus, for a specific material and environment choosing a sheltered exposure condition rather than unsheltered may, or may not, increase the corrosion rate. When the rain acidity is high the detrimental effect often dominates and the effect is usually quantified in terms of total acid load.

### 6.1.2 Dose-Response Functions

During the last decades a number of field exposure programmes have been performed. Most of the exposures have been confined to relatively small geographical areas and express the corrosion/degradation on a local or regional basis. Others can be classified as national exposure programmes, where the variation in meteorological parameters may be more extensive. Finally there are a few international exposure programmes, which cover extensive geographical areas. The weak point of most studies is usually the quality of environmental data, which is often inferior to that of the material degradation data. For many materials numerous dose-response functions have been reported. It is not the intention of the present report to give a full review of all those functions. Instead, recent functions for selected materials obtained within the UN ECE International Co-operative Programme on Effects on Materials, including Historic and Cultural Monuments (ICP Materials) is shortly presented and discussed for the purpose of performing cost calculations. A short description of the programme is also included. For further reference see Tidblad et al. (1998). One important task for the programme has been to estimate the relative contribution of dry and wet deposition to the degradation of materials. Therefore, and also because it makes sense from a mechanistic point of view, the dose-response relations are of the form where the corrosion attack, K, is described in terms of dry and wet deposition effects separated as additive terms

$$K = K_{dry} + K_{wet} \tag{6.1}$$

The dry deposition term is quantified in terms of the parameters $SO_2$, relative humidity and temperature whereas the wet deposition is quantified in terms of total amount of precipitation and precipitation acidity.

#### *6.1.2.1 ICP Materials*

The aim of ICP Materials is to perform a quantitative evaluation of the effect of sulphur pollutants in combination with $NO_x$ and other pollutants as well as climatic parameters on the atmospheric corrosion of important materials. This is achieved by measuring gaseous pollutants, precipitation and climate parameters at or nearby each test site and by evaluating the corrosion effects on the materials. A Task Force is organising the programme with Sweden as lead country and the Swedish Corrosion Institute serving as the Main research centre. Subcentres in different countries have been appointed, each responsible for their own group of materials. The materials included are:

- *Structural metals* including steel, weathering steel, zinc, aluminium, copper and bronze
- *Stone materials* including Portland limestone and White Mansfield dolomitic sandstone
- *Paint coatings* including coil coated steel with alkyd melamine, steel with alkyd paint, wood with alkyd paint system and wood with primer and acrylate
- *Electric contact materials* including nickel, copper, silver and tin as coupons, and Eurocard connectors of three different performance classes
- *Glass materials* including potash-lime-silica glasses M1 (sensitive) and M3 representative of medieval stained glass windows
- *Polymeric materials* including polyamide and polyethylene

Being the most extensive materials exposure programme, the results of ICP Materials not only confirm the corrosive effect of $SO_2$ but also enable quantification for a wide range of materials. For most unsheltered materials also the effect of wet deposition (acid precipitation) has been quantified and comprises the second most important contribution to the corrosion rate. For selected materials the effect of ozone ($O_3$) and ($NO_2$) have been demonstrated.

ICP Materials is still going on. A finalised part, on which the present description is based, is the extensive 8-year field exposure programme that was started in September 1987 and involved 39 exposure sites in 12 European countries and in the United States and Canada. Dose-response functions have been obtained for many of the included materials, some of which are described in the following. Table 6.1 shows all parameters included in the final dose-response relations. In general, care shall be taken when extrapolating the equations outside the intervals of environmental parameters used for their calculations.

**Table 6.1.** Parameters used in final ICP Materials dose-response functions including symbol, description, interval measured in the programme and unit. All parameters are expressed as annual averages.

| Symbol | Description | Interval | Unit |
|---|---|---|---|
| t | Time | 1-8 | year |
| T | Temperature | 2-19 | °C |
| Rh | Relative humidity | 56-86 | % |
| [$SO_2$] | $SO_2$ concentration | 1-83 | $\mu g/m^3$ |
| Rain | Rainfall | 327-2144 | mm |
| [H+] | H+ concentration | 0.0006-0.13 | mg/l |

#### 6.1.2.2
#### Stone Materials

Two stone materials have been exposed within ICP Materials, Portland limestone and white Mansfield sandstone. The function obtained for limestone is

$$R = (2.7\,[SO_2]^{0.48}\,e^{-0.018\,T} + 0.019\,Rain\,[H^+]) \cdot t^{0.96} \tag{6.2}$$

where R is the surface recession in $\mu$m, $[SO_2]$ is the $SO_2$ concentration in $\mu g/m^3$, T is the temperature in °C, t is the time in years, Rain is the amount of precipitation in mm/year and $[H^+]$ is the hydrogen ion concentration in precipitation in mg/l. The dose-response function was based on results from 100 observations and the $R^2$ value was 0.88. When performing cost calculations the maintenance frequency (1/t) is sought for a specified value of the surface recession, corresponding to a critical damage where maintenance action is required or desirable. Transforming Eq. 6.2 gives the function

$$1/t = [\,(2.7\,[SO_2]^{0.48}\,e^{-0.018\,T} + 0.019\,Rain\,[H^+]) / R]^{1/0.96} \tag{6.3}$$

For sandstone, the damage function is slightly more complicated

$$1/t = [\,(2.0\,[SO_2]^{0.52}\,e^{f(T)} + 0.028\,Rain\,[H^+]) / R]^{1/0.91} \tag{6.4}$$

where R is the surface recession in $\mu$m, $[SO_2]$ is the $SO_2$ concentration in $\mu g/m^3$, f(T) is a function of temperature in °C, equal to 0 when T is lower than 10 °C and -0.013(T-10) when T is higher than 10 °C, t is the time in years, Rain is the amount of precipitation in mm/year and $[H^+]$ is the hydrogen ion concentration in precipitation in mg/l. The dose-response function (R as a function of t and other parameters) was based on results from 101 observations and the $R^2$ value was 0.86.

It is in principle also possible to use the sandstone equation for other stone materials like rendering and mortar, however, with a higher degree of uncertainty, and probably underestimating the maintenance frequency. To apply it without modification to more resistant materials like granite or gneiss is not recommended.

#### 6.1.2.3
#### Zinc and Galvanised Steel

Zinc is one of several metals that were exposed in the UN/ECE materials programme and the dose-response function from this program proposed after 8 years of exposure is

$$ML = 1.4\,[SO_2]^{0.22}\,e^{0.018\,Rh}\,e^{f(T)}\,t^{0.85} + 0.029\,Rain\,[H^+]\,t \tag{6.5}$$

where ML is the mass loss in $g/m^2$, $[SO_2]$ is the $SO_2$ concentration in $\mu g/m^3$, Rh is the relative humidity in %, f(T) is a function of temperature in °C, equal to 0.062(T-10) when T is lower than 10 °C and -0.021(T-10) when T is higher than 10 °C, t is the time in years, Rain is the amount of precipitation in mm/year and $[H^+]$ is the hydrogen ion concentration in precipitation in mg/l. The dose-response function was based on results from 98 observations and the $R^2$ value was 0.84. For mathematical reasons it is not possible to transform Eq. 6.5 into an exact function that expresses 1/t as a function of all other variables. Instead, an estimate of 1/t ($1/t_e$) is proposed,

$$1/t_e = 0.14\,[SO_2]^{0.26}\,e^{0.021\,Rh}\,e^{f(T)} / R^{1\,18} + 0.0041\,Rain\,[H^+] / R \qquad (6.6)$$

where f(T) is a function of temperature in °C, equal to 0.073(T-10) when T is lower than 10 °C and -0.025(T-10) when T is higher than 10 °C. Eq. 6.6 has also been expressed in terms of thickness values, R ($\mu$m), instead of ML values ($g/m^2$) using the density for zinc (7.14). The reason for this is that the critical damage can be tied to zinc layer thickness values on different galvanised products. The estimate $1/t_e$ is always lower than the true value 1/t and the relative error never exceeds 6%. Formally, 1/t is bound by the conditions

$$1/t_e < 1/t \leq 1.06 / t_e \qquad (6.7)$$

### 6.1.2.4
### *Paint Coatings*

Both paint coatings on steel and galvanised steel were exposed within ICP Materials. The function for steel panel with alkyd is

$$(10\text{-ASTM}) = (0.033\,[SO_2] + 0.013\,Rh + f(T) + 0.0013\,Rain\,[H^+])t^{0.41} \qquad (6.8)$$

where ASTM is the degradation measured according to ASTM D 1150-55, 1987 giving a rating between 1 and 10 where 10 corresponds to an unexposed sample, $[SO_2]$ is the $SO_2$ concentration in $\mu g/m^3$, Rh is the relative humidity in %, f(T) is a function of temperature in °C, equal to 0.015(T-11) when T is lower than 11 °C and -0.15(T-11) when T is higher than 11 °C, t is the time in years, Rain is the amount of precipitation in mm/year and $[H^+]$ is the hydrogen ion concentration in precipitation in mg/l. The dose-response function was based on results from 139 observations and the $R^2$ value was 0.68. The function can be transformed into a usable damage function by using the established criterion that maintenance should occur when ASTM=5,

$$1/t = [\,(0.033\,[SO_2] + 0.013\,Rh + f(T) + 0.0013\,Rain\,[H^+]) / 5]^{1/0.41} \qquad (6.9)$$

Similarly, the function for coil coated galvanised steel with alkyd melamine is

$$1/t = [\ (0.0084\ [SO_2] + 0.015\ Rh + f(T) + 0.00082\ Rain) / 5]^{1/0.43} \qquad (6.10)$$

where f(T) is a function of temperature in °C, equal to 0.040(T-10) when T is lower than 10 °C and -0.064(T-10) when T is higher than 10 °C. The dose-response function was based on results from 138 observations and the $R^2$ value was 0.73.

## 6.2 Soiling of Building Materials

Soiling is the effect of particle deposition that results in a darkening of the surface and can be measured as a change in light reflectance. Dirty buildings are a normal appearance in all larger towns. Old churches are often almost black, although they are built of stones with fairly bright colours. Soiling is not only restricted to old buildings, green houses need to be cleaned, or solar cells have less output due to soling. The alteration of the visual appearance may be unacceptable even if the base material is virtually unaffected and the costs related to cleaning may be substantial. Furthermore, the substances constituting the soiling matter, i.e., carbon particles, may indirectly take part in the degradation process by acting as a catalyst for various chemical reactions, particularly for the conversion of $SO_2$ and $NO_x$ into sulphuric acid and nitric acids (Newby et al., 1991).

### 6.2.1 Soiling Mechanisms

The deposition of particles is characterised by the deposition velocity, similar to gas deposition. Particles deposit due to sedimentation, impaction, and diffusion depending on the size of the particles. Sedimentation is important for particles larger than a few micrometers and mainly occurs at horizontal surfaces. Particles in this range have a high deposition velocity (typically 1mm/s). Since the lifetime of these particles in the atmosphere is short, they are found on surfaces near the source. Particles smaller than a few micrometers, i.e., sub-micron particles, have a much lower deposition velocity (typically 0.05 mm/s) thus less particles will reach the surface. Since diffusion is the main factor, the deposition takes place at any surface. Furthermore, sub-micron particles contain soot, therefore although a less effective deposition occurs, the effect is well visible by the dirty appearance of the surface. Particles normally increase but may in some cases decrease the deterioration rates of materials, e.g., basic particles deposited on a surface may neutralise the effect of other pollutants, such as $SO_2$ (Tidblad and Kucera, 1998). This section will describe the quantification of soiling, which is only one effect that particles have on the degradation of materials.

## 6.2.2 Dose-Response Functions

All soiling dose-response functions include the concentration of particles in $\mu g/m^3$ as an explanatory variable. Since particles as such have a large variation in size and composition the particle concentration depends on the analytical technique. The following parameters have been used in the quantification of soiling (QUARG, 1993):

- Dark smoke (DS) refers to non-reflective particulate matter measured by the smoke stain measurement method. The technique measures the reflectance of particular matter collected on a filter paper compared to that of a calibration curve.
- Total suspended particulate (TSP) is measured as the weight increase of a filter paper collecting particles through a high volume air sampler. Although correlation factors have been given between DS and TSP, dark smoke measurements are, nowadays, not in any meaningful way comparable with gravimetric.
- Particulate elemental carbon (PEC) is also known as black carbon and graphitic carbon. Few measurements have been made of PEC levels. To identify the sources in the urban aerosol a possibility is to use PEC emission factors.
- $PM_{10}$ refers to particular matter less than 10 $\mu m$ aerodynamic diameter.

All these parameters have been used in dose-response functions but TSP and PEC are dominating. The preferred parameter can depend on the type of local pollution. The parameters mentioned here is only a list of parameters used so far and this does not exclude that other parameters are valuable for quantification of soiling effects.

The available dose-response functions are based on two types of models, the exponential model and the square root model.

The exponential model has a theoretical foundation. It expresses the reflectance, R, as

$$R = R_0 \cdot \exp\{-k_e \cdot C \cdot t\} \qquad (6.11)$$

where $R_0$ is the reflectance of an unexposed surface, normally set to 100%, $k_e$ is a constant, C is the particulate concentration in $\mu g/m^3$ and t is the time of exposure in years. Two different approaches both results in this form. The first approach, developed by Haynie (1986), is based on the assumption that the loss of reflectance is caused by the deposition of a monolayer of opaque particles over the surface and that the loss is directly proportional to the fraction of covered surface. Lanting (1986) considered an alternative approach and described the loss of reflectance in terms of the thickness of the deposited film. Table 6.2 shows a summary of exponential dose-response functions. In addition to the theoretical functions only one author has chosen to report his data in this form.

**Table 6.2.** Summary of dose-response functions based on the exponential model, Eq.(6.11).

| $k_c$ | Particle | Surface | Reference |
|---|---|---|---|
| 0.0085 | TSP[a] | Theoretical[c] | (Haynie, 1986) |
| 0.095 | PEC[b] | Theoretical[c] | (Lanting, 1986) |
| 0.13 | PEC[b] | Theoretical[d] | (Lanting, 1986) |
| 0.0092 | TSP[a] | white-painted wood in a road tunnel | (Hamilton and Mansfield, 1993) |
| 0.062 | PEC[b] | white-painted wood in a road tunnel | (Hamilton and Mansfield, 1993) |

a Total suspended particles
b Particulate elemental carbon
c Fraction of covered surface model
d Multilayer model

The square root model has only an empirical background and was first reported by Beloin and Haynie (1975), which gave the following expression for the reflectance, R,

$$R = R_0 - k_s (C \cdot t)^{1/2} \tag{6.12}$$

where $R_0$ is the reflectance of an unexposed surface, normally set to 100%, $k_s$ is a constant, C is the particulate concentration in $\mu g/m^3$ and t is the time of exposure in years. Butlin et al.,(1994) report functions for many materials, including coated yellow brick

$$R = 43.21 - 0.1133\ (C \cdot t)^{1/2} \tag{6.13}$$

where t is measured in months and C is expressed as total suspended particulate. Recalculating Eq. 6.13 for time measured in years and expressing the reflectance relative to that of the original gives

$$R = 43.21 - 0.39\ (C \cdot t)^{1/2} \approx 43\ (100 - 0.91\ (C \cdot t)^{1/2}) \tag{6.14}$$

Table 6.3 shows a summary of square root dose-response functions. Note especially that for Eq. 6.14 the compiled value is 0.91, i.e. that which gives the reflectance normalised to 100%.

Almost all functions given in Table 6.3 have the particle concentration expressed as TSP (total suspended particles). Exceptions are those of Pio et al. (1998) which are based on concentrations measured as PEC (particulate elemental carbon). In these cases alternative functions have been calculated based on the original functions and particulate measurements of TSP and PEC tabulated in Pio et al. (1998).

**Table 6.3.** Summary of dose-response functions based on the square root model, Eq. (6.12).

| $k_s$ | Particles[a] | Surface | | Reference |
|---|---|---|---|---|
| 1.4 | TSP[a] | Acrylic emulsion paint | unsheltered | (Beloin and Haynie, 1975) |
| 2.8 | TSP[a] | Painted wood | unsheltered | (Hamilton and Mansfield, 1993) |
| 4.2 | TSP[a] | Painted wood | sheltered | (Hamilton and Mansfield, 1993) |
| 1.1[c] | TSP[a] | Oil based paint | unsheltered | (Butlin et al., 1994) |
| 1.1[c] | TSP[a] | Tint based paint | unsheltered | (Butlin et al., 1994) |
| 2.2[c] | TSP[a] | Acrylic emulsion paint | sheltered | (Butlin et al., 1994) |
| 1.6[c] | TSP[a] | Acrylic emulsion paint | unsheltered | (Butlin et al., 1994) |
| 1.6[c] | TSP[a] | Shingles | unsheltered | (Butlin et al., 1994) |
| 0.91[c] | TSP[a] | Coated yellow brick | unsheltered | (Butlin et al., 1994) |
| 4.5 | PEC[b] | Painted wood | sheltered | (Pio et al., 1998) |
| 5.7 | PEC[b] | Portland stone | sheltered | (Pio et al., 1998) |
| 1.3 | TSP[a,d] | Painted wood | sheltered | (Pio et al., 1998) |
| 1.6 | TSP[a,d] | Portland stone | sheltered | (Pio et al., 1998) |

[a] Total suspended particles.
[b] Particulate elemental carbon.
[c] The values have been adjusted to $R_0$=100%.
[d] The values have been recalculated from a PEC equation using data measured at the site.

Both the exponential model (6.11) and the square root model (Eq. (6.12) can be transformed into exposure-response functions,

$$1/t = C \cdot k_e / \ln(R_0 / R) \tag{6.15}$$

and

$$1/t = C \cdot k_s^2 / (R_0 - R)^2 \tag{6.16}$$

respectively. Eqs. 6.15 and 6.16 both have the general form

$$1/t = C / d_{crit} \tag{6.17}$$

where $d_{crit}$ is the 'critical' dose $(C{\cdot}t)_{crit}$. In other words, after application of a suitable maintenance criterion $R=R_{crit}$, both functions result in the same damage function. According to Hamilton and Mansfield, (1993) and Pio et al. (1998) maintenance is triggered when $R_{crit}$=70%. Using the data of Pio et al. (1998), and this criterion the critical soiling doses for sheltered painted wood and limestone due to soiling becomes 533 and 352 year·$\mu$gTSP/m$^3$, respectively. For sheltered painted wood, Hamilton and Mansfield, (1993), the data gives alternative critical dose of 51 year·$\mu$gTSP/m$^3$, i.e. ten times lower. In view of the longer exposure period, 850 vs. 120 days, the Pio data is preferred, however, recognising that it may overestimate the maintenance time, i.e. underestimate the replacement frequency.

On the other hand, in areas with high particulate matter levels, individuals can possibly accept greater reductions in reflectance before cleaning due to the general impression of dirty buildings. When humans judge the soiling status of an object it is compared to a surface in the surrounding considered being white. In reality, this white surface may also be soiled to a lesser extent depending on the general pollution level. Therefore, the maintenance criterion R=70% may be modified.

Table 6.4 shows the result of these modified criterions, calculated for the exponential and square root model and based on two assumptions, one that can be considered reasonable ('almost white') and one extreme ('grey'). In practice, this means that the mentioned critical doses can also be higher than what can be expected when using R=70% and the Pio functions.

**Table 6.4.** Calculated critical doses (see Eq. 6.17) for polluted areas based on a higher tolerance to soiling due to individuals perceiving partially soiled surfaces as white.

| Surface perceived as white is | Critical dose<br>Exponential model | Square root model |
|---|---|---|
| white (reference case[a]) | $d_{crit} = \ln(1 / 0.7) / k_e$ | $d_{crit} = 900 / k_s^2$ |
| almost white (10-percentile of soiled objects) | 1.1 $d_{crit}$ | 1.6 $d_{crit}$ |
| grey (average soiling level) | 1.9 $d_{crit}$ | 3.5 $d_{crit}$ |

[a] See Table 6.2 and Table 6.3 for $k_e$ and $k_s$ values, respectively.

## 6.3 Combination of Degradation and Soiling Effects

The exposure conditions are of great importance when estimating impacts. Horizontal surfaces soil more rapidly than vertical surfaces. Sedimentation and impaction applies for particles larger than 1 $\mu$m and are orientation dependent whereas convective diffusion is the mechanism for deposition of sub-micron particles. Also, wind and rain easily remove coarse (> approximately 2 $\mu$m) particles while fine (< approximately 2 $\mu$m) particles adhere more strongly (Creighton et al., 1988). In a soiling study performed over a relatively long time (850 days) the unsheltered data was erratic. The study included soiling measurements on painted wood and stone tablets and also corrosion measurements on the stone tablets. On unsheltered painted wood the soiling was initially rapid but, for longer exposures, there was a cleaning effect and the reflectance increased with time. The unsheltered stone tablets showed an increase in reflectance and later a decrease, however, not reaching below the initial value. When attempting to correlate these changes with rain events, periods of agreement between observed and predicted values were as common as the opposite behaviour. In

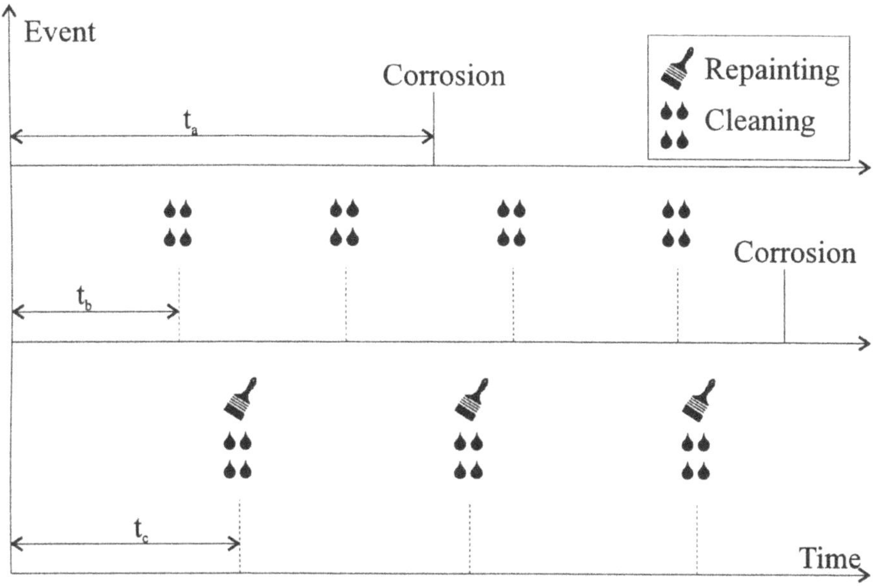


**Fig. 6.1.** Illustration of different maintenance practices showing a) no action, top b) cleaning at regular intervals, middle, and c) cleaning and repainting at regular intervals, bottom. For further discussion see text.

many occasions the rain does not remove the deposited particles but only rearranges them on the surface. For sheltered samples dose-response functions were obtained but not for unsheltered samples. Measurements of corrosion attack revealed that the unsheltered tablets decreased in weight whereas sheltered tablets increased in weight (Pio et al., 1998). The latter finding is consistent with observations within the ICP Materials program. The weight increases for sheltered stone samples are due to a combination of soiling and corrosion effects. In summary, it is possible to obtain dose-response functions for corrosion attack on unsheltered stones and paint coatings and soiling of sheltered samples. For corrosion of stone materials in sheltered position and for soiling of unsheltered objects functions are, however, at present stage not reliable.

Any building is subject to impact both by deterioration and by soiling and the decision to take a maintenance action may be due to deterioration or soiling or both. Different maintenance practices needs to be considered if to combine the results obtained from calculations on degradation and soiling of materials. Figure 6.1 shows an illustration of selected practices for a painted surface. In the first case no action is taken and corrosion occurs after a time, ta, with an associated maintenance frequency, $1/t_a$. In the second case cleaning is performed at regular intervals, which prolongs the time to corrosion, thereby reducing the cost. On the other hand, the cleaning cost at frequency $1/t_b$ is added to the total cost. In the last case repainting and cleaning is performed at regular intervals, practically eliminating the cost due to corrosion. Each of the practices a, b and c may be the most cost effective approach, depending on the material and application. In reality the used strategy may also be a mix of the three approaches.

In view of this, how should the maintenance frequencies, and associated costs, calculated from the degradation functions 6.3, 6.4, 6.6, 6.9 and 6.10, on one hand, and the critical soiling doses, on the other hand, be interpreted? The degradation functions are valid for unsheltered surfaces while the critical soiling doses are valid for sheltered surfaces. At present stage, however, the data on stock of materials at risk can not be divided into fractions corresponding to unsheltered and sheltered surfaces. Consider, for example, a facade consisting of partly unsheltered, partly sheltered areas. The decision to clean/paint may be taken either due unacceptably high corrosion in unsheltered areas or due to unacceptably high soiling of sheltered areas. Once the decision is taken, however, it is likely that the entire facade is cleaned/repainted regardless of the sheltering status of individual areas. Therefore the cost associated with the action should be multiplied with the total stock at risk area. The resulting total cost therefore represents one of two extreme cases. In the first it is assumed that all maintenance actions are taken as a direct result of corrosion damage while in the second it is assumed that all maintenance actions are taken as a direct results of soiling. It is not possible at present stage to combine these cost estimates.

To conclude, the following observations should be taken into account:

- New dose-response functions are available from ICP Materials, an international exposure programme, describing the degradation of a wide range of materials including metals, stone materials, paint coatings, electric contact materials, glass materials and polymeric materials.
- Selected functions from the programme, valid for unsheltered positions, have been adapted and exposure-response functions suitable for inclusion into EcoSense have been derived for limestone (Eq.6.3), sandstone (Eq. 6.4), zinc/galvanised steel (Eq. 6.6), painted steel (Eq. 6.9) and painted galvanised steel (Eq. 6.10). They include $SO_2$, temperature and relative humidity for quantification of dry deposition effects, and amount and acidity of precipitation for quantification of wet deposition effects.
- A critical review of dose-response functions for soiling to materials has been performed including additional analysis on strategies for selecting maintenance intervals. The functions quantify the loss of reflectance in terms of the particulate concentration times time of exposure, i.e. the particulate dose, which means that each exposure-response function can be summarised in the form of a critical dose. No other characteristics of the environment are included in the soiling functions.
- Critical soiling doses in sheltered conditions have been estimated to be 533 and 352 year·$\mu$gTSP/$m^3$ for painted wood and limestone, respectively (TSP = total suspended particulate). The uncertainties of these critical doses are high compared to the degradation functions and can be either significantly lower or higher.
- It is today not possible to combine cost estimates related to degradation and soiling of materials into a single cost estimate representative for the total impact to materials. Instead, the individual estimates should be regarded as illustrations pertaining to particular isolated maintenance practices.

# 7 Impacts on Terrestrial Ecosystems

## 7.1 Introduction

The impacts of transport on terrestrial ecosystems can be broadly divided into three different categories:

- Effects on the physical structure of the landscape, such as direct loss of land and habitat fragmentation, of the land used solely for transport
- Effects of management of the transport route and adjacent land, including cutting and clearing vegetation, salting and drainage
- Effects produced by use of the transport system, for example the impacts of atmospheric emissions from the transport vehicles and dispersal of organisms.

Only the localised, terrestrial impacts of transport have been considered here, that is where the impacts are concentrated by the use of fixed routes such as roads, rail lines and airports. There are other, more diffuse, impacts where routes are less well defined or not in direct contact with the earth, such as the impacts arising from emissions from aircraft during flight. Marine and freshwater transport systems, which have consequences on the terrestrial fringes through routine emissions, accidents and redistribution of organisms, have also not been considered here. The contribution of transport to changes that operate at a global scale, for example $CO_2$ emissions and climate change have not been considered.

## 7.2 Impacts Linked to Direct Effects on Ecosystem Structure

### 7.2.1 Loss of Habitat

All transport systems have land that is solely dedicated to them. Rail systems have permanent tracks with associated margins, air transport has airports, radar and radio stations, marine transport requires docks and freshwater transport uses canals or managed waterways. In terms of terrestrial habitats, road systems occupy the largest area, with tarmac carriageways and associated margins that together occupy a significant, measurable portion of the land surface. For example, in Britain, estimates of road area suggest that 1-2% of the land surface is affected (Table 7.1).

**Table 7.1.** Road area in Great Britain, in km$^2$

| Country | Motorway | A Roads | B Roads | Minor Roads |
|---|---|---|---|---|
| England | 105 | 401 | 178 | 991 |
| Wales | 5 | 46 | 26 | 136 |
| Scotland | 11 | 108 | 65 | 197 |
| Great Britain | 121 | 554 | 269 | 1323 |

This use of land for the road, or railway and linked infrastructure represents a loss of productive agricultural or forestry land, or of semi-natural habitats. Using the United Kingdom as an example, methods of quantifying the physical destruction/replacement of habitats as a result of road building was explored. The approach exploits GIS technology and requires digital databases of the road/rail system and of land cover or habitats for the area of study. In the exploratory study the focus lay on the motorway network of the UK, although this is only a small proportion of the total road surface in the UK it carries large volumes of traffic. The approach can be expanded to the whole road or rail system, providing that digital databases are available (see Sect. 7.5). To identify the land use and land cover/habitats that the motorways are replacing, a digital map of the motorway system was buffered 1 km on either side of the vector describing the road. The polygon produced was then used as a cutter to clip information from the ITE Land Cover Map (LCM) (Fuller et al., 1994); the LCM identifies 25 types of land cover to a spatial resolution of 25 m$^2$. It was then assumed that the motorway was constructed on land/habitats that was similar to that adjacent to it and has not significantly changed the surrounding land use. Table 7.2 contains the areas of the different cover types in the buffer zone; on the basis of the above assumption, it is possible to interpret the table as the land cover lost to the motorway.

The only four land cover types to exceed 10% of the total buffer area are arable land (tilled – 29%), two types of agricultural grassland (pasture and meadow – 35%) and suburban (16%). The values for these types exceed their proportion of cover for the whole of the UK (21%, 27% and 6% respectively). This should not be a surprise, as roads are not randomly located; their function being to link areas of population (suburban and urban areas). The centres of population are concentrated in the lowlands and roads generally follow the easiest routes between them, avoiding high and rugged land. The analysis shows a small proportion of the buffer strip contains sea, and although there may be occasional bridges over estuaries, this reflects an artefact of the analysis as motorways would replace land rather than water.

Both the spatial and categorical resolution of the satellite data (25 m$^2$ pixels and 25 categories) are too coarse to identify habitats that cover small areas that through their scarcity may have a high conservation value. To incorporate these into an analysis, higher resolution imagery (such as aerial photography) or field survey would be needed.

**Table 7.2.** Land cover within 1 km of Britain's motorways. Area is in km$^2$ and percent (%) is expressed as the proportion of the buffered area. Data from the satellite derived ITE Land Cover Map of Great Britain.

| Cover type | | Motorways | |
|---|---|---|---|
| | | Area | % |
| 0 | Unclassified | 94 | 1 |
| 1 | Sea | 21 | 0 |
| 2 | Inland water | 28 | 0 |
| 3 | Coastal bare ground | 6 | 0 |
| 4 | Saltmarsh | 5 | 0 |
| 5 | Lowland grass heath | 183 | 3 |
| 6 | Pasture amenity grass | 948 | 15 |
| 7 | Meadow/verge/unimproved grass | 1,295 | 20 |
| 8 | Marsh/rough grass | 73 | 1 |
| 9 | Montane/hill grass | 61 | 1 |
| 10 | Dwarf shrub/grass moorland | 15 | 0 |
| 11 | Upland dwarf shrub moorland | 6 | 0 |
| 12 | Bracken | 18 | 0 |
| 13 | Lowland heath | 30 | 0 |
| 14 | Scrub/orchard | 22 | 0 |
| 15 | Deciduous woodland | 319 | 5 |
| 16 | Coniferous woodland | 61 | 1 |
| 17 | Upland bog | 2 | 0 |
| 18 | Arable/tilled | 1,860 | 29 |
| 19 | Ruderal weeds | 26 | 0 |
| 20 | Suburban | 1,000 | 16 |
| 21 | Urban | 222 | 3 |
| 22 | Bare bround | 68 | 1 |
| 23 | Felled forest | 8 | 0 |
| 24 | Lowland peat bog | 7 | 0 |
| 25 | Dwarf shrub/grass heath | 42 | 1 |

A study in Switzerland has used a very similar approach to the above (Wittwer et al., 1998, Buser et al., 1999). Aerial photographs from two dates were used to identify the habitats lost to road building in the intervening period. The areas of recognisable habitats were digitised and each habitat was given a score for the 'quality' and range of the functions the area could fulfil. The areas of each type were then summed and thc total areas used to calculate an external cost through habitat loss.

If detailed spatial databases of soils or land capability are available, a similar approach can be used to determine the amount of land with different soils types and of different land capability classes, or 'quality' lost to road building and hence the land value.

A similar analysis to land lost under airports and airfields can be performed (Table 7.3). As a portion of the terrestrial surface, airfields take up an insignificant area, but they have high impacts because of the intensity of management and secondary effects due to needs for access.

**Table 7.3.** Land cover within 10 km of Heathrow and Manchester airports. Area is in $km^2$ and percent (%) is expressed as the proportion of the buffered area. Data from the satellite derived ITE Land Cover Map of Great Britain.

| Cover type | | Heathrow | | Manchester | |
|---|---|---|---|---|---|
| | | Area | % | Area | % |
| 0 | Unclassified | 2 | 1 | 2 | 1 |
| 1 | Sea | 0 | 0 | 0 | 0 |
| 2 | Inland water | 17 | 5 | 1 | 0 |
| 3 | Coastal bare ground | 0 | 0 | 0 | 0 |
| 4 | Saltmarsh | 0 | 0 | 0 | 0 |
| 5 | Lowland grass heath | 2 | 1 | 29 | 9 |
| 6 | Pasture amenity grass | 36 | 11 | 48 | 15 |
| 7 | Meadow/verge/unimproved grass | 61 | 19 | 77 | 24 |
| 8 | Marsh/rough grass | 2 | 1 | 3 | 1 |
| 9 | Montane/hill grass | 0 | 0 | 1 | 0 |
| 10 | Dwarf shrub/grass moorland | 0 | 0 | 0 | 0 |
| 11 | Upland dwarf shrub moorland | 0 | 0 | 0 | 0 |
| 12 | Bracken | 0 | 0 | 1 | 0 |
| 13 | Lowland heath | 0 | 0 | 9 | 3 |
| 14 | Scrub/orchard | 1 | 0 | 2 | 1 |
| 15 | Deciduous woodland | 7 | 2 | 18 | 6 |
| 16 | Coniferous woodland | 1 | 0 | 0 | 0 |
| 17 | Upland bog | 0 | 0 | 0 | 0 |
| 18 | Arable/tilled | 36 | 12 | 38 | 12 |
| 19 | Ruderal weeds | 5 | 2 | 0 | 0 |
| 20 | Suburban | 106 | 34 | 67 | 22 |
| 21 | Urban | 27 | 9 | 13 | 4 |
| 22 | Bare bround | 9 | 3 | 2 | 1 |
| 23 | Felled forest | 0 | 0 | 0 | 0 |
| 24 | Lowland peat bog | 0 | 0 | 0 | 0 |
| 25 | Dwarf shrub/grass heath | 0 | 0 | 3 | 1 |

### 7.2.2 Networks and Habitat Fragmentation

The division of a landscape into separate units by roads is known to produce isolated populations or meta-populations (Mader, 1984). The theory of island biogeography (MacArthur and Wilson, 1967) shows that smaller isolated populations have larger fluctuations than larger continuous populations and may become extinct. Roads or railways can act as simple barriers to animal movement and so reducing the potential for migration and colonisation; the local extinction of small populations may become permanent.

Although it is possible to model how the construction and use of roads will lead to the local extinction of species, it is impossible to produce general parameters to fit the extinction rate equations, as there are many confounding factors. One crucial factor is the starting point and time it will take to reach equilibrium. Terborgh (1974) described this process of relaxation and estimated that it may take 100 - 200 years for an isolated area of woodland to reach its new quota of species – a condition that is unlikely to be met in landscapes used by man.

The division of a population into smaller groups, especially if they remain isolated, can have effects on the genetic composition of the population. In Germany, the effects of isolation of common frog (Rana temporaria) populations by roads have been shown to produce reduced genetic heterozygosity and polymorphism (Reh and Seitz, 1990).

A Swiss study is currently exploring ways of quantifying habitat fragmentation due to road building and of assessing the impacts of the fragmentation (Buser, 1999). The study plans to use a series of sample sites, representative of the Swiss traffic system, in terms of traffic density and topographic setting. Approaches that can be used to ameliorate the impacts of roads on animal movements, e.g. bridges of tunnels, are also to be considered.

The road or rail network can equally be viewed in a more positive light. Rather than being a barrier isolating populations, the road and its associated verges and central reservation can be thought of as a new habitat and offering a corridor down which species can migrate. Although the tarmac carriageway is an inhospitable environment, the verges offer conditions of ample light (with regular mowing or selective removal of woody plants) and moisture. Verges are recognised as containing relatively high plant species richness, especially in areas of intensive agricultural management (Bennet, 1991, Milberg and Persson, 1994). The detailed management of roadside verges such as the mowing date and frequency, plays an important part in the determination and persistence of species.

Road density (i.e. length per unit area $km/km^2$) has been used as a broad index of the ecological effects of roads in a landscape (Forman and Herpsberger, 1996). A value of approximately 0.6 $km/km^2$ has been shown to be the limit for survival of large predators such as wolves and mountain lions in the United States (Forman and Herpsberger, 1996, Mech, 1989). The difficulty of using such an index is that it summarises a variety of interacting pressures including disturbance, habitat alteration, isolation and changes in species composition.

Applying the rules to a road network dataset for mainland Great Britain suggests that about a third of the $km^2$ would be capable of sustaining a population (Fig. 7.1). However, the biology of large mammals requires large home ranges and when contiguous squares are identified, only areas in Scotland can be seriously considered as offering any capability. The disturbance effect also influences the distribution of large raptors. With the exception of the Scottish islands, the distribution of the golden eagle (Aquila chrysaetos) is well reflected by the areas mapped (Gibbons et al., 1993).

The movement of exotic plant species along verges is widely accepted, but poorly documented. There are a few well-known examples (e.g. Wilcox, 1989, Ross, 1986), but the potential for redistribution of species should not be underestimated. There is no current method of quantification and evaluation of such an effect.

Similar assessments to the above could be carried out for rail networks or for airports. The loss of land to airports will be very small in national terms (for example see Table 7.3) but specific airport developments could lead to loss of habitats which are highly valued locally. In most new developments, measures are

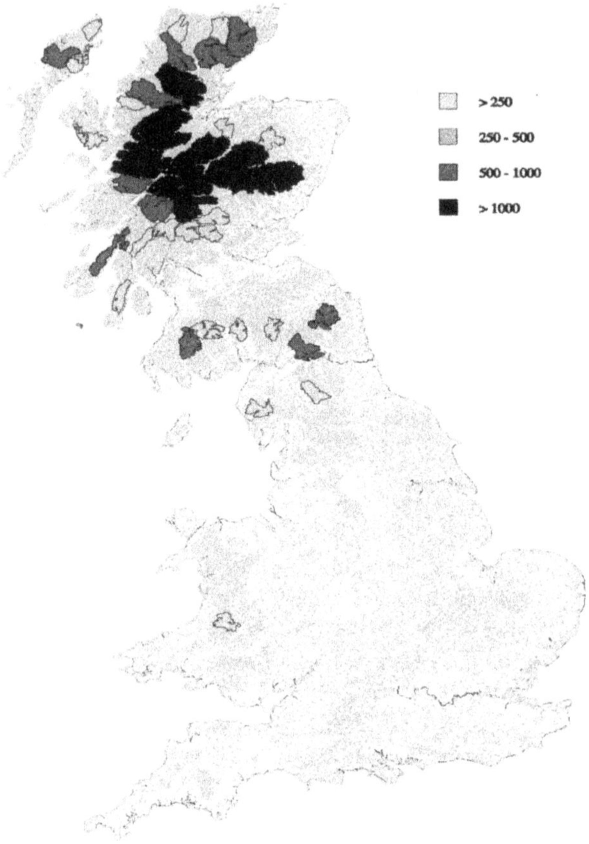


**Fig. 7.1.** The potential distribution of large predatory mammals in Great Britain as determined by the road network. Shaded unfragmented areas are shown in km$^2$

taken to restore or relocate the affected fragments of such habitats and these costs are internalised. Rail networks can occupy significant areas of land but in most countries, less than road networks. Land and habitat loss to these networks can be calculated using the approach outlined above for the UK motorway network. Railway networks can also lead to habitat fragmentation, impact upon animal movements and act as corridors for the spread of species.

### 7.2.3 Expenditure on Protection of Semi-Natural Habitats

Most roads were built before there was concern over the natural environment. Nowadays, new roads or re-routing generally requires complex impact assessment and monitoring. Mitigation and the no-net-loss principle is often used with lost habitat being compensated for by increasing the ecological value of a neighbouring site (Race and Fonesca, 1995) or relocating those fragments of high conservation value habitats likely to be destroyed by the transport development

(Buckley, 1989). The costs of such schemes, which are generally available from the relevant transport authorities, provide a means of valuing impacts on habitats. Similarly, the measures taken in some cases to ameliorate the impacts of roads on, for example animal movements by building tunnels and bridges, provide a value of part of the impacts. The latter costs are however, generally internalised.

## 7.3 Impacts Linked to Management of the Transport System

### 7.3.1 Roadside Verges and Rail Margins

Roadside verges and rail track margins are often managed to avoid vegetation growth overhanging the road margins, affecting visibility and producing a fire risk as biomass accumulates. In Britain legislation is in place to maintain visibility, consequently local authorities are committed to cutting and/or herbicide treatments. Concerns over changes in management regimes, most notably the increased use of herbicide, on the species composition have been expressed (Way, 1977) and effort is being made to maintain the biodiversity. Conversely, roadside verges are now increasingly recognised as refuges for plants and insects.

### 7.3.2 De-Icing

Ice on roads, rails and airport runways seriously limit safe movement. Most European states encounter frozen roads, although they are obviously most frequent at high latitudes or altitudes. The most common de-icing agent is sodium chloride which corrodes vehicles, bridges and the road surface (Obika et al., 1989, Vehovar et al., 1998), and may affect human health as a carcinogen and through contamination of drinking water supplies (Williams et al., 2000, Foster, 1993). The drift of the de-icing agent onto land adjacent to the treated area can impact on habitats and there are several reported effects on terrestrial ecosystems including the colonisation by salt tolerant plants usually found in salt marshes or inter-tidal areas that are frequently inundated with sea water. In the Netherlands three such species have colonised as far as 150 km from the sea (Aanen et al., 1991) and, in Britain in the first few centimetres of roadside verge vegetation species such as Puccinella distans can be commonly found (Scott and Davison, 1982). When combined with snow-plowing, airborne salt from de-icing has been shown to cause leaf injury to trees within 120 m of the road (Hofstra and Hall, 1971). Sodium also accumulates in the soil within the first 5 m from the road, where it alters soil structure and base cation status (NRC, 1997).

De-icing also strengthens the barrier function of the road for amphibians acting as a significant deterrent to crossing (Fahrig et al., 1995).

## 7.4 Impacts Linked to Use of the Transport System

As noted above, this section focuses mainly on the impacts resulting from gaseous and particulate pollutant emissions from vehicles during vehicle movements. The impacts of these emissions on terrestrial ecosystems are concentrated around the actual road or rail network. However, the magnitude of the effect is dependent on the volume of traffic, evenness of traffic flow throughout the year, landscape morphology and local climate. Pollutant emissions and disturbance are greater with higher traffic flow and both animals and plants may show increased sensitivity to the pollutants in different seasons or at different stages of the life cycle. Similarly, the dispersion and modification of traffic emissions varies significantly with local climatic conditions, with still air and temperature inversions contributing to seasonal smogs.

The zone of influence surrounding a traffic route reflects a concentration gradient that is determined by the quantity of the pollutant released and the form of the pollutant. There are a number of measurements of the size of corridor ranging from a few centimetres for elements such as lead to approximately 1000 m for more mobile pollutants and disturbance effects (Forman et al., 1997).

The main gaseous pollutants to be considered are carbon dioxide, nitrogen oxides, sulphur dioxide and ozone. Particulate pollutants will include heavy metals such as, in the past lead, zinc and, since the introduction of catalytic converters, platinum. Carbon dioxide emissions will contribute to global warming; these global scale impacts are considered elsewhere in this book and there will be no separate, local scale impacts of the carbon dioxide. The emissions of sulphur dioxide are small from the fuels used in motor transport and trains and unlikely to have a significant impact even adjacent to the highest density traffic routes. The largest pollutant gas emissions, other than the carbon dioxide are the nitrogen oxides. However, concentrations of these gases recorded adjacent to transport routes have not been large enough to cause direct phytotoxic effects on vegetation adjacent to the route. The enhanced, accumulated deposition of the nitrogen adjacent to the transport route could have significant impacts on sensitive habitats. In the case of the common agricultural and forest crops, the nitrogen deposition is more likely to act as a fertiliser although the additions will be small compared to those used in arable and intensive grass production. Some semi-natural habitats may be sensitive enough for adverse impacts to result and a methodology for exploring potential impacts is explored below. Ozone is not emitted directly from vehicle exhausts but nitrogen dioxide is a precursor for ozone formation. However, ozone concentrations will be depleted in the zone immediate adjacent to roads due to the reaction of nitrogen oxide with ambient ozone. Further ozone formation as a result of reactions involving nitrogen dioxide emitted by the vehicles will take place at a considerable distance downwind of the road.

## 7.4.1 Impacts on Semi-Natural Vegetation

A methodology based on the critical load concept has been explored to assess the possible impact of N deposition deriving from $NO_x$ emissions along road networks on semi-natural vegetation. The critical load is the maximum load of a pollutant that will not lead to adverse changes in the receptor, in this case semi-natural habitats. In the present approach the likelyhood of the critical load being exceeded was assessed, and hence damage resulting, as a result of the nitrogen deposition deriving from the motor vehicles added to ambient nitrogen deposition. The approach requires digital datasets of the road network, land cover or habitat distributions and current nitrogen deposition in the study area. For illustrative purposes, the approach has been applied to the motorway network in Great Britain.

A study by Bell and Ashenden (1997) suggests that a significant impact of traffic movements on nitrogen concentrations adjacent to roads is seen at annual average daily traffic flow (AADT) of more than 10,000 cars. As seen from Table 7.6 this effectively equates to motorways and busy trunk roads. We have therefore concentrated on UK motorways in this study. The roads in a digital road atlas were buffered, in a Geographical Information System (GIS), to produce a corridor of 1 km on either side of the motorway network. The satellite-generated map (LCM) (Fuller et al., 1994) described in Sect. 7.2.1 was then used to identify the land cover/habitats in the buffer strip. Critical load values for nitrogen have been defined for broad habitat types by the UN-(ECE, 1996). Table 7.4 also shows the allocation of the land cover types recorded in the LCM into the critical load classes. Using the table values assigned to the categories from the LCM it is then possible to translate the areas under different cover types into five standard critical load classes (Table 7.5) and produce a critical load map for the habitats in the buffer strip. None of the habitats in the buffer strip are in the two most sensitive classes. This critical load map for the buffer strip was then overlaid with data on the current nitrogen deposition to identify areas where the critical load is exceeded by the current deposition - no areas were exceeded.

The critical load for nitrogen produced by traffic is effectively the habitat value minus the background deposition. The figure will be a conservative estimate for the critical load as the background deposition will contain traffic emissions. In exceeded areas, the additional deposition from motor vehicles will increase the exceedance with increased likelihood of damage to the habitats. In areas of non-exceedance the gap between the critical load and current deposition will indicate whether or not the additional deposition resulting from emission from the motor vehicles will result in exceedance and therefore a risk of damage. Ideally, the next stage would involve the overlaying of data on the magnitude of the enhanced deposition resulting from the traffic movements. This would require data on the density of traffic for segments of the road system and a relationship between traffic density and nitrogen deposition. Such databases are not readily available.

**Table 7.4.** The critical loads for nitrogen for the land cover categories in the ITE Land Cover Map

| Habitat | Land Cover Map categories | Critical load |
|---|---|---|
| Arable | 18 Arable/tilled | - |
| Acid grassland | 8 Marsh/rough grass | 25 kg N/ha |
| | 9 Montane/hill grass | |
| | 10 Dwarf shrub/grass moorland | |
| | 19 Ruderal weeds | |
| | 23 Felled forest | |
| | 12 Bracken | |
| Sub alpine grassland | | 12.5 kg N/ha |
| Bogs | 17 Upland bog | 10 kg N/ha |
| | 24 Lowland peat bog | |
| Calcareous grassland | 6 Pasture amenity grass | 50 kg N/ha |
| | 7 Meadow/verge/unimproved grass | |
| Coastal | 1 Sea | - |
| | 3 Coastal bare ground | |
| | 4 Saltmarsh | |
| Freshwater | 2 Inland water | 10 kg N/ha |
| Heathland | 5 Lowland grass heath | 17 kg N/ha |
| | 11 Upland dwarf shrub moorland | |
| | 13 Lowland heath | |
| | 25 Dwarf shrub/grass heath | |
| Calluna vulgaris | | 15 kg N/ha |
| Arctic alpine heath | | 10 kg N/ha |
| Heathland on peat | | 10 kg N/ha |
| Urban/bare ground | 20 Suburban | - |
| | 21 Urban | |
| | 22 Bare ground | |
| Coniferous woodland | 16 Coniferous woodland | 12 kg N/ha |
| Deciduous woodland | 14 Scrub/orchard | 17 kg N/ha |
| | 15 Deciduous woodland | |

Our exploratory study indicates that, even added to the current total N atmospheric deposition values, the additional nitrogen produced by traffic is unlikely to push any of the habitats identified in the buffer zone, using the LCM data, into exceedance. However, a number of important points need to be taken into account. First it must be remembered that much of the British motorway network was established in the 1960s and the cumulative effect 40 years of deposition may well have resulted in changes, destroying any sensitive habitats. Second, the local elevation of deposition due to the traffic (even at high densities) is likely only to be noticeable over the first 100 m (Bell and Ashenden, 1997) and then will show a sharp decay curve. The atmospheric concentration is reflected in the nitrogen content of the vegetation as plants are known to take up both NO and $NO_2$ through their leaves. A transect crossing the M1 in England shows a doubling in concentration in plant nitrogen in the central reservation, but within 10-15 m the levels appear to be the same as background (Port and Thompson, 1980). The spatial resolution of the land cover map with its pixels of 25 m is too coarse to effectively describe the sensitivity. Thirdly, the broad categories of vegetation that could be derived from the LCM did not include the most sensitive semi-natural habitats occurring in the UK; while most of these habitats occur in remote areas some sensitive heathland and wetland habitats do occur in areas with a dense road network. Many of these habitats exist as small fragments that are not identified from satellite imagery. They may however, be identified if a similar approach was applied to a specific route and aerial photographs, backed up by field survey were used to characterise and map the habitats.

One consequence of increased nitrogen in vegetation is improved growth and reproduction in insect herbivores (McNeill and Southwood, 1978). Outbreaks of moths on vegetation adjacent to motorways have been attributed to elevated nitrogen concentrations in the vegetation (Port and Thompson, 1980) and in Switzerland, the frequency and scale of aphid (Aphis pomi) infestations on hawthorn (Cretagus monogyna) have been shown to be inversely correlated with distance from a motorway. They are also positively correlated to the high concentrations of free amino acids in the vegetation (Flückiger et al., 1978). The long term impacts of these effects on habitats are extremely difficult to quantify.

The effect of elevated nutrient nitrogen and the subsequent increased grazing can also lead to competitive replacement of species within a habitat (Beyschlag et al., 1996). The common plant species seen immediately adjacent to roadsides is clear evidence for the strong selection on those species.

It can be seen in Table 7.4 that most of the more sensitive habitats are wetlands, heathland or coniferous woodland. The literature shows a number of examples where the interaction between traffic exhaust fumes or salting have considerable effect on the habitat. However, it is important that these are specific examples and the values are then being used as generalisations. Freshwater habitats neighbouring roads and airports have often been damaged by other land use practices and are rarely capable of reflecting further damage.

A similar analysis to the above can be carried out for air traffic. The emissions during the main part of flights will contribute to global rather local impacts but the emissions at take off and landing are higher than those produced in steady flight and could give rise to local effects. A similar division of land cover around

**Table 7.5.** Critical loads for nitrogen for the habitats in the zone 1 km on either side of motorways in Great Britain

| Critical load | | Area | |
|---|---|---|---|
| Class | kg N/ha | $km^2$ | % |
| 1 | <= 2.8 | 0 | 0 |
| 2 | 2.8 - 7.0 | 0 | 0 |
| 3 | 7.0 - 14.0 | 97 | 2 |
| 4 | 14.0 - 28.0 | 803 | 13 |
| 5 | > 28.0 | 2242 | 35 |
| Sea | | 32 | 1 |
| Not affected | | 3149 | 50 |

**Table 7.6.** The annual average daily traffic flow (AADT) recorded at intersections of motorways in Great Britain

| Road Number | Motor Cycle | Car & Taxis | Bus & coaches | LGV | HGV Total | All Motor Vehicles | HGV % |
|---|---|---|---|---|---|---|---|
| M1 | 206 | 52388 | 542 | 5743 | 12096 | 70975 | 17 |
| M1 | 165 | 41242 | 398 | 5236 | 13316 | 60357 | 22 |
| M4 | 243 | 58247 | 437 | 3757 | 8035 | 70718 | 11 |
| M5 | 176 | 33109 | 399 | 3217 | 5437 | 42338 | 13 |
| M5 | 232 | 64354 | 405 | 7830 | 15188 | 88009 | 17 |
| M6 | 25 | 16374 | 263 | 1769 | 5376 | 23808 | 23 |
| M4 | 529 | 56867 | 472 | 7295 | 10374 | 75537 | 14 |
| M606 | 158 | 33904 | 216 | 4529 | 4740 | 43547 | 11 |
| M1 | 111 | 47160 | 346 | 5510 | 9906 | 63033 | 16 |
| M1 | 359 | 40028 | 764 | 5663 | 9322 | 56136 | 17 |
| Average | 220 | 44367 | 424 | 5055 | 9379 | 59446 | 16 |

Heathrow and Manchester airports can be seen in Table 7.3. The loss of land to airports is trivial, but in use they are similar to roads. De-icing management and drainage are known to cause local problems.

Even more than with the roads, the airports are built on land that would otherwise capable of supporting agriculture. They are near population centres and on flat open land. The level land can lead to their situation being near areas of wetland and open water which are generally more sensitive to pollutants. However, in the case of Heathrow much of the critical load class 3 area is sewage works and small lakes that are used for recreational fisheries.

Selection on the fitness of individuals to survive in transport corridors can occur both within and between species. One of the early examples of selection within species was the demonstration of lead tolerance in verge grasses. The ability of different tillers of grass (Festuca rubra) to grow in soils with elevated lead was shown to be highly correlated with the distance from the kerbside (Antonovics, 1971).

**Table 7.7.** Critical loads for nitrogen for the habitats in the zones of 10 km radius around Heathrow and Manchester airports.

| Class | Critical load kg N/ha | Heathrow km² | Heathrow % | Manchester km² | Manchester % |
|---|---|---|---|---|---|
| 1 | ≤2.8 | 0 | 0 | 0 | 0 |
| 2 | 2.8 - 7.0 | 0 | 0 | 0 | 0 |
| 3 | 7.0 - 14.0 | 17 | 6 | 1 | 0 |
| 4 | 14.0 - 28.0 | 18 | 6 | 64 | 21 |
| 5 | > 28.0 | 96 | 31 | 124 | 40 |
| Not affected | | 177 | 57 | 119 | 39 |

With the increased use of lead free petrol, lead tolerance may be declining, but other metals now need to be considered. Traces of platinum from catalytic converters are now appearing, even though it may be effectively inert, a better understanding of its deposition characteristics are needed. The range of heavy metals emitted by traffic may have important consequences in terms of selection of individuals and species, but in urban areas the deposition may be confounded by other sources.

The construction and management of rail tracks also creates a number of niches that are filled by specific plant species (Sargent, 1984).

### 7.4.2 Animal Mortality

Road-kills are now a greater human driven cause of death to vertebrates than hunting. In the Netherlands 159,000 mammals and 650,000 birds are killed each year (van der Zande et al., 1980). Even though amphibians tend to avoid roads for the reasons described above (see 7.3.2), roads by wetlands and ponds have the highest kill rates and have been suggested to be the highest mortality factor in amphibian dispersion (Ashley and Robinson, 1996, Fahrig et al., 1995). Even larger numbers of invertebrates are killed by fast traffic, but there are no studies of the consequences on national populations.

Although large numbers of animals are killed a study in Britain (Hodson, 1962, Hodson, 1966) suggested that none of the species are threatened by extinction by it. In countries such as Britain, where top predators have been eliminated, road deaths are replacing natural predation for species such as roe deer and are being monitored by deer control societies. Carrion produced by road-kills had produced a new niche that is being filled by species such as the corvids.

Road-deaths are involved directly in financial expenditure, in the deployment of traffic signs, the construction of tunnels and crossing points, and the expense of accidental damage to motor vehicles involved in accidents.

## 7.5 Conclusions and Further Information

Impacts of transport on (semi-) natural ecosystems were explored by a number of examples. Due to the site-specificity a general relationship between transport activity and impact cannot be derived. But results show that damages due to marginal changes can be expected to be only of minor importance. Even though, further work should be done to fill the existing gaps.

Reviews of the ecological effects of roads include (Forman and Alexander, 1998) and (Spellerberg, 1998). Digital datasets describing the transport networks across Europe can be found on the internet at http://www.maproom.psu.edu/dcw/. Land cover information information is available from a number of sources, CORINE data is described on

http://etc.satellus.se/the_data/webmap/eulc250m_L1_3_2.htm.

# 8 Economic Valuation, Sustainability Indicators and Alternative Assessment Techniques

## 8.1 Monetary Valuation: Mortality Risks

### 8.1.1 Background Methodology

The conventional approach for valuing mortality is based on the estimation of the willingness to pay for a change in the risk of death. This is converted into the 'value of a statistical life' (VSL or VOSL) by dividing the WTP by the change in risk. Thus, for example, if the estimated WTP is Euro 100 for a reduction in the risk of death of 1/10,000, the value of a statistical life is estimated at 100 * 10000, which equals one million Euro. The validity of this approach depends on a number of assumptions holding true, not least that there is 'linearity' between risk and payment. Whilst this is likely to be a reasonable assumption over a small range of the risk of death it will be indefensible for large ranges of risk levels.

### 8.1.2 Empirical Evidence

On the basis of the literature review reported in previous ExternE work (European Commission, 1995a and 1999) it has been established that a central range and an average for the Contingent Valuation Method (CVM) group of studies in Europe is Euro 2.9-3.9 million and Euro 3.4 million respectively. These figures are nested within an overall range of Euro 1.0 and 8.8 million covering all studies. No more recent studies have been undertaken which give grounds for adjusting these figures. This mean figure of Euro 3.4 million is therefore adopted as starting point for mortality valuation in this work, reflecting a belief that CVM measures can most comprehensively estimate the relevant preferences for mortality risks. It should be noted, however, that the mean estimate adopted has been derived from averaging unknown errors so that one cannot say what the bias may be. The level of uncertainty reflected in the ranges reported should also inform the interpretation of this data.

#### *8.1.2.1 Methodological and Empirical Applications to ExternE Context*

The values presented above have been derived by considering measures that reduce mortality risks in contexts other than environmental pollution, specifically in relation to improvements in both road safety and occupational safety. It could be questioned therefore whether it is appropriate to directly transfer these values to the context of the ExternE work.

To produce WTP values that are tailored to the ExternE context, the key questions over what types of adjustments to make may be formulated thus:

- Should the VSL values be adjusted for the fact that many of those affected are old and have lower life expectancy?
- Should some adjustment be made for their state of health?
- Should an adjustment be made for the risk context?
- Should some adjustment be made for a lapse of time between the exposure and the impact?

The ExternE team has followed through the argument that it seems unreasonable that the same value should apply in cases where life expectancy is 1 year as applies when it is 36 years; the presumption being that a lower value would apply for mortality effects where the loss of life expectancy is considerably shorter. In order to best quantify this effect, the ExternE work adopts a 'value of life years lost' (VLYL) approach. The conceptual justification for this is that premature death matters because life is shortened and the amount of the shortening is material. The theoretical models that underlie the derivation of the WTP for a change in the risk of death are sensitive to the survival probabilities that the individual faces at the time the valuation is made (see, for example, M.J. Moore and K. Viscusi (1988)). Hence a priori one would expect an empirical estimation of the WTP also to be sensitive to the amount by which life is shortened.

In previous work (European Commission, 1999) the value of a life year lost was derived from the VSL value of Euro 3.36 million (Year 2000 prices). The assumption was made that the observed VSL was the discounted present value of future years, allowing for the survival probabilities. In simple terms the following was assumed to hold:

$$VSL_a(\{P\},r) = VLYL.\sum_{i=a}^{i=T} P_i(1+r)^{-i} \tag{8.1}$$

Where a is the age of the person whose VSL has been estimated, $P_i$ is the conditional probability of survival in year i, having survived to year i-1. T is the upper age bound and r is the discount rate. The above formula assumes that VLYL is independent of age (though it can straightforwardly be modified to allow for the possibility that VSL is age-dependent). This assumption will not in general be valid, but is made as a simplifying one that allows us to get an initial value for the kind of changes in survival probabilities that can be expected in the area of air pollution.

Estimates of survival probabilities for the EU population are available from Eurostat. The male probabilities have been used to estimate VLYL for individuals aged from 35 to 45, which are presented in Table 8.1. The rate of discount has been taken in the range 0 to 3%. The mid-point estimates which are adopted for the "standard" acute VLYL are therefore: Euro 104,760 (0% discount rate); and 165,700 Euro (3% discount rate).

**Table 8.1.** VLYL for different discount rates (Euro, 2000). Corresponding value of VSL is 3.36 million Euro.

| Discount Rate | VOLY with Age 35 | VOLY with Age 45 |
|---|---|---|
| 0% | 89,900 | 119,300 |
| 3% | 150,840 | 180,130 |

The empirical evidence on the value of VLYL is limited. There are two studies that have looked empirically at the risk of mortality in terms of VLYL, e.g. Moore and Viscusi (1988). They derive VLYL from econometric models, where expected lives years lost appear as a specific variable. With a VSL of Euro 6.9 million they derive implicit values of an additional year of life of between Euro 180,000 and Euro 210,000 (2.6% to 3.1% of VSL).

A more recent study is that of Johannesson and Johansson (1996) who use the contingent valuation method to look at the WTP of different respondents, aged 18-69, for a device that will increase life expectancy by one year at age 75. A sample of the results obtained are reported in Table 8.2.

These figures are generally much lower than those of Moore and Viscusi (1988). An implicit discount rate can be computed from the values and this comes out at between 0.4 and 1.3%, which is much lower than the rates estimated in other studies (Cropper, 1994, Viscusi and Moore, 1989). Moreover, the average age of respondents was 42 years. Hence, the additional life year in question was, on average, about 44 years in the future. Whilst questions as to the robustness of the survey method has cast serious doubt on the accuracy of these estimates it may be valuable to consider their use in sensitivity analysis.

**Table 8.2.** WTP (Euro, 2000) for 1 Year of Life at Age 75 and Corresponding Values for 1 Year of Life Immediately

| Age of Payment | WTP for 1 Life Year at 75 | WTP for 1 Life Year Now (3% Discount rate) |
|---|---|---|
| 18-34 | 1,596 | 6,834 |
| 25-51 | 2,019 | 6,026 |
| 52-69 | 2,317 | 3,555 |

Source: (Johannesson and Johannesson, 1996)

### *8.1.2.2 Treatment of Age Dependency*

It was argued that the assumption of a VLYL that is independent of age is not defensible and is important in this context because some of the studies, and much of the clinical evidence, suggest that pollution disproportionately affects the elderly. Hence other data should be considered, and, if necessary primary studies be conducted to collect the relevant information. One criticism of a constant VLYL is that it is not consistent with findings that the VSL does not decline uniformly with age. One such pattern is illustrated in Table 8.3.

Indeed, on the basis of this, and other data, a UK government working group (Stationary Office, 1999) recommends applying adjustment factors to the baseline WTP-based VSL. In recognition of the fact that epidemiological studies only report data on the whole population at risk over 65, the group suggests using an adjustment factor of 70%, weighted to take into account of the average distribution of ages within the over 65 age group in the general population.

This age-dependency relationship was investigated further. The fact that VSL takes this shape as a function of age in fact suggests that there are two forces at work. First, individuals become more risk averse as they grow older and second that they face a reduced life expectancy. The first would raise the VSL and the second would reduce it. A simple reduced form equation might then take the form:

$$VSL = A.AGE^{\alpha} LE^{\beta} \tag{8.2}$$

Where LE is life expectancy, and $\alpha$ and $\beta$ are parameters of the VSL function. This would imply that the valuations of different persons would be age dependent but they would also be dependent on life expectancy and would allow us to replace the 'normal' life expectancy with the reduced life expectancy that is characteristic of those affected by air pollution.

The above equation was estimated for data for European males, using a VSL age profile as given in Jones-Lee (1989) and this is given in Table 8.4.

Assuming that for particulate pollution mainly persons over 65 are affected, the corresponding value for mortality would be between Euro 38,970 and Euro 84,450, which is lower than the range suggested by the simple ExternE approach. However, these low figures are not inconsistent with the estimates from other studies, as reported above.

**Table 8.3.** Mean Estimates of VSL for Different Ages as a Percentage of VSL at Age 40

| | | | | | | | | | | | | |
|---|---|---|---|---|---|---|---|---|---|---|---|---|
| Age | 20 | 25 | 30 | 35 | 40 | 45 | 50 | 55 | 60 | 65 | 70 | 75 |
| VSL at age 40 | 68 | 79 | 88 | 95 | 100 | 103 | 104 | 102 | 99 | 94 | 86 | 77 |

Source: (Jones-Lee and al, 1985)

**Table 8.4.** Estimates of the VSL for a Person with 0.75 Years to Live (Euro 2000)

| | | | | | | | | | |
|---|---|---|---|---|---|---|---|---|---|
| Age | 30 | 40 | 50 | 60 | 65 | 70 | 75 | 80 | 85 |
| VSL | 3894 | 9610 | 18298 | 30952 | 38971 | 48250 | 58773 | 70917 | 84454 |

To conclude, the regression analysis presented above suggests that two separate effects can be identified when examining the evidence of VSL dependency on age: a life expectancy effect and a pure age effect. The ExternE approach explicitly recognises the importance of life expectancy by using measures of VLYL, rather than VSL. The evidence given above implies then that if an adjustment is to be made to the VLYLs on the basis of age, it should increase with age. At present the empirical evidence does not appear sufficient to adjust the VLYLs in relation to age, one way or another.

### 8.1.2.3
### Pure Health Impairment

An important assumption underpinning the analysis above is that the person at risk is in an 'average' state of health for someone of his or her age group. Clinical evidence would suggest, however, that individuals who die prematurely from air pollution already are in a state of severely impaired health. If a person's quality of life is severely impaired this may effect his or her WTP for a reduction in the risk of death. The question therefore arises as to whether it is appropriate to adjust the VLYL to account for impaired health (or quality of life).

There is little empirical evidence however, substantiating a significant positive correlation between WTP and the quality of life that individuals expect during an additional year of life, although health service professionals do use a 'Quality Adjusted Life Years' (or QALY) approach in their decision making process.

Work has been undertaken in the UK, (Stationary Office, 1999), which examines the effect of assuming that WTP is proportional to an index of health-related quality of life. Taking the loss of one year, for example, the suggested quality of life adjustment factor is 53% ( ± 1 SD: 26% to 92%). However, the quality of life measure used in the study (from COPD) may not be representative of all types of respiratory illness. Also, there is no evidence as to the extent to which the quality of life might depress WTP; it is quite possible that WTP might change more or less than proportionately to changes in the quality of life. There is therefore a high degree of uncertainty associated with such an adjustment, which is exacerbated by other methodological problems highlighted in previous ExternE work. This suggests that no adjustment should be made to VLYL in this regard at present.

### 8.1.2.4
### The Context of Air Pollution Risks

Factors influencing the nature of the risk itself may affect individuals' WTP for avoiding particular types of risks. Such factors include:

- The type of health effect (e.g. people may dread a chronic risk more than an acute risk).
- The risk context (e.g. evidence shows that people regard involuntary risks, over which they have no control, as worse than others).
- The futurity of the health effect (e.g. effects which happen today are generally worse then those which occur tomorrow).

Whilst there are very good reasons to believe that the risk context is likely to be important in determining the WTP values in question, it is too soon to make such an adjustment for policy purposes. Empirical evidence of this effect is therefore required as a matter of urgency.

#### *8.1.2.5*
#### *Chronic Mortality*

A model is outlined that account for chronic effects – i.e. that once exposed, impacts can occur with a latency that is variable, but could be as much as 50 years. Once the impact is underway, survival probabilities are altered to an extent that is undetermined. Although there are great uncertainties, for representative populations, the number of years of life lost was estimated, as a result of an increment in the hazard in year i, in each future year. Call this $YOLL_i$. In addition the total number of years of life lost in the population ($YOLL_{tot}$) was estimated for such a case. The value to be attached to a case of chronic mortality can then be calculated as $VOLY_{chronic}$. The formula proposed for this is:

$$VOLY^r_{chronic} = \sum_{i=1}^{i=T} \frac{YOLL_i}{YOLL_{tot}} * \frac{VOLY^r}{(1+r)^{i-1}} \tag{8.3}$$

There will be different values for chronic illnesses, depending on the values obtained for $YOLL_i$ and $YOLL_{tot}$, as well as on the value of the discount rate. The numbers derived for chronic mortality in the case of PM exposure are: 104,760 Euro (0% discount rate) and 96,500 Euro (3% discount rate).

It is important to remember that for cases with a long latency period it will be appropriate to add a cost for the period of pain and suffering in addition to the values of life years lost. The presentation of morbidity values below facilitates an appropriate addition.

#### *8.1.2.6*
#### *Impacts of Latency on the VSL*

If exposure to air pollution today causes the risk of death to increase T years from now the WTP to avoid that risk is not likely to be the same as that associated with an increase in the risk of death now. (This discussion does not apply to acute effects, which tend to be realised immediately following the precipitating pollution event. The same cannot be said of chronic and latent effects, which result in death after the onset of exposure to the pollutant). The accepted way to deal with such latency effects is to discount future risks. Hence, if the WTP for an immediate reduction in risk is £X, then the WTP for a reduction in a risk with a latency period of T years is $X(1+r)^{-T}$ . The key question, of course, is what value should r take? This is discussed in a later section but it is recommended to use 0%, 1% and 3%.

### 8.1.2.7 Conclusion on the WTP Unit of Measurement for Mortality Risk Valuation

The discussion above reveals that, both in empirical and methodological terms, the valuation of mortality risks in the ExternE context is in its infancy. Although recent developments in the EU and US support a move towards using VLYLs, more work needs to be done before a first set of estimates are made. For the present, it is recommended that Equation 8.1 be used to estimate the costs of air pollution via mortality for all cases, using the values of VLYL as given in Table 8.1, and replacing the normal conditional survival probabilities with the probabilities associated with the particular case. Note, however, that the exposure-response function for chronic effects of air pollution implicitly covers reduced life expectancy from acute effects as well as chronic. Investigation of the sensitivities of different approaches suggests that quantification based solely on the recommendations made above for chronic effects will give an acceptably accurate result.

The methodological advantages of using a VLYL measure are first that it allows greater flexibility in valuation, and second that it is one that clinicians are more comfortable in estimating in that it brings the WTP approach closer to the QALY approach. In addition, it is more amenable to valuing chronic mortality as it is easier to express chronic impacts in terms of life years lost.

Critics of the VLYL approach (NERA and CASPAR, 1997) have argued that the VLYL approach is invalid because it assumes a constant value of a life year over the remaining lifetime, and the evidence suggests that life year values vary with age (among other factors). It is correct to say that VLYL values should not be constant and independent of age, but that does not make the method invalid. It is perfectly possible to develop values for life years that vary with age, and indeed the simple model presented above gives a way of doing so. The other issue raised by critics of the VLYL approach is that it is invalid to add up life years, i.e. saving one life year for two people is not equal to saving two life years for one person. It is correct to say that the adding up of life years cannot be carried out in a simplistic fashion. The value of future years is less than that of present years on account of the discount factor and any application of the VLYL method has to take account of that. As far as adding up values across individuals, there are problems associated with such a procedure but they are no more or less than adding up VSL values across individuals.

In general it is not possible to derive a VLYL value that is independent of age and that declines at a constant discount rate over the remaining lifetime (The normal assumption is that VLYLs decline over time at a constant discount rate and it is this assumption that is challenged). As noted above, it is possible to allow for different VLYLs for people of different ages. The use of a constant discount rate for 'adding up' the WTP for a change in the present risk of death is a simplification, as is the use of VSLs that are independent of risk. It is worth checking how good a simplification it is, but the fact that it may not be consistent with a theoretical model that itself is based on unrealistic assumptions about individual rationality is not a serious criticism of the method. As long as individuals have a coherent notion of life years and a WTP for life years, it is possible to make empirical estimates of these categories. Indeed that is exactly what some researchers have done and others are presently doing. Hence in our opinion, the use of VLYL as a measure of risk is both valid and useful as a means of valuing changes in mortality risk.

### 8.1.3 Measurement of the Cost Per Life Saved (CPLS) / Cost Per Life Year Saved (CPLYS)

It is well known that both the VSL and VLYL figures are higher than the values that are implicitly or explicitly used by policy makers in EU countries, and represented by the resource cost of life (year) saved. For example, values relevant to the ExternE context that have been identified are given in Table 8.5.

For clarification, the methodology of this measurement is outlined below before drawing conclusions as to how this disparity should be interpreted in ExternE.

The concept of a cost per life saved is deceptively simple. It takes a given intervention and measures the costs associated with that intervention. It also measures the number of persons whose death is delayed as a result of the measure. Dividing the cost by the number of lives saved gives the cost per life saved.

In practice there are a lot of issues that have to be resolved in constructing such a measure. First, costs are incurred over a period of time and have to be converted into a present value. This requires a discount rate, which has to be chosen. Second, estimates of lives 'saved' require data on the effects of the programme over a

**Table 8.5.** Values of CPLS used in different interventions (1998 prices)

| Source | Date | Central Value (Euro) 1999 | Comments |
|---|---|---|---|
| UK Department of Environment, Transport and the Regions | 1995 | 1,376,400 | Value of prevention per fatality, based on WTP and other considerations |
| U.K. National Radiological Protection Board (NRPB) | 1993 | 1,322,150 – 2,115,750 | NRPB recommends a value of Euro 30,000 per man sievert for public exposure and Euro 75,000 for occupational exposure. Values are based on crude extrapolation of life years lost. |
| U.K. Railtrack | 1995 | 1,444,600 – 3,888,950 | From Railtrack Railway Safety Case |
| U.K. Department of Health | 1997 | 640,500 | Based on a CPLYS of Euro 15,000 for a healthy life year, a discount rate of 1% and age of 35 years. |
| Norway, Ministry of Finance | 1998 | 1,410,000 | Used for evaluation of public projects |
| Norway, Directorate for public Roads | 1995 | 2,162,854 | Only sector to have explicit value |
| Germany, Ministry of transport | 1997 | 814,488 | Uses human capital approach |
| Netherlands, Ministry of Health | 1998 | 15,000 | Note: CPLYS value |

period of time on the death rates of a base population. At best this is an imprecise exercise, with problems of estimation and interpretation of data.

The conventional response of economists to this discrepancy between implicit and WTP values is to argue that mortality risk valuation research should inform the implicit values adopted by policy makers. The counter-argument, however, in this context is that the WTP values currently proposed are so much greater (an order of magnitude) that the level of health expenditure implied would not be countenanced by a public administration. In addition, it can be argued that the current level of expenditure is implicitly approved by the population through the democratic process i.e. that the administration is elected, in the countries considered, on the basis of a commitment to individual public services.

One way to reconcile the discrepancy between the values considered is to accept that the WTP values are subject to many measurement errors and uncertainties so that they are in fact over-estimated by a factor of 10. Statistical analysis on existing data supports this possibility. Therefore, in view of this high level of uncertainty, and the reasonably open and democratic process by which investment in health is decided, it could be argued that both the scientific/economic WTP estimates and the revealed preference estimates should be used in the damage estimation until further empirical evidence is forthcoming.

## 8.2 Discount Rates

Discounting is the practice of placing lower numerical values on future benefits and costs as compared to present benefits and costs The reasons why individuals, and society as a whole, might wish to discount future costs and benefits have been described in detail in European Commission (1995a). It was noted that there remains considerable debate about the appropriate rates to be used in the context of the environment and environmental decision making. This debate has been reflected in on-going discussions within the ExternE project team. Our conclusions on appropriate rates to use in the ExternE project context are outlined below.

The two main contenders for the choice of the social discount rate are the individual rate of time preference, based on the rate of individual trade-off between present and future benefits and costs, the opportunity cost of capital, based on the marginal productivity of capital, and the social rate of time preference, based on ethical considerations about the rate at which present and future costs should be traded off.

There is no simple solution to the problem of which discount rational to use and it has been at the centre of the debate about inter-temporal valuation for a long time, and in many other contexts. In general, here the view is taken that since the type of decisions which the ExternE project is interested in helping with are those taken by public bodies on behalf of society, for public policy it is of interest how the social time preference (discount) rate should be derived. Therefore, the individual components and their values have been briefly reviewed below.

The social time preference rate attempts to measure the rate at which social welfare or utility of consumption for society falls over time. Clearly this will depend on the rate of pure time preference, on how fast consumption grows and, in turn, on how fast utility falls as consumption grows. It can be shown that the social rate of time preference is (for details see Ray (1984)):

$$i = z + ng \tag{8.4}$$

where z is the rate of pure time preference (often described as being due to impatience), g is the rate of growth of real consumption per capita, and n is the percentage fall in the additional utility derived from each percentage increase in consumption. (n is referred to as the 'elasticity of the marginal utility of consumption'). With no growth in per capita consumption therefore, the social rate of time preference would be equal to the private rate, z.

The role of pure time preference, z, has been questioned particularly in inter-generational assessments. In terms of personal preferences, no one appears to deny the impatience principle and its implication of a positive individual discount rate. However, arguments exist against permitting pure time preference to influence social discount rates, i.e., the rates used in connection with collective decisions. For example, it has been argued that public policy should reflect collective, not private interests (Sen, 1982). There are indeed strong arguments for paternalism though there is no agreement on this issue. To the extent that the value judgement underlying liberal economic tradition, i.e. that individual preferences should count for social decisions, continues to have weight – not least in the welfare theoretic methodology on which other economic valuations used in ExternE are based –the retention of two values for z is suggested; 0 and 1.5.

The per capita income growth rate, g, is clearly very variable. However, the long term sustainable growth rate of 1.5% seems to the project team to be reasonable. It also coincides with the assumption used by other bodies involved in the development of environmental policies that have inter-generational impacts (e.g. IPCC (1995)). It should be noted that in the case where there are explicit assumptions about economic growth, differentiated by time period and economic region, it is therefore valid to use these assumptions in determining the appropriate discount rate.

A value of 1 for the elasticity of the marginal utility of consumption component, n, has been widely adopted to date. However, it is recognised in the project team that critiques of consumerism have argued that this value may be too high in wealthier countries. To acknowledge this as a possibility the use of a second value of 0 is suggested.

From the discussions on the composition of the discount rate values of 0% and 3% are derived. These are comprised in the following way:

| z | g | n | i |
|---|---|---|---|
| 0 | 1.5 | 0 | 0 |
| 1.5 | 1.5 | 1 | 3 |

For additional sensitivity analysis it is recommended that an intermediate value of 1% discount rate be adopted.

## 8.3 Revised Morbidity Values

In the two years that have passed since Navrud (1997) there have been three new European projects using the Contingent Valuation (CV) method for valuing morbidity impacts from air pollution, one in Helsinki (Finland), one in Strasbourg (France) and Kehl (Germany), and a 5-country study (Amsterdam (NL) Lisbon (P), London and surroundings(UK), Oslo (NO) and Vigo (ES)). In the following the results of each of these three studies will be reviewed.

### 8.3.1 The Helsinki Study

Otterström et al. (1998) conducted a CV mail survey of 2000 households in Helsinki to estimate the willingness-to-pay (WTP) to prevent eight symptoms: Symptom day, Asthma attack day, Restricted Activity Day (RAD), Emergency Room Visit (ERV), Hospital Admission (HA) case, Children's cough day). The respondents were asked to state their WTP to avoid the welfare loss (pain and loss of free time). Medical care costs and the value of lost labour time were added to come up with the unit costs shown in Table 8.6.

**Table 8.6.** Economic values of eight morbidity symptoms. The unit estimates includes individuals' welfare loss (based on a CV study in Helsinki), medical care costs and the value of lost labour time. 1996-FIM converted to Euro (1 FIM = 0.16 Euro)

| Morbidity symptom | Unit Value (Euro) | Percentage medical care costs and value of lost labour (%) |
|---|---|---|
| Symptom day | 15 | 35 |
| Asthma attack day | 22 | 17 |
| Restricted Activity Day (RAD) | 90 | 63 |
| Emergency Room Visit (ERV) | 200 | 80 |
| Hospital Admission (HA) case | 1679 | 85 |
| Children's cough day | 34 | 19 |

Source: Otterström et al. (1998)

### 8.3.2 The 5-Country Study

The most extensive of the three new European morbidity valuation studies is a 5 country CV-study valuing six morbidity episodes in the Netherlands, Norway, Portugal, Spain and the UK (CSERGE et al., 1999). Table 8.7 gives the results (when no context was given, i.e. the respondents were just asked to state their WTP to avoid the episodes and were not told that the episodes could be caused by air pollution).

#### 8.3.2.1
#### *Study Description*

The same contingent valuation (CV) questionnaire (and same survey mode; in-person interviews) was carried out in Portugal (Lisbon), the Netherlands (Amsterdam), Norway (Oslo), Spain (Vigo) and the UK (London and surroundings) to estimate the willingness to pay (WTP) to avoid certain health effects that are associated with air pollution. Air pollution is an example of a contextual feature of the study and policy sites, i.e. it links the WTP to a health effect arising from a specified context. One issue is whether context affects WTP, e.g. whether people are WTP more or less for a reduction in what is effectively the same health effect if the context is air pollution rather than, say, some other form of pollution or risk. To test for contextual effects, the CV studies were carried out in a context free mode, i.e. by seeking WTP for changes in given health effects only. The basic idea is that context free WTP provides a 'core' value that might be transferred, with other factors then being introduced to modify that core value for a given policy site.

#### 8.3.2.2
#### *Results*

The WTP results from the five CV surveys are given in Table 8.7. The parametric results are shown since these tend to be more conservative.

The results suggest some consistent patterns (CSERGE et al., 1999):

- WTP is highest for the longer lasting episodes, as would be expected
- Spain and Norway have the highest WTP for all episodes
- UK and Netherlands have the lowest WTP
- Portugal shows more variation, being the country with the highest WTP for 'eyes' and a low WTP for 'cough'

The estimates were subject to tests for statistical significance, i.e. to see if the different valuations reflect more than random error. These tests suggest that Norway and Spain tend to have significantly higher WTP to avoid ill-health episodes and the UK has significantly lower WTP.

**Table 8.7.** Unadjusted WTP for health endpoints in the 5-country study (UK£ per episode (in 1997-prices ). UK£1 = 1.53 Euro).

| | Pooled 5 countries | Netherlands | Norway | Portugal | Spain | UK |
|---|---|---|---|---|---|---|
| Hospital | 306 | 283 | 301 | 300 | 426 | 164 |
| Casualty | 158 | 128 | 239 | 185 | 146 | 131 |
| Bed | 97 | 71 | 119 | 88 | 113 | 83 |
| Cough | 27 | 28 | 36 | 28 | 39 | 20 |
| Eyes | 35 | 40 | 31 | 70 | 53 | 14 |
| Stomach | 35 | | | 61 | | 26 |

Source: CSERGE et al. (1999); Note: parametric estimates only, rounded to nearest £

## 8.3.3 Testing for Benefit Transfer (BT) across Countries

It is estimated that any given estimate transferred to a policy site would over or understate the 'true' value by 36-44%. Interestingly, the error increases when the 'functional' transfer is made, i.e. when socio-economic factors are taken into account (CSERGE et al., 1999).

The previous results suggest that transfer of average results across EU countries involves an error of about one third if no account is taken of socio-economic differences, and between one third and one half if adjustments are made for socio-economic differences. What is being transferred are context-free mean WTP estimates.

### *8.3.3.1 Does Context Matter?*

A further CV study was conducted in each of the five countries, this time citing air pollution as the context. The 'without context' results are then compared to the 'with context' results and tests of statistical significance were applied. In all cases, bar two from Portugal, there was no statistically significant difference between the with and without context answers. Context does not therefore affect WTP. In the two cases from Portugal, the 'with context' WTP was lower than the without context WTP.

The absence of contextual effects is encouraging for policy purposes since, subject to the errors in site-to-site transfer, it suggests that values can be transferred across contexts where the source of the problem changes. (How far this extends to very different risk contexts, e.g. radiation risks, has not, however, been tested here).

A more detailed study on context was conducted for the UK. Here the 'with' and 'without' context cases were supplemented by a third survey in which the 'with context' scenario contained potential policy measures to reduce air pollution. Thus, the context is expanded to include an added dimension. In this case context did affect WTP, with WTP being significantly higher in the 'with policy' case.

### *8.3.3.2 Sequencing*

An issue widely debated in the contingent valuation literature is whether WTP answers are sensitive to the order in which questions are asked. Varying the order of questions was found to have no significant effect on WTP. Again, this is reassuring for policy implementation of BT (CSERGE et al., 1999).

## 8.3.4 The Strasbourg and Kehl Study

Rozan (1999) performed a CV study in two cities, Strasbourg (France) and Kehl (Germany). She used a CV scenario where respondents were asked to state their willingness to pay (WTP) for a 30-40 % reduction in the risk of contracting light respiratory symptoms (eye irritation, headache, shortness of breath, coughing etc.). However, the change in the number of days the respondents would experience the different symptoms was not specified in the CV scenario. Therefore, unit values for the different symptoms cannot be estimated.

Rozan (1999) also tested for the impact of telling the respondents that the reduction in morbidity symptoms was due to a public health program versus reductions in air pollution, and found a significantly higher WTP for the subsample that was told that the reduction in health effects was due to an air pollution program. This result is consistent with the results from the UK test of the "with context" - option with policy elements in the 5-country study.

### 8.3.5 New European Estimates of Productivity Loss and Health Services Costs

CSERGE et al. (1999) also provide a list of Productivity Loss from absenteeism (Table 8.8) and Health Service costs of going to Emergency rooms and hospitalisation in selected EU-countries (Table 8.9). These Costs of Illness (COI) have to be added to the individual welfare loss estimates presented in table 8.7 to provide the economic estimates.

**Table 8.8.** Estimated per diem productivity costs from absenteeism in the European Union (ECU; 1997 prices).

| Country | Per diem productivity cost in ECU |
|---|---|
| Belgium | - |
| Germany | 68.46 |
| Greece | 18.06 |
| Spain | 46.03 |
| France | - |
| Ireland | 48.64 |
| Italy | 53.56 |
| Luxembourg | 121.11 |
| Netherlands | 61.34 |
| Austria | - |
| Portugal | 23.30 |
| Finland | - |
| Sweden | 54.07 |
| United Kingdom | 53.35 |
| European Union Average | 54.97 |

Source CSERGE et al. (1999)

**Table 8.9.** Health service costs (Local currencies)

| Country | Emergency Room – cost per visit | Hospitalisation – cost per inpatient day |
|---|---|---|
| Belgium | - | BF 8000 (1995 prices) |
| France | FF 150 (1996 prices) | FF 2020 (1994 prices) |
| Germany | DM 37.8 (1996 prices) | DM 515.47 (1995 prices) |
| Italy | L 31075 (1994 prices) | - |
| Netherlands | - | G 706 (1995 prices) |
| Spain | P 14174 (1995 prices) | - |
| United Kingdom | £ 52.24 (1997 prices) | £187 (1996 prices) |

Source: CSERGE et al. (1999)

### 8.3.6 Suggested Revision of Morbidity Estimates

Since the 5-country study is the most comprehensive study in terms of European countries covered and is a well designed CV survey, it is suggested that most weight should be put on the results from the 5-country study (CSERGE et al., 1999) when updating the list of economic values used in ExternE. This study most probably produces the most representative values for the EU as a whole. However, for those morbidity symptoms that are not covered in this study, it will be necessary to rely on the results from the other two studies.

One cough day is estimated to be 41 Euro/day. The same value should be applied to minor RAD (restricted activity day) and symptom day (note that this is probably a low estimate for a symptom day as one day with mildly, red watering, itchy eyes and runny nose (EYES) is valued at 53.5 Euro.

For RAD the BED value can be used, which is 148 Euro. Since this value is to avoid an episode lasting three days, the estimate has to be divided by three. To this should be added the EU average per diem productivity loss. Thus, one RAD can be valued at 104 Euro (i.e. 49 + 55 Euro).

For one case of ERV (Emergency Room Visits) the CASUALITY episode estimate of 242 Euro could for simplicity be used for all three ERV estimates in ExternE (ERV for croup pre-school child, COPD and asthma). The episode assumes 4 hours in the Emergency room and 5 days at home in bed. To the estimate above one should add the estimated productivity loss for 5 days in bed, which is 275 Euro. The health service costs of an Emergency Room visit should also be added (i.e. 80 Euro). Thus, the economic value of ERV is 597 Euro per case.

For a case of HA (Hospital Admission), the HOSPITAL episode estimate can be used as starting point. This episode is described as admission in hospital for treatment of respiratory distress, and assumes that the patient stays in hospital for treatment for 3 days followed by 5 days in bed. The individual welfare loss of this episode was valued at 468 Euro in the CV study. In addition there will be productivity loss for 8 days of 440 Euro and costs of hospitalisation for three days at 858 Euro. This gives a total economic estimate of 1766 Euro/case for HA from respiratory distress. For the enpoints considered, i.e. respiratory HA, congestive heart failure HA, and cerebrovascular HA, adjustments were made according to productivity losses and hospital costs This resulted in values of 4,320, 3,260 and 16,730 Euro respectively. Otterström et al. (1998) found an estimate of a general HA episode caused by air pollution of 1679 Euro. Thus, this recent empirical evidence indicates that the economic value for HA, at least from respiratory HA, should be lower that the estimate currently used in ExternE.

Otterström et al. (1998) found an economic estimate of an asthma attack day in Helsinki of 22 Euro (including medical care costs and value of lost labour, which accounts for 17 % of this estimate). In comparison Navrud (1998) found a WTP to avoid a day of asthma attack (excluding medical care and lost productivity costs) of 67, 139 and 295 Euro per case for adult non-asthmatics, adult asthmatics and asthma attack among the respondents ´own children, respectively. These were the values for the sub-sample of respondents that were asked to express their WTP to avoid one additional day of asthma attack (in addition to what they had experienced the last 12 months). The corresponding asthma attack daily values for the sub-sample that were asked to value 14 additional days were 14, 15 and 42, respectively. These values correspond better with the observed value from the Helsinki study. If a value for an asthma attack day among those that have asthma is required, it would be recommended using the results from those that have asthma, and use the marginal value from the 14-days sample in Navrud (1998), i.e. 15 Euro per day. However, productivity loss of one day should be added, which is 55 Euro per day. Thus, an asthma attack (for adults that have asthma ) should be valued at 70 Euro/day.

## 8.4 Sustainability Indicators

### 8.4.1 Introduction

The aim of the sustainability indicator task in this phase of the ExternE project is to develop indicators that measure progress towards sustainable development, but also a way to handle the impact that are currently not valued, in a consistent and transparent way. The impact pathway damage function approach of ExternE builds on the weak sustainability approach, which does not take account of any thresholds in critical natural assets or limits to the substitutability between natural resources and produced assets. Therefore this approach, which is based on monetary indicators, has to be supplemented with a set of strong sustainability indicators, which are measured in physical units.

These physical indicators should show how far we are from clearly defined sustainability goals, and the progress made towards these goals. In addition they should adhere to the OECD criteria for good indicators, and the ExternE criteria of transparency and consistency. The critical levels concept satisfies these criteria best. It was considered how the critical loads/levels concept can be used as strong sustainability indicators. Two main types of sustainability indicators were looked at, which both measure the distance to a sustainability goal: i) the ecological approach of critical loads for acidification damages to ecosystems using indicators like "exceedance area" and "relative exceedance weighted area" (the latter also takes into account the extent of the exceedance), and ii) the engineering approach of life cycle analysis (LCA), using the Ecoscarcity index as the indicator.

These two strong sustainability approaches were compared with the weak sustainability approach of economic valuation. It was tried to develop links between indicators of these two sustainability approaches. This was done by performing a more detailed analysis of a Contingent Valuation (CV) study of increased fish populations in Norway due to reduced exceedance of critical loads for sulphur (S) (Navrud, 1999), and by analysing an empirical study, using both CV and the LCA method ECO (Ecological Scarcity Method), to value the net environmental impacts from re-injection of produced water from oil drilling into reservoirs in the North Sea (which reduce marine ecosystem damages, but increases greenhouse gas emissions due increased use of energy in the re-injection process (1998). These two case studies provide more insight into the development of an algorithm between the "distance to sustainability" - indicators of Critical loads and ECO, and monetary valuation. The results are encouraging, but there are still many problems to overcome, and it is premature to generalise from these case studies.

### 8.4.2 Linking Strong and Weak Sustainability Indicators

The analysis in Navrud (1999) is based on a CV survey (in-person interviews) of a representative sample of 1000 households in Norway. In the CV scenario description maps showing the change in Norwegian fish stocks as a result of a full implementation of the second international sulphur protocol are presented. These maps are based on calculated changes in critical load exceedance for sulphur as a result of this protocol, and dose-response functions between critical load exceedance and the state of the fish stocks. Thus, from this study economic values for exceedance of critical loads can be calculated. The results can be presented in different units. Previously they were presented as weight-based units (value per tonne of sulphur, and per tonne of sulphur exceeding the critical load), but for benefit transfer purposes an area-based unit corrected for differences in population density between Norway (13.5 inhabitants per km$^2$) and EU (115.4 inhabitants per km$^2$) was developed. Taking into account differences in population density, an EU-wide estimate of 69.2-131.6 million Euro / km$^2$ / year can be derived. This is the economic value of avoiding fish losses from exceedance of critical loads for acid deposition (sulphur and nitrogen).

### 8.4.3 Comparison of the LCA-Method ECO and Monetary Valuation Method CV

Knudsen et al. (1998) conducted a small CV survey (in-person interviews) of 50 randomly selected households in Stavanger, Norway The respondents were asked to value the net environmental impacts from re-injection into the reservoir of produced water from oil drilling in the North Sea (which reduce marine ecosystem damages from bio-accumulation of emissions of PAH and heavy metals, but also increases greenhouse gas emissions). At the same time they also used the Ecoscarcity index of the Life Cycle Analysis (LCA) method Ecological Scarcity Method (ECO, also named BUWAL). The Ecoscarcity index measure the value of reduced emission of a substance as the distance to a sustainability goal, either a goal based on natural science (e.g. critical loads) or a goal set by politicians (e.g. national goal to reduce emission of some pollutant by X %). While the CV study valued the aggregate net environmental impact directly, the ECO method used estimates of ecopoints for one unit of the different substances emitted (ecopoint values provided by the National Pollution Control Authority (SFT) in Norway, which were based on a literature review of the LCA literature), multiply by the number of units emitted of each substance and aggregate over the different substances. However, since ecopints estimates for all substances in produced water do not exist, an aggregate ecopoint for produced water was used with the OSPAR convention of stabilising the amount of produced water as the sustainability goal. Ecopoints was then summed for the following substances; produced water, $CO_2$-equivalents, $NO_x$, NMVOC, CO and $SO_2$ Net reduction in ecopoints from re-injection of produced water was estimated at 6.92 x 1010. Then ecopoints were converted to monetary estimates, by dividing the ecopoint estimate for 1 tonne of sulphur by the monetary estimate for environmental acidification damages from 1 tonne of sulphur that SFT uses (and which is based on a combination of previous environmental valuation studies and expert assessment). 33.6 million ecopoints corresponded to an economic value 9.800 NOK. This conversion implicitly assumes a constant environmental damage cost per ecopoint. In this way the ECO results can be compared with the results from the CV survey on the same monetary scale. Knudsen et al. (1998) showed that the CV study gave only somewhat higher (28 % higher) economic value than ECO. This relatively close level of correspondence between strong and weak sustainability indicators is encouraging, but one should be careful in drawing conclusions since there are many uncertainties in both approaches, and the method of comparison of results that was used.

### 8.4.4 Decision Rules

To conclude, it seems like it is difficult to use critical loads as a general strong sustainability indicator, since critical loads only exists for a few substances and a few ecosystems. The LCA method ECO has a similar problem as clear sustainability goals for many substances do not exist. Even if the comparison of CV and ECO methods in Knudsen et al. (1998) gave quite similar results, it is premature to generalise from this one study. Thus, for the time being it is suggested that the decision rule described below should be applied for those impacts that do not have economic values, when applying the methodology of this study in cost-benefit analysis (CBA) of new EU directives and strategies (where environmental and health impacts would be benefits, rather than costs)

A CBA of a new EU directive which will reduce air pollution (and thus create environmental and health benefits). It is assumed that the costs (C) of implementing the directive is known, and that the benefits are calculated for the impacts where economic values exists ($B_1$).

1. If there are positive net benefits of the valued impacts ($NB_1 = B_1 - C$), unvalued benefits ($B_2$) will just add to the net benefits, and strengthen the conclusion that the directive have gross benefits ($B_1 + B_2$ ) larger than gross costs (C), and should be implemented
2. If there are negative net benefits of the valued impacts ($NB_1 = B_1 - C$), it depends on the size of the unvalued benefits ($B_2$) whether the recommendation would be to implement the directive or not.

   a) If $B_2 > NB_1$, the directive have gross benefits ($B_1 + B_2$ ) larger than gross costs (C), and should be implemented

   b) If $B_2 < NB_1$, the directive have gross benefits ($B_1 + B_2$ ) smaller than gross costs (C), and should not be implemented

(This decision rule is also called the Krutilla-Fischer algorithm)

$NB_1$ should be converted from annual net benefits to a value per household in the affected population, which should be compared to the scarce valuation literature on the unvalued impacts (using existing databases of valuation studies) and crude benefit transfer techniques (i.e. expert assessments - based on CV practitioners experiences from previous CV studies of similar impacts to those considered), to decide whether a) or b) applies.

## 8.5 Evaluating "Non-Marginal" Changes

### 8.5.1 Introduction

The valuation methodology currently employed within ExternE assumes that the impact of any policy intervention under consideration is "marginal", and therefore the prices of affected goods and services do not change. Subject to this assumption, the benefit/cost of an intervention may be valued by multiplying the anticipated change in the quantity demanded by the appropriate, revealed or stated, price. In some cases however, a policy intervention may result in "non-marginal" impacts in the targeted market(s), which may in turn change the equilibrium price. We are now faced with the dilemma of which price to use in the policy appraisal – the initial price or the price that prevails subsequent to the implementation of the intervention? Moreover, depending on the nature of interrelationships between markets, a price change in the targeted market may disrupt price and quantity equilibria throughout the economy. A further question therefore arises - how many markets must be considered to derive an accurate measure of the "true" welfare effect of the policy?

In short, when a policy has a significant ("non-marginal") effect and prices change as a result, it must be resorted to the relevant supply and demand curves in order to attach an appropriate "social" valuation to the resulting changes in output or inputs.

The cost-benefit literature distinguishes three situations in which prices change as a result of policy interventions; the analytical treatment of each case is different. The three cases are:

1. Situations where price changes in the directly affected (targeted) market.
2. Situations where prices change in indirectly affected (related) markets.
3. Situations where price and quantity equilibria change throughout the economy.

Welfare changes in the first case are evaluated within a partial equilibrium framework. The second case may also be dealt with within a partial equilibrium framework, albeit an "extended" one. This may take the form of a sectoral analysis. The third case, in which prices potentially change throughout economy, must be examined within a general equilibrium framework. Each of the these cases will be considered in turn.

### 8.5.2 Valuing a Change in Welfare: The Case of a Single Affected Market

It is demonstrated that if an intervention yields "non-marginal" changes thereby rendering the ceteris paribus assumption invalid, the simply "multiplication" approach may no longer represent an appropriate approach to valuation. However, for small price changes the error is negligible. The original ExternE approach, identified in the formula above is justified by assuming that the demand curve is horizontal.

The next task is to identify situations relevant to ExternE in which "prices" change, and identify the appropriate valuation framework. Two special cases can be identified immediately, where the simple "multiplication" approach may still be used even in the presence of "non-marginal" changes in output. These are:

- If the demand curve for a good is perfectly elastic and the anticipated change in air quality affects the supply-side of the market.
- If the supply curve is perfectly elastic, and the anticipated change in air quality affects the demand-side of the market.

In general, the magnitude of the elasticity of supply for any particular commodity depends on: (1) the elasticity of the supply of the inputs used in its production; and (2) the amount of time elapsed since the price change. If a primary input is highly elastic, for example, then the supply of the product will also be highly elastic. Conversely, the supply of a product will be inelastic if the supply of a primary input is also inelastic. In terms of the length of time that has elapsed after a price change affects the responsiveness of the quantity supplied, as the time frame increases (e.g. from momentary to short-term to long-term) the more elastic the supply curve becomes..

Of course, this is only the case when the affected market is considered in isolation. If these assumptions cannot be made, then it is necessary to estimate the parameters of the supply and demand functions.

## 8.5.3 The Indirect Effects of Price Changes

### *8.5.3.1 Introduction*

In a market economy, a policy intervention that has a significant affect on a single, or a few related markets, may have indirect effects that are much more widespread. Due to mutual interdependence among markets, the direct impact can be dispersed from market to market, throughout the economy. This section begins by examining the key factors that may result in policy disrupting price and quantity equilibria in indirectly affected markets. We then look at the possibility of using "extended" partial equilibrium analysis to value the welfare implications of such an intervention. In terms of valuation, we are now moving away from the simple "multiplication" approach towards sectoral analysis (e.g. modelling the agricultural sector or transport sector).

### *8.5.3.2 Why is it Necessary to Look Beyond the Directly Targeted Market*

As illustrated in the previous sections, partial equilibrium analysis concentrates on the determination of an equilibrium price and quantity in a given market. Moreover, the market is considered in isolation, i.e. it is viewed as largely self-contained and independent of other markets. For example, an analysis of an increase in the price of passenger car trips on the quantity of car trips demanded, using supply and demand curves for passenger car travel, is a partial equilibrium analysis. The supply and demand curves which underpin the analysis are drawn on the assumption of given, and unchanging prices in other, related markets. As a consequence, the analysis can ignore all other markets and concentrate solely on the market for car trips. Formally, this is accomplished through the now familiar "other things being equal", or ceteris paribus, assumption.

Markets are, however, interrelated. For example, changes in the market for passenger car trips, affect the bus market. In turn, changes in the bus market affect the market for car trips. In some situations the "fixity" assumptions in the ceteris paribus condition may be reasonable. An increase in the price of car trips, for example, is unlikely to have a measurable effect on the equilibrium price of computers. An increase in the price of car trips, however, could conceivably have a significant effect on the demand for bus trips. In this case, the partial equilibrium assumption that the price of bus trips does not change could be seriously inadequate. The interrelationships between markets may therefore cause inaccuracies when the simplifying assumptions of partial equilibrium analysis are employed.

Consider the market for car trips and the market for bus trips. As these two goods are close substitutes, interdependence on the demand-side is expected.

Consequently, an increase in the price of car trips would be expected to increase the demand for bus trips; likewise, an increase in the price of bus trips would be expected to shift the demand curve for car trips upward (recall the previous discussion on the determinants of demand).

The above analysis neglects two important consequences of the mutual interdependence between the car and bus trip markets: spill-over effects; and feedback effects.

The change in the car trip market will have a spill-over effect on the bus trip market, which disrupts the equilibrium in that market As the two goods are substitutes in this case, an increase in the price of car trips will shift the demand curve for bus trips upward. The magnitude of the shift will depend on the cross elasticity of demand. The closer the two goods are as substitutes, the higher the cross elasticity, and the larger the rise in the number of bus trips demanded as a result of an increase in the price of car trips.

There would be little cause for concern if spill-over effects in other markets were the only effect neglected by partial equilibrium analysis. Partial equilibrium analysis, however, ignores a second, important effect. The induced change in the market for bus trips has a feedback effect on the market for car trips. This effect is not considered in a partial equilibrium analysis of the directly targeted market since the initial increase in the price of car trips induced the price of bus trips to rise. Consequently, it is necessary to retrace the demand curve for car trips.

The above example considered interrelationships between only two markets. Moreover, the relationships were restricted to the demand-side of the market. In reality, a change in one market, albeit a relatively important one, may disrupt the equilibria in tens of markets, and be affected by changes in those markets. Markets may also be related on the supply-side. The pricing of inputs in one market, for example, may affect other markets that utilise those, or closely related inputs. As with the demand curves, the position of the supply curve equally depends on what happens in other markets.

#### *8.5.3.3 Partial Equilibrium Analysis in Several Markets*

A question that is frequently encountered in practical applications of economic analysis to environmental policy appraisal, is "how many markets must be examined to determine an adequate measure of the welfare implications of a particular policy intervention?" This is an important issue in environmental policy appraisal, as most environmental interventions change conditions in markets other than the one directly targeted.

Conventional analyses of environmental interventions, and that employed in ExternE to date (under the assumption that changes in output are "marginal" and prices therefore remain fixed), tend to concentrate on the market directly affected by the intervention. Nonetheless, in many instances this still provides an accurate and complete measure of the welfare change. At the same time however, there is no guarantee that this will be the case. It is therefore critical to understand the conditions under which it is necessary to go beyond the specifically targeted market to ensure that the measurement of the resultant welfare effects is complete.

#### *8.5.3.4*
#### *Propositions for Extending Partial Equilibrium Analysis*

Presented below are some propositions that identify the conditions under which it is advisable to extend the analysis beyond the specifically targeted market. It must be said however, that the results of an "extended" partial equilibrium analysis may still differ form the correct general equilibrium results.

These propositions, which represent the consensus view of the cost-benefit literature, are (it is not practical herein to illustrate these proposition; detailed examples are provided in Sugden and Williams (1978), Arnold (1995) and Zerbe and Dively (1994); NB if welfare effects are determined allowing for substitution and income effects, then these propositions apply directly to general equilibrium analysis):

- When there are price changes in indirectly affected markets with no distortions, there is no net welfare change and these markets can therefore by neglected. What we have is a pecuniary external effect – changes in consumer surplus in the indirectly affected market are approximately offset by changes in producers surplus.
- When distortions exist in an indirectly affected market and quantities change as a result of the intervention in the targeted market, these markets ought to be examined. (The welfare effect is approximated by multiplying the size of the distortion per unit by the change in the number of units.)
- When distortions exist in an indirectly affected market and prices change as a result of the intervention in the targeted market, these markets ought to be examined. (The welfare effect is approximated by multiplying the average value of the price distortion by the by the change in quantity.)

Therefore, depending on the questions to be answered by the economic analysis, an extended partial equilibrium analysis, or a full general equilibrium analysis – outlined below - may be required.

## 8.6
## General Equilibrium Analysis

### 8.6.1
### Introduction

In this section general equilibrium analysis is considered, where, in contrast to the partial equilibrium frameworks considered above, all determinants of prices are variable, and the analysis focuses on the simultaneous determination of equilibrium in all markets, although at a more aggregated level. Consequently, in general equilibrium analysis the interdependence of the prices and quantities of products and inputs in the economy are explicitly taken into account.

Consider a policy of increasing "charges" to good hauliers for using major interregional highways. (Furthermore, assume that the policy will increase the "price" of freight haulage.) Obviously, the intervention directly affects HGV operators. If we are only interested in the welfare implications in this market, then a simple partial equilibrium analysis is sufficient. However, confronted with increased costs, hauliers may decide to shift freight onto the rail system (if available). At the same time, reduced HGV usage of the highway network may induce passenger car usage to increase. The direct effect in the freight haulage market may therefore have indirect welfare effects in other, related markets. In this case, an "extended" partial equilibrium or sectoral analysis may be required to capture the main welfare effects of the policy.

Some sectors of the economy are heavily reliant on road transport (e.g. retailers of food and other domestic products). It is thus possible that this policy may affect economic conditions in these sectors. Ultimately, these effects may be felt by households, if the prices of consumables change. In this case, it will be necessary to adopt an economy-wide approach, using general equilibrium analysis.

To measure the full welfare change in this case, more sophisticated appraisal techniques are now needed. Techniques which are capable of modelling the linkages between different economic agents in the economy, and therefore quantifying the "general equilibrium effects" of a policy intervention, include computable general equilibrium models (CGE) and, to a lesser extent, input-output models. Each will briefly be considered in turn.

A caution may also be useful at this point: it is wrong to think of general equilibrium analysis as a complete analysis; no analysis is capable of taking into account all the complex interdependencies in the economy. General equilibrium analysis is simply a more complete framework for evaluating policy interventions than partial equilibrium analysis. (In this sense, partial equilibrium analysis may be viewed as one way of simplifying a system of general economic equilibrium.) The real difference between the two forms of analysis is therefore really one of degree. A summary framework that identifies the appropriate use of these forms of analysis is given in Fig. 8.1.

## 8.6.2 Computable General Equilibrium (CGE) Models

GE models essentially simulate markets for production factors, products, etc. with systems of equations specifying supply and demand behaviour across all markets. There are many examples of CGE models, each "tailor-built" with a specific purpose in mind and a recent summary is given in OECD, (1997). Several applied GE models have also been specifically designed to assess the overall economic impact of addressing the enhanced greenhouse effect.

Limitations of GE Models: The fact that GE modules start at the 'top', i.e. with a representation of what should happen if the economy in question conformed to the assumptions of the model, rather than the more traditional 'bottom up' approach that takes a set of observations relating to what is actually happening, some commentators take the view that these models are too abstract for the real world (European Commission, 1996). In common with input-output models, the inherent complexity of GE models means that the amount of time and effort required to collect the basic data, and build a suitable model, is often prohibitive.

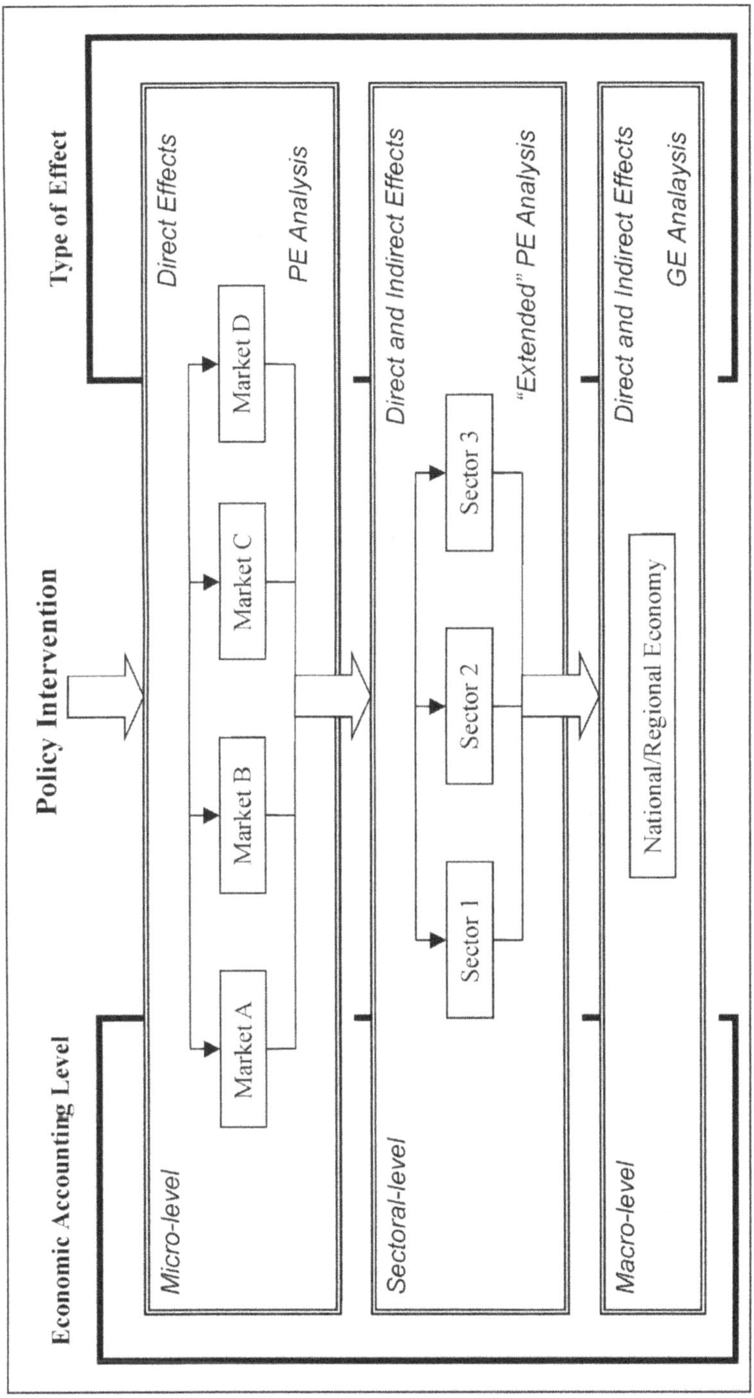


**Fig. 8.1.** Accounting Framework for Assessing Welfare Effects of Policy Interventions (Source: Adapted from FSO (unpublished))

## 8.6.3 Input-Output Models

Wassily Leontief developed the first set of basic input-output tables in 1936. By 1941, Leontief had produced input-output tables for the US economy for 1919 and 1929. Input-output analysis is based on the fact that in modern economic systems linkages exist between activities. Basically, each production activity acts both as a "supplier" and a "buyer". Input-output analysis is essentially a method of systematically quantifying the linkages between various sectors in an economy.

### *8.6.3.1 Examples of Input-Output Models*

National and regional economic-environmental input-output models have been built for several countries. The primary purpose of such models is to forecast residual discharges, mainly air emissions. Examples include Proops et al. (1996), De Haan (1996) and Koppl et al. (1996). In addition to forecasting discharges of residuals, input-output models can be used to assess both the direct and indirect effect of controlling flows of residuals from economic activities. The cost of installing new abatement technology, for example, could be added to the primary input costs of an input-output model. In a non-environmental context, input-output models have been applied in various transport-related studies to examine the wider impacts (i.e. direct and indirect effects) of investment in infrastructure on a regional or national economy (see Table 5.3 in EC, (1996) for a summary of these studies).

### *8.6.3.2 Limitations of Economic-Environmental Input-Output Models*

An obvious drawback of input-output models is therefore the amount of time and effort required to collect the basic data and build a suitable model. The assumptions underlying the construction and operation of input-output (environmental quality) models are also subject to criticism. For example, it is not always the case that fixed coefficients (assumed in input-output models) accurately describes real production relationships or environmental quality effects, especially when non-marginal changes in output are anticipated. (Note, however, that these models can be extended to variable coefficients). With the above problems, the huge number of variables considered, and the resulting complexity, input-output models can be cumbersome and frustrating for the decision maker.

Nonetheless, one of the questions we are trying to answer when faced with "general equilibrium effects" is, what are the channels by which a change in policy directed at one market affects behaviour in other markets? Input-output tables can prove very useful in answering this question, and therefore could be used as a part of an initial screening exercise to identify what markets need to be examined, and possibly, whether some form of general equilibrium analysis is necessary.

## 8.6.4 When to Use General Equilibrium Analysis

General equilibrium analysis, as used by economists, typically involves studying the relationships between a few, highly aggregated sectors. Consequently, detail is sacrificed for completeness. By contrast, in partial equilibrium analysis one can take into account many specific factors underlying the workings of an individual market; completeness is sacrificed for detail. The choice between "bottom-up" and "top-down" modelling approaches requires a similar trade-off to be made; the former is typical of partial equilibrium analysis, whereas the latter is typical of general equilibrium analysis.

## 8.6.5 Recommendations

It is recommended that the way in which the welfare effects of each policy option considered by ExternE are estimated, should be determined on an individual basis since there are no hard and fast guidelines on this issue.

Nevertheless, some broad conclusions can be drawn. They are:

- Partial equilibrium analysis, as utilised by ExternE to date, is appropriate as long as either the supply or demand curve for the directly affected market can be characterised as being more or less flat, and no other markets are affected. This rule also holds if other markets are affected but they too have highly elastic supply or demand curves.
- Extended partial equilibrium analysis is likely to be appropriate if distortions exist in an indirectly affected market and prices and/or quantities change. This will be most necessary where the effects are sizeable and the directly targeted market is highly integrated with other markets or sectors of the economy. This is applicable to the agricultural, transport and health sectors in ExternE policy analysis.
- General equilibrium analysis is appropriate when there are strong inter-dependencies between the directly affected sector and the indirectly affected sector, and the effects result in a significant percentage change in resource costs. This rule is likely to be true if the policy impacts on a number of sectors that provide inputs to final products e.g. transport and energy.
- Extended partial equilibrium analysis or general equilibrium analysis is therefore more likely to be needed if (a) the policy necessitates structural adjustment of the economy and (b), the policy is applied across a wide geographical area (i.e. national or international).
- Any recommendation made regarding the appropriate form of analysis also needs to consider the greater uncertainties that are inherent in more complex modelling forms that reduce the accuracy of the results. It also needs to be borne in mind that resource cost demands also rise with more complex modelling requirements.

## 8.7 Incorporating Non-Monetised Impacts into the Decision Making Process

### 8.7.1 Introduction

Policy appraisals address a minimum of two alternatives, but typically focus on three to five alternatives. A two-alternative appraisal usually represents a choice between implementing a project or policy intervention versus "doing nothing" (i.e. a baseline scenario). Each alternative to be assessed may encompass a wide range of considerations, for example, success in meeting defined needs and identified objectives, financial costs and benefits, environmental impacts and public preferences. Since these considerations influence the choice among options they are frequently referred to as decision factors (in the language of decision analysis or decision theory, decision factors are referred to as "attributes"; the two terms are used interchangeably herein). Project or policy appraisal essentially attempts to collect information on the impacts of alternative actions on priority decision factors. This information alone however, is not necessarily sufficient to make a decision; particularly when some project impacts are expressed in money terms, while others are reported in "physical" units (e.g. mass, area, volume, etc).

In order to provide a clear basis for selecting among alternatives, all decision factors relevant to the set of proposed alternatives therefore need to be presented in a comparative form; ideally in similar units of "value" or "worth". Given current economic valuation state-of-the-art, it is not possible to express all priority impacts in money terms, at least with an acceptable level of uncertainty. Therefore, it is desirable to seek a systematic method of expressing the priority impacts of each alternative in a common form so as to facilitate the aggregation of impacts to ultimately arrive at some measure of overall project "value". The challenge is thus "to adequately capture and express in sufficient detail the full range of impacts that need to be considered by the decision maker without overwhelming that individual or individuals with either the sheer amount of information, the diversity of information, or the lack of comparability of information" (SETAC, 1993).

The purpose of this section is therefore to consider ways in which the ExternE results may be combined with data on other decision factors, which are not expressed in money terms, but are equally important determinants of policy selection. It should be emphasised at the outset that the intention is not to identify a single methodology to adopt in ExternE. Rather, the intention is reflect on the main options available to decision makers whom might wish to use the ExternE results in a multi-attribute decision making context. Before considering the possibilities for multi-attribute decision-making we give some general guidance on the use of avoidance costs in the absence of willingness to pay monetary values.

## 8.7.2 Avoidance Costs

Impacts may be taken into account without using willingness to pay monetary measures by the use of a measure of the costs that are, or would be, incurred in avoiding environmental damage. This is a more narrowly defined mechanism in terms of its scope of coverage than that allowed under the weighting systems described above - in that it can only apply to those impacts that can be avoided by economic measures. Avoidance costs may be taken to include costs of mitigation measures, preventative costs and replacement costs. These can be seen as a proxy for WTP values in that the value that people place on preserving their environment is inferred from what they are prepared to spend to prevent its degradation or to restore its original state after it has been damaged. However, this measure takes for granted that the expenditure is worth making. There is no guarantee that just because the resource costs of avoiding the damage are known they will be justifiable.

Avoidance costs are also less than satisfactory as WTP proxies since they ignore consumer surplus. Thus, they will only provide minimum estimates. Further, replacement costs will be minimum values unless full restitution of the environment is possible after damage. It is more likely that many environmental effects will not be fully perceived, will arise in the long term, or will not be capable of being restored.

However, avoidance costs do serve to give a form of social weighting in that they reflect the resource costs that will be incurred in avoiding the environmental damage or replacing/renewing the environmental asset. In this context, a recent EC research project, GreenStamp, has explored a more ambitious use of avoidance costs, whereby they are used to measure the economic costs associated with moving the economy to an environmentally sustainable development path. General equilibrium modelling is therefore used to track this economic transformation. The preference for avoidance costs in that research project over the use of a measure of welfare loss is as a result of the uncertainties that are associated with some WTP measures of environmental damage.

As the EC DG Research GARPII research project comments, however, it should be noted that the methodological approach required to compute empirical estimates of the avoidance costs raises certain issues. The availability of statistics on costs that can be aggregated from the firm (micro) level to the branch (meso) level required in the compilation of national income statistics has proved in practice to be very limited. Indeed the Green Stamp project included an empirical application of the approach for various compounds of nitrogen, using German data which found that there were in fact "no usable cost statistics available in Germany." It reports that "a lot of intermediate steps related to many uncertainties and assumptions are necessary to come to a common classification level of both the physical and economic side". It is likely that these difficulties will be greatly exacerbated when, in bringing about volume reductions, an entirely new resource-mix exists in the economy. Clearly, in order to successfully model future GDP scenarios the cost parameters are of crucial importance, yet these are highly uncertain at the present time.

The GreenStamp approach is an extension and alternative to the conventional way in which avoidance costs have been used – namely as proxies for WTP. To the extent that policy makers need to have an assessment of both the costs and benefits of a potential policy, the emphasis on the cost of supplying environmental services is complementary to the work of ExternE. For the reasons outlined above and in more detail in the GARP research project, however, it is suggested that avoidance costs should be used as proxies for WTP with care in the ExternE context, and as a complement rather than a substitute in the overall methodology.

### 8.7.3 Conceptual Framework for Attribute Decision Making

#### *8.7.3.1 Generic Approaches Trade-Off Analysis*

To achieve a systematic approach to policy selection in an environmental context, it is desirable to use trade-off analysis. Typically, trade-off analysis involves the comparison of a set of alternatives relative to a set of decision factors and there are several approaches which can be used to make such trade-offs These approaches, or decision techniques, may be sorted into two categories: qualitative procedures and analytical procedures. The former do not analytically identify the preferred alternative from the set of proposed alternatives. In the main they rely on the qualitative judgements of the analyst in differentiating among alternatives. Analytical procedures, in contrast, uniformly process data on the consequences of alternatives to identify the preferred alternative(s). There are three main groups of analytical procedures: (1) conjunctive rules; (2) rating, scaling and/or ranking; and (3) rating, scaling and/or ranking plus weighting.

Each group of analytical procedures should be used in accordance with the generic decision making objective to be achieved. These are:

- To eliminate non-feasible alternatives, thereby identifying "legitimate" alternatives for further evaluation.
- To eliminate dominated or inferior alternatives. That is, to compare feasible alternatives with each other in order to identify, and eliminate, poorer alternatives.
- To select a preferred alternative by ranking the alternatives in order to identify the "optimum" or "balanced" alternative. Compromise programming, MAUA and AHP are widely used to this end in environmental decision problems. The former suffers from the fact that it selects a "balanced" alternative rather than an "optimum" one. The combination of ExternE results with other, non-monetised impacts within a MAUA framework is theoretically defensible and logical. Moreover, MAUA provides a mechanism for explicitly incorporating the uncertainty of impacts and the preferences of policy makers into environmental decision making.

In general, when moving from simple qualitative approaches to analytical approaches involving weighting-scaling, the technical complexity of the method, input requirements, demands on the decision maker, etc. increase significantly whilst the approach becomes more "informative". The most useful techniques strike a good balance between input demands, transparency and complexity, and their overall contribution to the decision making process. In the context of ExternE, the compatibility of the technique with environmental cost-benefit analysis is also crucial.

If the objective is to select the "optimum" alternative, then we are restricted to the third group of decision approaches. Subject to the arguments outlined above, some form of MAUA would be our preliminary choice for a decision tool. Ideally, one could establish a generic utility function for relevant priority attributes, and the corresponding scaling constants.

### *8.7.3.2 Example of MAUA Use in the ExternE Context*

An example problem may be in selecting between two development projects, one of which reduces river-flow. The set of attributes applicable to the specific environmental issue need to be identified. In this example a range of environmental, socio-economic and financial attributes may be relevant. One environmental impact that is considered in more detail below is salmonid damages. Whilst many of the attributes are "objective" by nature; i.e. there already exists a commonly understood scale (or numeraire) for that attribute and its levels (x) are objectively measurable. There are attributes, however, for which no objective scale (or numeraire) exists, e.g. several environmental impacts and other attributes falling under the heading socio-economic factors. In such cases, a "subjective" index must be constructed. Subjective indices are also required when attributes attempt to capture many detailed impacts, i.e. at a highly aggregate level.

### *8.7.3.3 Constructing the Multi-Attribute Utility Functions*

Once the set of attributes (X) has been established, a measurement scale (index) established for each member of X and ranges of possible impacts determined, the next step in the MAUA is to construct an appropriate multi-attribute utility function. Under certain conditions, (preferential independence and utility dependence (Keeney and Raiffa, 1976), the multi-attribute utility function may be expressed in the simpler functional form of *m* one-attribute utility functions.

Subject to certain generally accepted axioms of rational decision making, the preferences of the decision maker may be reflected by a utility value $U(x_i)$ satisfying the following properties (Winkler, 1972, Neumann and Morgenstern, 1947):

- If $x_i' > x_i'''$, then $U(x_i') > U(x_i''')$; and if $x_i' \approx x_i'''$, then $U(x_i') = U(x_i''')$.
- If $(x_i'; x_i''')p$ represents an uncertain outcome (or "lottery") with a probability p of receiving consequence $x_i'$ and (1-p) of receiving $x_i'''$, and a certain outcome $x_i''$ is equally as desirable as the "lottery", then $U(x_i'') = pU(x_i') + (1-p)U(x_i''')$. By convention, $x_i'$ refers to the most desirable consequence of $x_i$, $x_i'''$ to the least desirable, and $x_i''$ to a consequence between $x_i'$ and $x_i'''$.

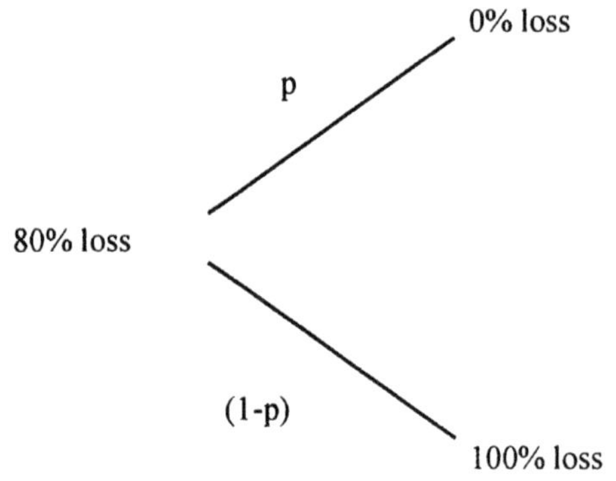


**Fig. 8.2.** Representation of example "lottery".

The former property states that if, for example, we consider $x_i''$ = 80 percent loss of salmonids, the second property implies that:

80 percent loss of salmonids ~ ($x_i'$ = zero percent, $x_i'''$ = 100 percent)p.

This "lottery" may be represented diagrammatically as in Fig. 8.2. In this case, the utility value associated with a consequence of 80 percent salmonid loss is given by:

$$U(80\%) = pU(0\%) + (1-p)U(100\%) \tag{8.5}$$

Without violating any of the above two properties, the utility of the "best" and "worse" consequences can be assigned values of 1.0 and 0.0, respectively. Consequently, U(80%)=p(1.0)+(1-p) (0.0) = p. What is the appropriate value of p?

For the purpose of illustration, the utility assessment forms used by Collins and Glysson (1980) will be considered: other elicitation approaches involve the use of questionnaires, of various degrees of sophistication. The numbers in the right two columns of Table 8.10 represent possible p, (1-p) combinations for the "best" and "worse" possible consequences, respectively. The preferences of the decision maker for different "lotteries" are recorded in the two centre columns. For example, the bottom left check in Table 8.10 indicates that the preference of the decision maker is for the certain consequence of a 80 % loss of salmonids, as opposed to an uncertain situation where the probability of no losses is zero and 100 percent losses is 1.0. That is, 80% > (0%, 100%) 0.0.

Clearly, one would expect the decision maker to exhibit the above preferences for p = zero and p = 1.0. However, for other p and (1-p) combinations, the preferences of the decision maker are less clear. These are ascertained by asking the decision maker to place checks up the left hand column and down the right hand column, until a point of indifference is reached between the certain consequence and the uncertain consequence. A completed form may look like the one shown in Table 8.10. In this case, indifference is achieved where p = 0.5, or 80% ~ (0%, 100%) 0.5.

**Table 8.10.** Completed Utility Assessment Form: Salmonid Damages

| Attribute: | Salmonid Damages Percentage Loss | Evaluator: Date: | Ms. Smith 26th June 1999 | |
|---|---|---|---|---|
| Certain Outcome Level ($x_i''$) | Preference | Uncertain Outcome Preference | Level (%) $x_i' = 0$ | $x_i''' = 100$ |
| | | yes | 1.0 | 0.0 |
| | | yes | 0.9 | 0.1 |
| | | yes | 0.8 | 0.2 |
| | | yes | 0.7 | 0.3 |
| 80% | | yes | 0.6 | 0.4 |
| | ~ | ~ | 0.5 | 0.5 |
| | yes | | 0.4 | 0.6 |
| | yes | | 0.3 | 0.7 |
| | yes | | 0.2 | 0.8 |
| | yes | | 0.1 | 0.9 |
| | yes | | 0.0 | 1.0 |

Source: Adapted from Collins and Glysson (1980)

The above outlined assessment procedure can be used to establish the utility value assigned to various levels of $x_i''$ between $x_i'$ and $x_i'''$, for several decision makers. A utility function can then be fitted to the set of values generated.

### *8.7.3.4 Evaluating the Scaling Constants*

The first step in combining utility functions is to evaluate the scaling constants for each attribute. These scaling constants, which will be denoted by $k_i$ , reflect the relative contribution of each attribute in set X to the overall multi-attribute utility function. In other words, they account for the preference of the decision maker regarding, for example, levels of public health effects versus noise pollution versus investment expenditure versus salmonid losses, and so on. Evaluating the scaling constants is undertaken in two stages. The first stage involves ranking the members of set X in order of importance; the second stage involves measuring the magnitude of the each scaling constant.

The next stage is to establish the actual magnitude of the scaling constants; this involves assessing specific trade-offs between attributes. A scaling constant $k_i$ can be assessed for attribute $X_i$ by looking for the value $p_i$ such that:

$$x_i', x_j''' \sim \left(x_i', x_j'; x_i''', x_j'''\right) p_i \qquad (8.6)$$

where $x_j = (x_1, ..., x_i\text{-}1, x_i\text{+}1, ..., x_m)$ and $0.0 < p_i < 1.0$. As before, the objective of the exercise is to establish the point of indifference between the left and right hand side of the relationship. This is accomplished by following the procedure outlined above for single attribute utility functions. The point of indifference yields the scaling constant.

### 8.7.3.5 Specifying the Multi-attribute Utility Function

The single attribute utility functions and the attribute scaling values can now be combined in utility function U(A) can be written for m attributes in either an additive form or in a multiplicative form. In general, the multiplicative form is the appropriate form to use when interrelationships exist between the attributes, and the additive form can be used when the attributes are independent of each other. Finally, the single attribute utility functions, scaling constants and, if desirable, probability distributions, are input into the appropriate multi-attribute utility model. The (expected) total utility of each alternative in set A can then be determined, and the alternative with the highest utility selected – assuming the decision rule is to maximise utility.

***Derivation of weights.*** The central question remaining relates to the problem of how weights are derived for the assessment of different environmental impacts. So far it has simply been asserted that the "decision maker" will work through the series of exercises outlined above to produce each single attribute utility function and then to decide the ranking and magnitude of the scaling constants between different attributes. However, the legitimacy of the weights used in the type of public policy processes envisaged for the ExternE results demands that the decision maker is well defined and accepted as appropriate in the society.

In the application of monetary willingness to pay values, the ExternE project is implicitly adopting the weights that are provided by the real or hypothetical expenditure decisions made by individuals, and aggregated across the general population. The weightings derived therefore have a basis in democratic principles, but are further differentiated since the monetary values reflect the strength of individual preferences. An implicit assumption is that the income and wealth distribution within the society is optimal.

Where willingness to pay for an environmental good/service is not, or cannot be, identified the mechanism for deriving weights is defined using either a public survey, the Delphi (expert survey) technique or from a survey of interest-group priority action areas. The three methods have obvious strengths and weaknesses concerning degree of representation; degree of knowledge and cost. It is suggested above that in the immediate ExternE context the Delphi method could be applied to a case study - based on the technical expertise that resides in the project team.

***The derivation of implicit weights.*** The decision-maker, however formulated, will reveal points of indifference between different combinations of attributes. Once this point of indifference has been quantified this information can be manipulated to derive the weights implicit in the trade-offs made. For example, it may be possible for an intermediate system of weights to be derived from the existing monetary and physical information relating to the attributes that would result from different policy measures. This weighting system essentially derives ratios between physical and monetary units (though in theory the two sets of units could both be in physical units) so that one can be compared with the other.

# 9 Global Warming

## 9.1 Introduction

Marginal costs of the emissions of a selected number of radiatively-active gases are presented, three uniformly-mixed gases – carbon dioxide, methane, nitrous oxide – and two region-specific gases – nitrogen (from aircraft) and sulphur, which influence ozone and sulphate aerosol concentrations, respectively. Eyre et al. (1997) extensively discuss the literature on climate change impact estimation and valuation, and estimate the marginal costs of $CO_2$, $CH_4$ and $N_2O$, including substantial sensitivity and uncertainty analyses (cf. (Tol, 1999a, Pearce et al., 1996)). This analysis does not repeat Eyre et al. (1997), but rather complements it by adding a third model (FUND2.0), adding region-specific gases, and by presenting an alternative accounting framework.

As the uncertainties involved in estimating damage costs for grenhouse gases are considerable, estimates based on the avoidance cost approach are presented as well.

## 9.2 Models

Three models are used in this analysis, viz. the Open Framework for Economic Valuation of Climate Change (OF), developed by Downing et al. (1995, 1996), and two versions of the Climate Framework for Uncertainty, Negotiation and Distribution (FUND), developed by Tol (1995, 1996) and Tol (1999b), respectively. The models are briefly presented here.

### 9.2.1 Open Framework

The Open Framework for Economic Valuation of Climate Change (Downing et al., 1995, Downing et al., 1996) follows a sequence of steps in calculating economic damages. A reference scenario, based on the IPCC 1992a scenario, is used to project economic conditions, sensitivity to climatic variations and the climate forcing of global GHG emissions.

Global-average temperature change and sea level rise are calculated by the 1995 version of MAGICC (Wigley et al., 1993). MAGICC is a relatively simple upwelling-diffusion, energy balance climate model that distinguishes between land and ocean and between hemispheres. In all cases, MAGICC's default parameters are used.

Spatial scenarios of climate change are based on the $2xCO_2$ equilibrium run of the general circulation model experiment from the Goddard Institute of Space Sciences (GISS). The GCM scenario was scaled to the global-average temperature projection from MAGICC. This results in a time-dependent climate change scenario consistent with the assumptions of the global emissions. The spatial pattern of anomalies from the GISS equilibrium scenario is retained, however.

Simple impact models are run for the current climate and for the scenario of climate change. The baseline climate is based on the 0.5 degree latitude by longitude climatology of Cramer and Leemans (1991). Climate parameters for the baseline are mean monthly temperature and precipitation for the period of record. These methods of creating scenarios are common in climate change impact assessment (see Viner and Hulme, 1994). The impact variables include heating and cooling degree days, agricultural suitability, and water balance. They are extracted to provide country-average values.

The model calculates costs for seven time slices from 1990 to 2100. However, the spatial impacts are only calculated for 1990 (the base year) and 2100, scaling the simple impact indicators for intervening time periods according to changes in the global mean temperature.

For coastal impacts (coastal protection, loss of wetlands, loss of drylands, migration and biodiversity), agriculture, energy demand and water, country-level economic impacts are derived from the reference projections, simple impact models, and climate sensitivity-cost equations. The equations vary in form, depending on the availability of previous sectoral studies. Most include changes in supply (from the simple impact models), projected prices (or value of production) and the sensitivity of demand to changes in supply or price. The impacts of natural disasters are only calculated at a global-average level.

The country-level direct costs are summed to a global total. An additional global cost, representing higher order, non-market effects, is calculated as a multiplier on the net damages (i.e. not including the positive impacts of climate change). This is done to provide an estimate of the total cost of climate change, rather than only the costs that have been captured by the sectoral models. The used scalar corresponds to the ratio of non-market to market costs in other models (such as FUND), but is still a subjective estimate of the potential damages to "other sectors".

### 9.2.2 FUND 1.6

The Climate Framework for Uncertainty, Negotiation and Distribution (FUND) is a model that closes the loop population – economy – technology – greenhouse gas emissions – atmospheric composition – climate – climate change impacts – emission abatement. Using simple representations of these components, the model runs in time steps of one year from 1950 to 2200, for nine major world regions. The fact that FUND is able to perform a cost-benefit analysis -- with multiple actors and under uncertainty – implies that the impact module is subject to strict demands on computational speed.

FUND was developed to compare the impacts of climate change against the impacts of greenhouse gas emission abatement. Here, the main interest is in the climate change impact module. This is described in Tol (1995, 1996), and applied to estimate the marginal costs of greenhouse gas emissions in Eyre et al. (1998) and Tol (1999a).

A standard five-box carbon cycle model (cf. Hammitt et al., 1992) is used for carbon dioxide concentrations in the atmosphere. The influence of methane and nitrous oxide emissions on concentrations geometrically declines over time, with life-times according to Schimel et al. (1996). Other human disturbances of climate are omitted. Changes in radiative forcing follow from Shine et al. (1990). Radiative forcing drives the equilibrium change in the global mean temperature, to which actual temperature geometrically converges. Actual temperature determines equilibrium sea level rise, to which actual sea level rise geometrically converges. Equilibrium sensitivities and convergence rates are calibrated to the typical outcomes of simple climate models (cf. Kattenberg et al., 1996).

Impact of climate change $I_t$ at time t is modelled as either:

$$I_t = \alpha_t W_t + \beta_t W_t^2 \tag{9.1}$$

or

$$I_t = \alpha_t \Delta W_t + \beta_t \Delta W_t^2 + \rho I_{t-1} \tag{9.2}$$

with W an appropriate climate variable, and $\alpha$, $\beta$ and $\gamma$ parameters. W may be the global mean surface air temperature, or the global mean sea level. The parameters $\alpha$ and $\beta$ differ per impact category, and depend on agricultural production, per capita income, and urbanisation (cf. Tol, 1996). Impact I is measured either in percentage of Gross Domestic Product, or in percentage of population. In the latter case, climate change induced mortality is valued at 200 times the per capita income in the relevant region at the relevant time for each casualty.

### 9.2.3 FUND 2.0

The impact module of FUND1.6 reflects the insights into the effects of climate change of the first half of the 1990s, as laid down in the Second Assessment Report of the IPCC (Pearce et al., 1996, Watson et al., 1996). These insights have substantially changed since. Whereas earlier studies emphasised the negative impacts of climate change, later studies increasingly found positive aspects as well (Mendelsohn and Neumann, 1999), for example in energy consumption and agriculture. Other developments include the extension of studies to new sectors and new countries, better inclusion of adaptation, better integration of sectors, and the addition of more dynamics. These changes are reflected in the climate change impact module of FUND2.0, which is a complete revision of the impact module of FUND1.6. FUND2.0 is described in Tol (1999b). There are four reasons why version 2.0 is placed alongside version 1.6, rather than replacing it:

- FUND2.0 is currently being peer-reviewed whereas FUND1.6 was published in the scientific literature;
- FUND1.6 is known, inter alia from previous ExternE applications;
- the revision of FUND2.0 was so extensive that it is almost a completely different model; and
- although FUND1.6, reflecting earlier impact literature, may be too pessimistic about climate change, FUND2.0, reflecting more recent literature, may be too optimistic.

## 9.3 Conventions for Valuation and Reporting

### 9.3.1 Discounting

Climate change is a problem of the long term. Discounting is therefore of utter importance (cf. Arrow et al., 1996). There are four reasons to discount the future:

1. *Impatience, or myopia.* People prefer current consumption over later consumption.
2. *Economic growth.* If people are richer in the future, a dollar now has greater (relative) value than a dollar later.
3. *Changing relative prices.* Certain impacts, such as on human health, may well be valued higher in the future.
4. *Uncertainty.* Because future consumption is less certain, it is worth less.

Arguments 1, 2, and 4 call for a positive discount rate (i.e., one that places more weight on the near than on the far future). Argument 3 may call for a negative discount rate.

Arguments 3 and 4 are improper arguments for discounting. Changes in relative prices are better explicitly modelled than included in the discount rate, primarily for reasons of transparency. Similarly, it is better to include uncertainty explicitly, rather than to short cut it through tinkering with the discount rate. The Open Framework and FUND model changes in relative prices. FUND includes uncertainty, the OF reports low, middle and high estimates.

The discount rate thus consists of the first two elements only. In the standard neo-classical formulation the discount rate r follows from $r = \rho + \eta g$ with $\rho$ the pure rate of time preference, $\eta$ the consumption elasticity of marginal utility, and g the growth rate of per capita consumption. Here, $\eta=1$, that is, utility is proportional to the natural logarithm of consumption. The growth rate of per capita consumption g is assumed to equal the growth rate of per capita income. The pure rate of time preference $\rho$ varies between 0%, 1%, and 3% per year. 0% is preferred on moral grounds (Koopmans, 1967, Ramsey, 1928)) and in line with sustainability ((Broome, 1992, Brown, 1997)). 3% is observed on markets (Nordhaus, 1994). 1% is an arbitrary value in between, that will be used for the central estimates below.

**Table 9.1.** Assumed regional per capita income growth rates.[a]

| | FUND1.6[b] | | | FUND2.0 | | | Open Framework | | |
|---|---|---|---|---|---|---|---|---|---|
| | avg. | min. | max. | avg. | min. | max. | avg. | min. | max. |
| OECD-A | 1.2 | 0.6 | 3.0 | 1.2 | 0.7 | 3.0 | 1.6 | 1.1 | 2.4 |
| OECD-E | 1.3 | 0.6 | 2.5 | 1.2 | 0.9 | 2.4 | 1.6 | 1.2 | 2.4 |
| OECD-P | 1.3 | 1.0 | 3.5 | 1.4 | 1.0 | 3.5 | 1.6 | 1.1 | 2.4 |
| CEE&fSU | 1.5 | -0.5 | 4.3 | -0.1 | -2.4 | 3.7 | 1.4 | -0.9 | 2.8 |
| ME | 1.4 | 0.5 | 2.2 | 1.6 | 0.8 | 2.4 | 2.0 | 1.6 | 2.3 |
| LA | 1.6 | 0.9 | 2.5 | 1.7 | 0.9 | 2.7 | 2.7 | 1.8 | 3.0 |
| S&SEA | 2.0 | 0.8 | 3.1 | 2.1 | 0.9 | 3.2 | 2.5 | 1.5 | 2.8 |
| CPA | 2.8 | 1.3 | 5.0 | 2.9 | 1.3 | 5.0 | 3.5 | 3.0 | 4.5 |
| AFR | 1.6 | -0.0 | 3.3 | 1.3 | -0.0 | 3.0 | 2.0 | 1.6 | 2.3 |
| World | 1.4 | 1.0 | 1.8 | 1.4 | 1.0 | 1.8 | 2.2 | 1.6 | 2.4 |

[a] The table shows average, minimum and maximum growth rates per decade for the period 1990-2100.

[b] FUND allows for different growth rates as part of its uncertainty analysis, assuming a Normal distribution with a coefficient of variation of 20%.

Table 9.1 displays the assumed growth rates per region. Although all analyses are based on the IS92a scenario, the different models had to make different assumptions to fit this to their different spatial and temporal resolutions. In addition, the feedback of climate change on economic and population growth is different across the models.

## 9.3.2 Aggregation

Impacts of climate change are valued at national or regional prices in both the Open Framework and FUND. When aggregated, this may be objectionable because, for example, a life lost in Bangladesh counts less than a life lost in Germany. Altering values, however, would undermine the theoretical foundations of monetisation and yield inconsistencies between national and international environmental policies (Fankhauser et al., 1998). Therefore, the aggregation of monetised impacts is changed, using so-called equity weights (Fankhauser et al., 1997).

The framework of Fankhauser et al. (1997) assumes a global decision maker. The audience of the ExternE project, however, is a European decision maker. The economic theory of how one decision maker values impacts in other than her own country is underdeveloped. An ad hoc solution is therefore chosen, inspired by the equity-weights of Fankhauser et al. (1997). Four alternatives are reported here:

1. EU impacts only, with EU values ('EU only');
2. 1, plus impacts in other regions with local values ('regional values');
3. 1, plus impacts in other regions with globally averaged values ('world average');
4. 1, plus impacts in other regions with EU values ('EU values').

**Table 9.2.** The marginal costs of carbon dioxide emissions (in \$/t C).[a]

| | EU only | regional values | world average | EU values |
|---|---|---|---|---|
| FUND1.6 | | | | |
| 0% PRTP | 2.2 | 38.9 | 109.5 | 454.3 |
| 1% PRTP | 1.7 | 26.1 | 73.8 | 302.7 |
| 3% PRTP | 0.8 | 12.3 | 37.0 | 150.3 |
| FUND2.0 | | | | |
| 0% PRTP | 7.2 | 19.7 | 27.5 | 96.3 |
| 1% PRTP | 3.1 | 3.5 | 12.5 | 45.5 |
| 3% PRTP | 0.0 | -6.8 | 1.3 | 6.7 |
| Open Framework | | | | |
| 0% PRTP | 7.6 | 74.5 | 104.0 | 295.9 |
| 1% PRTP | 4.6 | 45.8 | 64.0 | 182.1 |
| 3% PRTP | 1.7 | 16.3 | 22.8 | 64.9 |

[a] Emissions are in the period 2000-2009. Costs are discounted to 2000. Time horizon is 2100. Scenario is IS92a. Morbidity risks are valued based on the value of a statistical life.

**Table 9.3.** The marginal costs of nitrous oxide emissions (in \$/t$N_2O$).a The marginal costs of nitrous oxide emissions (in \$/t $N_2O$).[a]

| | EU only | regional values | world average | EU values |
|---|---|---|---|---|
| FUND1.6 | | | | |
| 0% PRTP | 291.3 | 4,562.9 | 11,547.1 | 46,645.9 |
| 1% PRTP | 187.7 | 2,687.4 | 7,079.0 | 28,769.5 |
| 3% PRTP | 80.4 | 1,123.0 | 3,234.0 | 13,155.2 |
| FUND2.0 | | | | |
| 0% PRTP | 727.4 | 2,350.6 | 2,646.0 | 9,077.0 |
| 1% PRTP | 337.1 | 782.1 | 1,266.7 | 4,446.1 |
| 3% PRTP | 51.3 | -269.5 | 231.4 | 912.3 |
| Open Framework | | | | |
| 0% PRTP | 1,155.0 | 11,384.7 | 15,898.6 | 45,239.1 |
| 1% PRTP | 673.2 | 6,635.9 | 9,267.0 | 26,369.0 |
| 3% PRTP | 210.8 | 2,077.8 | 2,901.6 | 8,256.4 |

[a] Emissions are in the period 2000-2009. Costs are discounted to 2000. Time horizon is 2100. Scenario is IS92a. Morbidity risks are valued based on the value of a statistical life.

The alternatives are ordered in terms of estimated severity of the marginal cost estimates. Alternative 1 ignores impacts outside the EU, closest to realpolitik. Alternative 2 is based on the expressed willingness-to-pay of people outside of the EU, but suffers from the welfare-economic inconsistencies induced by aggregating money measures over people with very different incomes. Alternative 3 argues from the perspective of a benevolent world leader. Alternative 4 argues from the

**Table 9.4.** The marginal costs of methane emissions (in \$/t $CH_4$).[a]

| | EU only | regional values | world average | EU values |
|---|---|---|---|---|
| FUND1.6 | | | | |
| 0% PRTP | 1.7 | 93.3 | 453.1 | 1,959.8 |
| 1% PRTP | 6.0 | 110.4 | 407.3 | 1,670.4 |
| 3% PRTP | 5.0 | 78.4 | 274.3 | 1,086.5 |
| FUND2.0 | | | | |
| 0% PRTP | 5.0 | -89.7 | 41.6 | 183.4 |
| 1% PRTP | -5.1 | -116.5 | -0.3 | 33.4 |
| 3% PRTP | -12.0 | -118.8 | -29.8 | -75.0 |
| Open Framework | | | | |
| 0% PRTP | 25.9 | 255.6 | 357.0 | 1,015.8 |
| 1% PRTP | 23.6 | 232.7 | 325.0 | 924.8 |
| 3% PRTP | 14.1 | 139.2 | 194.5 | 553.2 |

[a] Emissions are in the period 2000-2009. Costs are discounted to 2000. Time horizon is 2100. Scenario is IS92a. Morbidity risks are valued based on the value of a statistical life.

**Table 9.5.** The marginal costs of greenhouse emissions according to FUND2.0.[a]

| | EU only | regional values | world average | EU values |
|---|---|---|---|---|
| Carbon dioxide (\$/tC) | | | | |
| 0% PRTP | 1.2 | 6.1 | 15.1 | 60.1 |
| 1% PRTP | 0.7 | 5.1 | 8.9 | 35.2 |
| 3% PRTP | 0.3 | 4.1 | 3.8 | 14.9 |
| Nitrous oxide (\$/t$N_2O$) | | | | |
| 0% PRTP | 102.1 | 413.9 | 1,319.0 | 5,242.1 |
| 1% PRTP | 59.7 | 320.0 | 78.3 | 2,967.7 |
| 3% PRTP | 24.3 | 250.8 | 296.5 | 1,163.6 |
| Methane (\$/t$CH_4$) | | | | |
| 0% PRTP | 4.9 | 68.2 | 65.5 | 257.0 |
| 1% PRTP | 3.4 | 60.9 | 44.9 | 175.0 |
| 3% PRTP | 1.9 | 49.2 | 24.9 | 95.7 |

[a] Emissions are in the period 2000-2009. Costs are discounted to 2000. Time horizon is 2100. Scenario is IS92a. Morbidity risks are valued based on the value of a year life lost.

moral perspective of a European decision maker inclined to Kant. We reject alternative 2 as internally inconsistent (but do report the outcomes), and use alternative 3 – in between alternatives 1 and 4 – for the central estimates below.

Alternatives 3 and 4 are approximated to avoid expensive reprogramming of FUND. Impacts in other regions are valued at local levels. In aggregating, impacts are weighted with the ratio of world and EU per capita income,

**Table 9.6.** Sectoral breakdown of the estimated marginal costs of carbon dioxide (cf.Table 9.5).

| | EU only | regional values | world values | EU values |
|---|---|---|---|---|
| 0% | | | | |
| Water | 15.4% | 178.0% | 71.6% | 77.2% |
| Forestry | -1.2% | -0.9% | -0.5% | -0.5% |
| Energy | 60.5% | -34.7% | 31.3% | 28.1% |
| Agriculture | 22.2% | 1.8% | -0.4% | -1.9% |
| Coast | 2.8% | 3.2% | 2.6% | 2.5% |
| Migration | 0.2% | 0.1% | 0.1% | 0.1% |
| Species | 0.0% | -0.1% | 0.0% | 0.0% |
| Mortality | 0.0% | -47.5% | -4.6% | -5.5% |
| 1% | | | | |
| Water | 15.3% | 129.2% | 68.9% | 74.5% |
| Forestry | -1.2% | -0.7% | -0.6% | -0.5% |
| Energy | 63.2% | -28.9% | 30.9% | 27.6% |
| Agriculture | 19.2% | 1.0% | -0.9% | -2.2% |
| Coast | 3.2% | 2.8% | 3.3% | 3.2% |
| Migration | 0.2% | 0.1% | 0.1% | 0.1% |
| Species | 0.0% | -0.1% | 0.0% | 0.0% |
| Mortality | 0.0% | -3.5% | -1.7% | -2.7% |
| 3% | | | | |
| Water | 15.0% | 66.9% | 62.7% | 68.0% |
| Forestry | -1.2% | -0.4% | -0.6% | -0.6% |
| Energy | 67.9% | -19.0% | 27.6% | 24.3% |
| Agriculture | 14.2% | 0.3% | -1.7% | -2.7% |
| Coast | 3.9% | 2.0% | 4.8% | 4.7% |
| Migration | 0.3% | 0.1% | 0.1% | 0.1% |
| Species | -0.1% | -0.1% | 0.0% | 0.0% |
| Mortality | 0.0% | 50.1% | 7.2% | 6.3% |

respectively, over regional per capita income. Open Framework estimates are approximated using the period, sector and region impacts for slightly different reporting conventions.

## 9.4 Carbon Dioxide, Methane and Nitrous Oxide

Table 9.2 presents the marginal costs of carbon dioxide emissions. Patterns are clear and obvious. The more weight one places at impacts outside the European Union, the higher the marginal cost. The higher the discount rate, the lower the marginal cost. The regional and temporal patterns of the three models are quite different. FUND1.6 finds lowest marginal costs for the EU, FUND2.0 highest, and the Open Framework lies somewhere in between. This pattern changes if more weight is placed on the rest of the world, depending on the discount rate.

**Table 9.7.** The uncertainty about the marginal costs of carbon dioxide (in $/tC) according to FUND2.0.[a]

| | EU only | regional values | world average | EU values |
|---|---|---|---|---|
| 0% PRTP | | | | |
| Central est. | 1.2 | 6.1 | 15.1 | 60.1 |
| Mean | 1.4 | 7.9 | 19.8 | 78.1 |
| Median | 1.2 | 7.4 | 15.9 | 63.5 |
| Std. Dev. | 0.9 | 4.9 | 14.8 | 56.2 |
| Geom. Mean[b] | 1.2 | n.a. | 16.7 | 66.1 |
| Geom. Std. Dev.[c] | 1.8 | n.a. | 1.8 | 1.7 |
| 1% PRTP | | | | |
| Central est. | 0.7 | 5.1 | 8.9 | 35.2 |
| Mean | 0.9 | 4.8 | 11.1 | 43.9 |
| Median | 0.7 | 4.0 | 9.8 | 38.1 |
| Std. Dev. | 0.5 | 3.2 | 7.6 | 29.1 |
| Geom. Mean | 0.7 | n.a. | 9.6 | 38.0 |
| Geom. Std. Dev. | 1.8 | n.a. | 1.7 | 1.7 |
| 3% PRTP | | | | |
| Central est. | 0.3 | 4.1 | 3.8 | 14.9 |
| Mean | 0.4 | 2.3 | 4.3 | 16.7 |
| Median | 0.3 | 2.9 | 4.0 | 15.4 |
| Std. Dev. | 0.2 | 2.2 | 2.5 | 9.4 |
| Geom. Mean | 0.3 | n.a. | 3.8 | 14.9 |
| Geom. Std. Dev. | 1.8 | n.a. | 1.6 | 1.6 |

[a] Emissions are in the period 2000-2009. Costs are discounted to 2000. Time horizon is 2100. Scenario is IS92a. Morbidity risks are valued based on the value of a year life lost. The statistics are based on a Monte Carlo experiment with 1,000 runs. The assumed probability density functions are given in Tol and Downing (2000).

[b] The geometric mean is the exponent of the mean of the natural logarithms of the observations. 'n.a.' stands for 'not applicable' because there are negative marginal costs in the sample.

[c] The geometric standard deviation is the exponent of the standard deviation of the natural logarithm of the observations. If the observations are lognormally distributed, the 95% confidence intervals ranges from the geometric mean divided by twice the geometric standard deviation to the geometric mean times twice the geometric standard deviation.

Table 9.3 displays the results for nitrous oxide. The patterns are similar to those for carbon dioxide. However, FUND2.0 finds a positive marginal impact for a 3% discount rate for regional values: Short-term benefits to relatively rich regions outside the EU dominate the global estimate. Examples are the positive impact of climate change on agriculture in the USA and China.

Table 9.4 presents the marginal cost estimates for methane. The Open Framework now finds the highest marginal costs to the EU. FUND2.0 finds marginal benefits in a number of cases, reflecting its assumption that climate change is, in the short run, good for a number of sectors and countries.

Table 9.5 presents the results of FUND2.0 if mortality is valued not on the basis of a 'value of a statistical life' (200 times per capita income) but rather at a

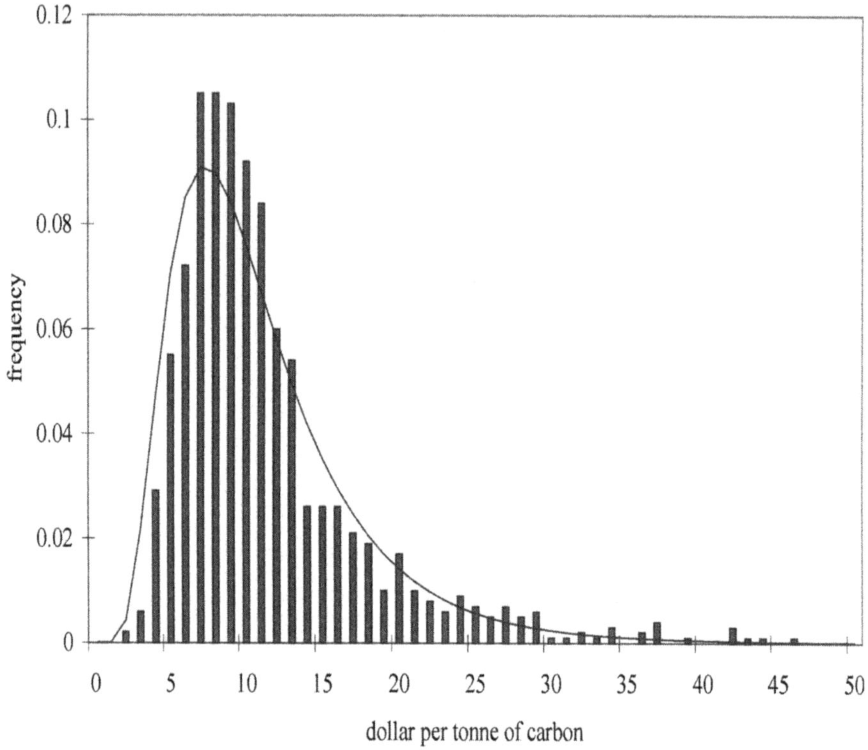


**Fig. 9.1.** Uncertainty about the marginal costs of carbon dioxide emissions for a 1% PRTP and world average values. Source: FUND2.0; see Table 9.7.

value of a life year lost' (10 times per capita income, per year). Cardiovascular and respiratory diseases, for which relatively little life-time is lost, dominate in the rich regions, do YOLL methodology leads to substantially lower marginal costs than does the VSL methodology. The gap closes if more weight is placed on developing countries, where vector-borne diseases, for which more life-time is lost, dominate.

The sectoral breakdown of the marginal costs of carbon dioxide is given in Table 9.6

## 9.5 Uncertainty

Uncertainties abound in climate change. The uncertainties about the impact of climate change are estimated in Tol (1999b, 1999a). These are confounded by uncertainties about the scenarios (which expand through time) and about the workings of the climate system. The exact specification can be found in Tol and Downing (2000). The uncertainty analysis is restricted to parametric uncertainty. The uncertainties reflect the ranges found in the literature. Thus, the uncertainty calculated below is a lower bound to the 'true' uncertainty. This particularly holds for the uncertainties about the impacts of climate change. The literature on that is thin (excepting agriculture), and ranges are therefore narrow.

Table 9.7 displays the results for carbon dioxide for a Monte Carlo analysis with 1,000 runs. All results are based on the 'years of life lost' methodology for morbidity risks. Figure 9.1 depicts the uncertainty about the marginal costs of carbon dioxide emissions for a 1% PRTP and world average values. The uncertainty is large and right-skewed. The probability density can be reasonably approximated with a lognormal distribution (the line in Fig. 9.1). In case regional values are used, the marginal costs may be negative (i.e., benefits). In this case, the positive impacts of climate change on the USA, the former Soviet Union and Japan may dominate the negative impacts on other regions.

**Table 9.8.** Uncertainty about the marginal costs of carbon dioxide emissions (in \$/tC) with a time horizon up to 2200, according to FUND2.0.[a]

| | EU only | regional values | world values | EU values |
|---|---|---|---|---|
| 0% PRTP | | | | |
| Central est. | 1.9 | 20.0 | 79.4 | 244.8 |
| Mean | 2.7 | 30.4 | 1849.8 | 3552.3 |
| Median | 2.2 | 22.8 | 123.4 | 350.2 |
| Std. Dev. | 1.6 | 21.2 | 7052.1 | 13010.6 |
| Geom. Mean[b] | 2.2 | 25.0 | 182.2 | 486.4 |
| Geom. Std. Dev.[c] | 1.8 | 1.8 | 6.1 | 5.3 |
| 1% PRTP | | | | |
| Central est. | 0.9 | 8.3 | 21.3 | 71.5 |
| Mean | 1.1 | 9.5 | 274.8 | 552.9 |
| Median | 1.0 | 8.8 | 28.5 | 89.1 |
| Std. Dev. | 0.6 | 4.0 | 981.9 | 1835.4 |
| Geom. Mean | 1.0 | 8.7 | 43.3 | 126.5 |
| Geom. Std. Dev. | 1.7 | 1.5 | 4.6 | 3.8 |
| 3% PRTP | | | | |
| Central est. | 0.3 | 4.3 | 4.4 | 16.8 |
| Mean | 0.4 | 2.7 | 10.4 | 29.1 |
| Median | 0.3 | 3.0 | 4.7 | 17.5 |
| Std. Dev. | 0.2 | 1.8 | 21.8 | 44.1 |
| Geom. Mean | 0.3 | n.a. | 5.7 | 20.0 |
| Geom. Std. Dev. | 1.6 | n.a. | 2.2 | 1.9 |

[a] Emissions are in the period 2000-2009. Costs are discounted to 2000. Time horizon is 2200. Scenario is IS92a. Morbidity risks are valued based on the value of a year life lost. Statistics are based on a Monte Carlo experiment with 1,000 runs. Results are trimmed for the top and bottom 5%. The assumed probability density functions are given in Tol and Downing (2000).

[b] The geometric mean is the exponent of the mean of the natural logarithms of the observations. 'n.a.' stands for 'not applicable' because there are negative marginal costs in the sample.

[c] The geometric standard deviation is the exponent of the standard deviation of the natural logarithm of the observations. If the observations are lognormally distributed, the 95% confidence intervals ranges from the geometric mean divided by twice the geometric standard deviation to the geometric mean times twice the geometric standard deviation.

**Table 9.9.** The uncertainty about the marginal costs of nitrous oxide (in $\$/tN_2O$) according to FUND2.0.[a]

| | EU only | regional values | world average | EU values |
|---|---|---|---|---|
| 0% PRTP | | | | |
| Central est. | 102.1 | 413.9 | 1,319.0 | 5,242.1 |
| Mean | 126.6 | 672.9 | 1,782.6 | 7,010.6 |
| Median | 110.1 | 594.8 | 1,405.5 | 5,613.2 |
| Std. Dev. | 79.2 | 452.0 | 1,383.7 | 5,213.0 |
| Geom. Mean[b] | 106.4 | n.a. | 1,480.6 | 5,872.8 |
| Geom. Std. Dev.[c] | 1.8 | n.a. | 1.8 | 1.8 |
| 1% PRTP | | | | |
| Central est. | 59.7 | 320.0 | 748.3 | 2,967.7 |
| Mean | 73.1 | 437.8 | 970.9 | 3,826.5 |
| Median | 63.5 | 377.9 | 786.8 | 3,129.8 |
| Std. Dev. | 45.1 | 260.8 | 794.4 | 2,639.9 |
| Geom. Mean | 61.8 | n.a. | 825.4 | 3,276.1 |
| Geom. Std. Dev. | 1.8 | n.a. | 1.7 | 1.7 |
| 3% PRTP | | | | |
| Central est. | 24.3 | 250.8 | 269.5 | 1,163.6 |
| Mean | 29.2 | 169.4 | 348.3 | 1,366.6 |
| Median | 25.7 | 186.2 | 301.3 | 1,181.3 |
| Std. Dev. | 17.7 | 154.5 | 210.7 | 806.3 |
| Geom. Mean | 24.9 | n.a. | 307.5 | 1,212.6 |
| Geom. Std. Dev. | 1.8 | n.a. | 1.6 | 1.6 |

[a] Emissions are in the period 2000-2009. Costs are discounted to 2000. Time horizon is 2100. Scenario is IS92a. Morbidity risks are valued based on the value of a year life lost. The statistics are based on a Monte Carlo experiment with 1,000 runs. The assumed probability density functions are given in Tol and Downing (2000).

[b] The geometric mean is the exponent of the mean of the natural logarithms of the observations. 'n.a.' stands for 'not applicable' because there are negative marginal costs in the sample.

[c] The geometric standard deviation is the exponent of the standard deviation of the natural logarithm of the observations. If the observations are lognormally distributed, the 95% confidence intervals ranges from the geometric mean divided by twice the geometric standard deviation to the geometric mean times twice the geometric standard deviation.

As a sensitivity analysis, and to emphasise the 'real' uncertainties, Table 9.8 displays what happens if the time horizon is extended from 2100 to 2200. The uncertainties get really large in the 22nd century, partly because of the uncertainty about the scenarios and partly because parameter uncertainties accumulate over time. The result is that it cannot be excluded that some economies collapse, for climate change or other reasons. If per capita income sharply declines, the discount rate becomes negative, and small changes in impacts are amplified in the net present value. This effect can be quite dramatic, and substantially increase the estimated marginal cost. However, the model was not designed for such extreme scenarios, so this result is a model artefact rather than a genuine finding. Comparing the central estimates of Table 9.7 and Table 9.8, it can be seen that

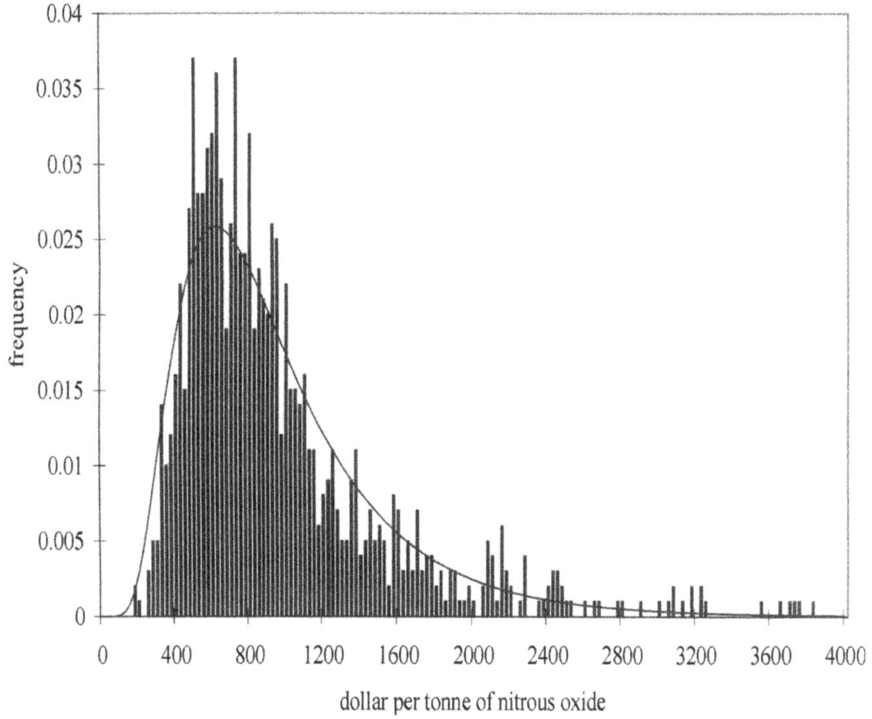


**Fig. 9.2.** Uncertainty about the marginal costs of nitrous oxide emissions for a 1% PRTP and world average values. Source: FUND2.0; see Table 9.9.

there are substantial impacts of emissions in the period 2000-2009 in the 22nd century. The outcomes of the Monte Carlo experiment are trimmed, that is, the highest and lowest 5% are removed. Nevertheless, an even larger increase in the mean estimate can be observed, and a substantial increase in the uncertainty, particularly for 'world values' and 'EU values' and a low discount rate.

Table 9.9 and Fig. 9.2 display the results for nitrous oxide. The findings are similar as those for carbon dioxide in Table 9.7.

Table 9.10 and Fig. 9.3 display the results for methane. The uncertainty is large and right-skewed, but the lognormal approximation is not very good.

## 9.6 Ozone and Sulphate Aerosols

The methodology to estimate the costs of region-specific climate change is described in Grewe and Tol (1999). Geographical climate change patterns resulting from regional nitrogen and sulphur emissions are derived from the DLR atmospheric chemistry model and the MPIM general circulation model (for

**Table 9.10.** The uncertainty about the marginal costs of methane (in $\$/tCH_4$) according to FUND2.0.[a]

| | EU only | regional values | world average | EU values |
|---|---|---|---|---|
| 0% PRTP | | | | |
| Central est. | 4.9 | 68.2 | 65.5 | 257.0 |
| Mean | 6.8 | 43.6 | 88.0 | 345.2 |
| Median | 5.9 | 51.1 | 75.1 | 294.8 |
| Std. Dev. | 3.8 | 35.1 | 53.5 | 204.5 |
| Geom. Mean[b] | 5.8 | n.a. | 77.2 | 304.4 |
| Geom. Std. Dev.[c] | 1.7 | n.a. | 1.6 | 1.6 |
| 1% PRTP | | | | |
| Central est. | 3.4 | 60.9 | 44.9 | 175.0 |
| Mean | 4.6 | 31.3 | 55.5 | 216.6 |
| Median | 2.6 | 37.3 | 49.5 | 193.3 |
| Std. Dev. | 1.4 | 30.6 | 30.6 | 116.7 |
| Geom. Mean | 3.9 | n.a. | 49.5 | 194.3 |
| Geom. Std. Dev. | 1.7 | n.a. | 1.6 | 1.6 |
| 3% PRTP | | | | |
| Central est. | 1.9 | 49.2 | 24.9 | 95.7 |
| Mean | 2.4 | 19.3 | 26.4 | 101.5 |
| Median | 2.1 | 21.9 | 25.3 | 97.6 |
| Std. Dev. | 1.4 | 23.2 | 12.5 | 47.0 |
| Geom. Mean | 2.1 | n.a. | 24.0 | 92.8 |
| Geom. Std. Dev. | 1.7 | n.a. | 1.5 | 1.5 |

[a] Emissions are in the period 2000-2009. Costs are discounted to 2000. Time horizon is 2100. Scenario is IS92a. Morbidity risks are valued based on the value of a year life lost. The statistics are based on a Monte Carlo experiment with 1,000 runs. The assumed probability density functions are given in Tol and Downing (2000).

[b] The geometric mean is the exponent of the mean of the natural logarithms of the observations. 'n.a.' stands for 'not applicable' because there are negative marginal costs in the sample.

[c] The geometric standard deviation is the exponent of the standard deviation of the natural logarithm of the observations. If the observations are lognormally distributed, the 95% confidence intervals ranges from the geometric mean divided by twice the geometric standard deviation to the geometric mean times twice the geometric standard deviation.

nitrogen and ozone) and from the UIUC atmospheric chemistry and general circulation model (for sulphur and sulphate aerosols). Changes in the regional climate patterns are imposed on FUND1.6 so as to estimate the regional damages.

Table 9.11 presents the estimated marginal costs of nitrogen emissions from aircraft flying over Europe. The effects on Europe are positive at first, and negative latter. The pulse of nitrogen emissions causes temporary warming, which brings the positive effects of climate change on European agriculture forward. Although the direct climatic effect ends with the pulse, the rate of warming is slightly lower afterwards. This brings costs to Europe, because its cold-related deaths fall less. Nitrogen emissions over Europe also warm the rest of the world. Here, generally negative effects ensue.

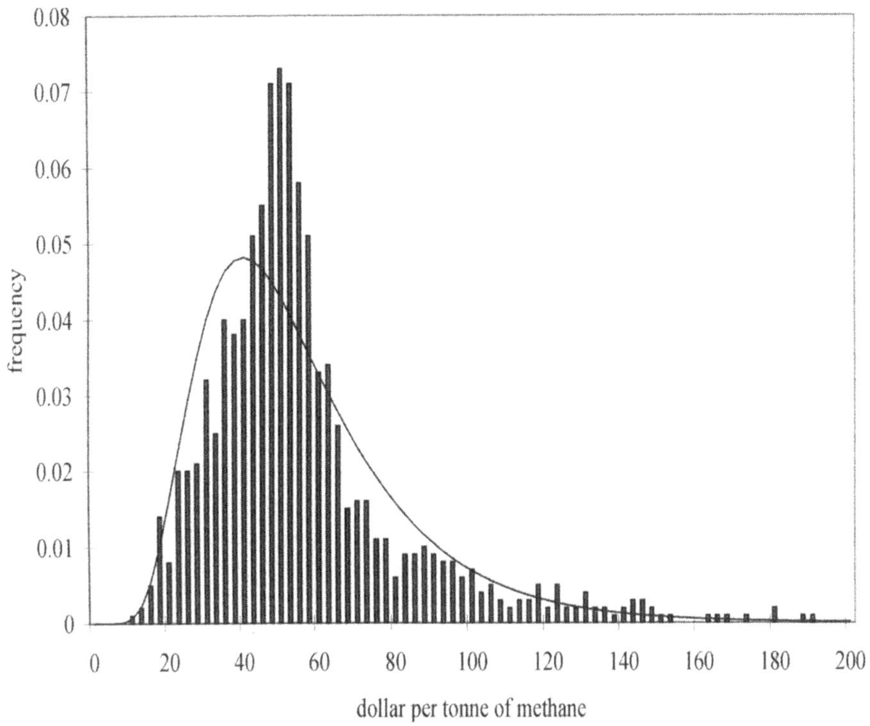


**Fig. 9.3.** Uncertainty about the marginal costs of methane emissions for a 1% PRTP and world average values. Source: FUND2.0; see Table 9.10.

**Table 9.11.** The marginal costs of nitrogen emissions from European aircraft ($/kgN).

| | EU only | regional values | world average | EU values |
|---|---|---|---|---|
| FUND1.6 | | | | |
| 0% PRTP | -5.5 | 53.1 | 359.9 | 1,270.2 |
| 1% PRTP | -2.5 | 52.8 | 337.0 | 1,232.2 |
| 3% PRTP | 0.8 | 62.2 | 330.7 | 1,233.5 |

Notes: Emissions are in the period 2000-2009. Costs are discounted to 2000. Scenario is FUND.

**Table 9.12.** The marginal costs of sulphur emissions from Europe ($/kgS).

| | EU only | regional values | world average | EU values |
|---|---|---|---|---|
| FUND1.6 | | | | |
| 0% PRTP | 1.2 | -2.0 | -11.3 | -35.8 |
| 1% PRTP | 0.6 | -1.4 | -9.8 | -34.1 |
| 3% PRTP | 0.0 | -1.5 | -9.6 | -35.4 |

Notes: Emissions are in the period 2000-2009. Costs are discounted to 2000. Scenario is FUND.

Table 9.12 displays the marginal costs of sulphur emissions in Europe. Sulphate aerosols have a short-lived cooling effect, so the pattern of the marginal costs of sulphur emissions is opposite to the pattern of the marginal costs of nitrogen emissions.

## 9.7 Summary of Damage Costs

The marginal costs of greenhouse gases, viz. carbon dioxide, methane, and nitrous oxide, were reassessed and the marginal costs of sulphur emissions (via sulphate aerosols) and nitrogen emissions from aircraft (via ozone.) were estimated. The recommended values are given in Table 9.13.

The discounting and valuation procedures for marginal cost estimation were refined, but the estimates for the three greenhouse gases do not substantially differ from those in Eyre et al. (1998), Pearce et al. (1996), and Tol (1999a).

The currently recommended values (Table 9.13) are lower than those of Eyre et al. (1998), though, reflecting the more optimistic tone of recent impact literature. The older models, FUND1.6 and the Open Framework, are unable to estimate marginal costs on the basis of the value of a life year lost. With comparable valuation methods, these models estimate the marginal costs of $CO_2$ and $N_2O$ to be a factor 5-7 higher than does FUND2.0.

It should be noted that with the inclusion of new insights into the impacts of climate change, it can no longer be excluded that marginal costs are negative, particularly for methane. The sign of the costs is model and region dependent.

Despite their short life-time, the marginal costs of nitrogen and sulphur emissions are relatively large, primarily because they are not much discounted.

**Table 9.13.** Recommended marginal costs for ExternE.[a]

| | Minimum[b] | Low[c] | Central estimate[d] | High[c] | Maximum[b] |
|---|---|---|---|---|---|
| $CO_2$ (€/t$CO_2$)[e] | 0.1 | 1.4 | 2.4 | 4.1 | 16.4 |
| $N_2O$ (€/t$N_2O$)[e] | 24.3 | 440.2 | 748.3 | 1,272.1 | 5,242.1 |
| $CH_4$ (€/t$CH_4$)[e] | 1.9 | 28.2 | 44.9 | 71.5 | 257.0 |
| N (€/kgN)[f] | -5.5 | 198.2 | 337.0 | 527.9 | 1,270.2 |
| S (€/kgS)[f] | -35.8 | -16.6 | -9.8 | -5.8 | 0.0 |

[a] Emissions are in the period 2000-2009. Costs are discounted to 2000.

[b] Minimum and maximum are as in Tables 9.5, 9.11 and 9.12.

[c] High and low approximately span the 67% confidence interval.

[d] PRTP equals 1%. Values are world averages.

[e] Model is FUND2.0. Time horizon is 2100. Scenario is IS92a. Morbidity risks are valued based on the value of a life year lost. Note that the marginal costs of carbon dioxide are here expressed per tonne of $CO_2$, rather than per tonne of carbon as in earlier tables.

[f] Model is FUND1.6. Time horizon is 2200. Scenario is FUND. Morbidity risks are valued based on the value of a statistical life. Uncertainty is based on an assumed geometric standard deviation of 1.7, in line with $CO_2$, $N_2O$ and $CH_4$. Nitrogen emissions are from aircraft only.

The results presented here should not be taken as final estimates. The impacts covered by the models used are only a fraction (of unknown size) of all climate change impacts. Particularly, large scale disruptions, such as a breakdown of North Atlantic Deep Water formation or a collapse of the West-Antarctic Ice Sheet, are excluded from the analysis. The methodologies to estimate climate change impacts in a different future remain weak. Adaptation is not included in its full complexity. Valuation of impacts is still troublesome, particularly for nature and health. Our knowledge of atmospheric chemistry and climate has substantial gaps. The estimates reflect our current best knowledge, and indicate a stimulating research agenda.

## 9.8 Avoidance Costs

With regard to the rather broad and uncertain range of results coming from the analysis of the damage costs, an alternative for estimating costs per unit of greenhouse gas is the avoidance cost or shadow value approach. This approach uses given emission reduction targets and calculates the marginal avoidance costs (i.e. the highest costs per tonne of avoided $CO_{2equiv.}$) to reach this target. Again, uncertainties are involved here, as especially the generation of the necessary avoidance cost curve is difficult and assumptions about the future development of emissions and reduction measures and their costs have to be made. To give an example: Fahl et. al. (1999) estimate marginal avoidance costs for Germany to reach the Kyoto target to 19 Euro per t of $CO_{2equiv.}$; this is equivalent to ca. 4.6 Eurocent per litre of gasoline and 5.1 Eurocent per litre of diesel. Duerinck (2000) estimates marginal costs for $CO_2$ abatement for meeting the Kyoto targets in Belgium as 25 Euro per tonne of $CO_2$. Criqui and Viguier (2000) report marginal costs of reaching the Kyoto target for the EU of 37 US$ per tonne of $CO_2$ avoided. With international emission trading this amount could be reduced to ca. 14 $ / t of $CO_2$. Capros and Mantzos (2000) give marginal abatement costs for the EU of 38 Euro (1990 prices) per tonne of $CO_2$ (without emission trading).

These costs are obviously somewhat higher than the damage costs calculated in the previous sections. One might argue, that the damage costs only contain costs of impacts, that are reasonably well known and understood, whereas there might be further impacts not known today, so that with regard to the precautionary principle somewhat higher reduction targets and marginal reduction costs might be justified.

Looking further in the future, more stringent reductions than the Kyoto aims are assumed to be necessary to reach sustainability. Based on a reduction target of 50% in 2030 compared to 1990 INFRAS/IWW (2000) use avoidance costs of 135 Euro per t of $CO_2$; however one could argue, that this reduction target is not or not yet accepted.

In the following, for most of the results damage costs (as given in Sect. 9.7, with a central estimate of 2.4 Euro per t of $CO_2$) are used. However, for some sensitivity analyses, also avoidance costs of 19 Euro per t of $CO_2$ are taken.

# 10 Treatment of Up- and Downstream Processes

## 10.1 Life Cycle Assessment (LCA) and Impact Pathway Analysis

In the spirit of life cycle assessment (LCA) one should account for all the environmental consequences, from cradle to grave, of a decision about a system, process or product (e.g. the choice between nuclear and coal fired power plants, or the choice between plastic and glass beverage containers). For example, the replacement of conventional cars by electric ones would eliminate air pollution from cars at the expense of increased pollution from power plants; in addition there would be increased pollution from the manufacture of batteries and their eventual disposal. Electricity production, and manufacture and disposal of batteries, are processes that occur upstream or downstream of the main process, i.e. driving the car.

→ real impacts for each stage (site specific)

**Goal** evaluate the entire matrix

| **Steps of impact pathway analysis** → <br> **Stage of fuel chain** ↓ | Emission | Dispersion | Exposure-response function | Economic valuation |
|---|---|---|---|---|
| Fuel extraction | | | | |
| Fuel transport | | | | |
| Power plant | | | | |
| Transmission of electricity | | | | |
| Management of wastes | | | | |

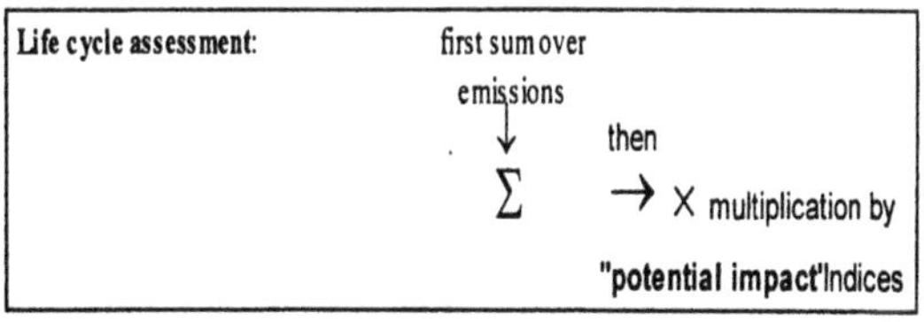


**Fig. 10.1.** Relation between impact pathway analysis and current practice of most LCA, illustrated for the example of electricity production (from Spadaro and Rabl 1999).

However, before ExternE most life cycle assessments have been more or less limited to an inventory of resource use and emissions (Heijungs et al. 1992, SETAC 1992). They have not gone far with the calculation of impacts, if they have attempted to quantify impacts at all. For example one of the principal guides for LCA, Heijungs et al. (1992), proposes general site independent toxicity indices ("classification factors") for each pollutant. For air pollution these factors are calculated with the assumption of uniform mixing in the entire atmosphere of the earth. The reason for this drastic approximation lies in the emphasis on a precise accounting of all material flows, direct and induced (i.e. upstream or downstream). Since induced flows occur at many geographically different points under a variety of different conditions, it is not feasible to keep track of the local details of the emissions. However, neglect of such site dependence is justified only for greenhouse gases and other pollutants with long residence times; it is totally unrealistic for particulate matter, $NO_x$, $O_3$ and $SO_2$: a tonne of diesel particles emitted in the ocean is quite harmless, emitted in a big city it has severe health effects. A correct assessment requires an impact pathway analysis (IPA) of the type carried out here. Fig. 10.1. illustrates the relation between IPA and LCA.

## 10.2 Simplified Damage Estimates for LCA

Unfortunately a full site-specific impact pathway analysis is far too complex to be practical for LCA, quite apart from the fact that in most cases the sites of the upstream or downstream processes are not even known. Clearly a compromise is needed for the analysis of upstream or downstream processes, a compromise that involves typical damage estimates that are approximately right for typical processes.

Two possible approaches for obtaining typical damage estimates have been explored in the current phase of ExternE. The results are essentially the same for industrial processes in the EU. One approach is based on the multi-source version of EcoSense and the CORINAIR data base for emissions. The average damage cost per kg of a pollutant is calculated by calculating the total damage in Europe with and without emissions from the sector of interest (e.g. industry) and dividing by the total emission of that pollutant from that sector.

The other approach is the "uniform world model" (UWM) of Curtiss and Rabl (1996), a formula with five factors that follows from conservation of mass under certain simplifying assumptions. This model has the simple form

$$\text{Damage cost} = P\ f_{cr}\ \rho\ Q/k \tag{10.1}$$

where

| | | |
|---|---|---|
| Q | = | emission [kg/yr] |
| K | = | depletion velocity [ms/s] of pollutant |
| $\rho$ | = | average receptor density |
| $f_{er}$ | = | slope of E-R function |

P = unit cost of end point of E-R function [e.g. Euro/YOLL for mortality].

The generalisation to secondary pollutants is straightforward.

Spadaro and Rabl (1999) have verified the relevance of UWM by comparing it with site-specific EcoSense results for a wide range of sites, including all the ones evaluated by ExternE. The agreement is excellent; for example, the damage costs for all coal fired power plants in France are reproduced within a few %. Correction factors for site type (urban vs rural) and stack height have also been developed (Spadaro, 1999).

Since many LCA practitioners object to monetary valuation and prefer instead to express impacts in physical units, the respective valuations of health impacts (disability adjusted life years DALYs vs. ECUs) were compared; the results are in good agreement for the relative ranking (Spadaro, 1999).

## 10.3 The Uniform World Model

To explain the UWM (Curtiss and Rabl 1996) let us write the damage D due to an emission Q as an integral over land area

$$D = f_{er} \int dx \int dy \; \rho(x)\, c(x) \qquad (10.2)$$

where

$\rho(x)$ = density of receptors at point x = (x,y),

$c(x)$ = concentration increase at x due to Q, and

$f_{er}$ = slope of dose-response or exposure-response (E-R) function.

The slope of the E-R function states the incremental number of cases (e.g hospitalisations) per concentration increment. It is assumed that $f_{er}$ has a constant value independent of x.

It is instructive to relate the concentration c(x) to the depletion rate of the pollutant. There are essentially three mechanisms by which an air pollutant can disappear from the atmosphere (Seinfeld and Pandis, 1998):

1. dry deposition (uptake at the earth's surface by soil, water or vegetation)
2. wet deposition (absorption into droplets, removed by precipitation),
3. decay or transformation (e.g. transformation of $SO_2$ to $(NH_4)_2SO_4$).

When evaluating the damage of a primary pollutant, this pollutant is no longer counted in the equation once it has been transformed; rather from that point on a different exposure-response function comes into play for the secondary pollutant.

The dry deposition rate is proportional to the concentration c(x) at the earth's surface, and it is customarily written in the form

$$F_{dry}(x) = v_{dry}\, c(x) \qquad (10.3)$$

where

$F_{dry}(x)$ = deposition flux [in kg/(m²·s)], and
$v_{dry}$ = dry deposition velocity [m/s].

Wet deposition and decay or transformation can likewise be characterised in terms of fluxes $F_{wet}(x)$ and $F_{trans}(x)$, defined as the rate at which the pollutant is removed by these mechanisms per m2 (horizontal surface area) and per second. Now let us define a "depletion velocity" k(x) as ratio of the total depletion flux

$$F(x) = F_{dry}(x) + F_{wet}(x) + F_{trans}(x) \quad (10.4)$$

and the surface concentration c(x) as

$$k(x) = F(x)/c(x), \quad (10.5)$$

the units of k being m/s. Using F(x) and k(x) the damage can be written in the form

$$D = f_{er} \int dx \int dy\, \rho(x)\; F(x)/k(x)\,. \quad (10.6)$$

Let us now consider a situation where k(x) are $\rho(x)$ are independent of x. With uniform receptor density $\rho(x) = \rho_{uni}$ and uniform depletion velocity $k(x) = k_{uni}$, the integral in Eq.10.6 is simply

$$D = D_{uni} = f_{er}\, \rho_{uni}\, Q/k_{uni} \quad (10.7)$$

because, averaged over time, the surface integral of the depletion flux equals the emission

$$Q = \int dx \int dy\; F(x) \quad (10.8)$$

by conservation of matter. Eq.10.7 (and its generalisation for secondary pollutants, Eq.10.12) will be called "uniform world model".

As an illustration $D_{uni}$ is calculated for acute mortality due to $SO_2$, with $f_{er}$ = 5.34E-06 YOLL/(pers·yr·µg/m³) (Sunyer et al. 1996). By fitting EcoSense dispersion data a depletion velocity k = 0.0073 m/s was found for $SO_2$. Inserting these numbers into Eq.10.7 with the regional average population density $\rho$ = 8.0E-5 person/m² (see Sect. 10.4) leads to Q = 106 kg/yr = 3.17E07 $\mu$g/s

$$D_{uni} = (5.34 \times 10^{-6}\ \text{YOLL} / (\text{person yr}\ \mu\text{g/m}^3) \times 8.0 \times 10^{-5}\ \text{person/m}^2) / 0.0073\ \text{m/s} \times 3.17 \times 10^{7}\ \mu\text{g/s} = 1.86\ \text{YOLL/yr} \quad (10.9)$$

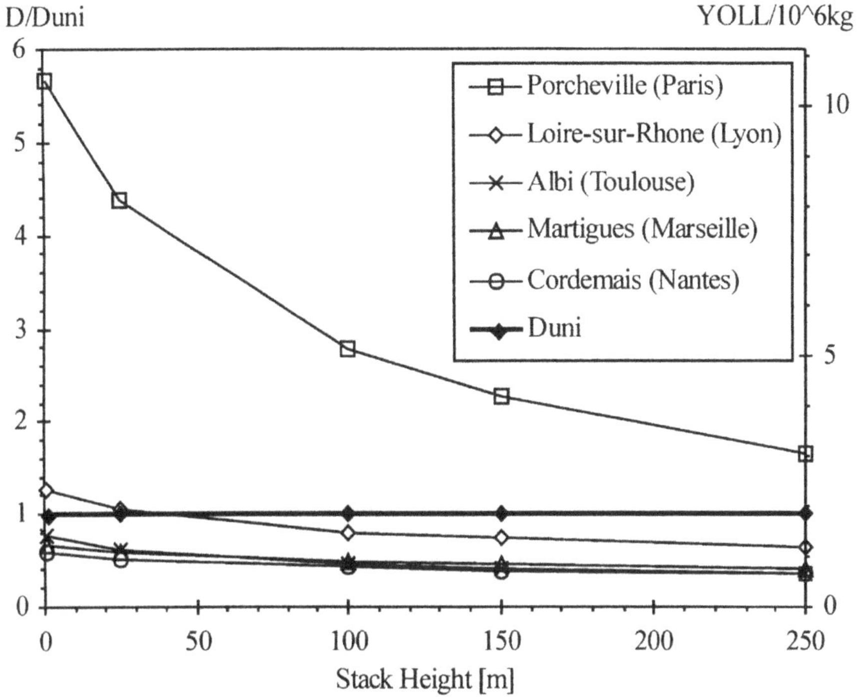


**Fig. 10.2.** An example of dependence on site and on height of source for a primary pollutant with linear E-R function: damage D from $SO_2$ emissions, for five sites in France, in units of $D_{uni}$ for uniform world model Eq.10.7 (the nearest big city, 25 to 50 km away, is indicated in parentheses). Scale on right indicates YOLL/yr (acute mortality) from a plant with emission 106 kg/yr. Plume rise for typical conditions is accounted for.

This is shown as the horizontal line in Fig. 10.2. It lies right in the middle of the curves for the five sites.

Even though the assumption $k(x) = k_{uni}$ may not appear very realistic, especially near a point source, the sensitivity to deviations from uniformity is surprisingly small. The reason is that much of the total impact occurs in regions sufficiently far from the source where the pollutant is fairly well mixed vertically in the planetary boundary layer, and variations of k(x) are not too large. Emission height dependence and deviations from uniformity are most pronounced when the receptors are concentrated near the source. A source near Paris seems like a fairly extreme example because France is highly centralised, with 19% of its population in Greater Paris. Cordemais, in a semi-rural zone on the Atlantic cost, is an opposite extreme.

Thus it is plausible that these results are fairly representative and that the simple "uniform world model" of Eq.10.7 can be a useful first estimate. If one wants typical results for public policy, without being able to evaluate each and every site, $D_{uni}$ seems as good a choice as any - and it has the advantage of being simple and transparent.

## 10.4 Generalisation to Secondary Pollutants

The generalisation of Eq.10.7 for the damage $D_2$ due to a secondary pollutant is

$$D_{2uni} = f_{er2} \times \rho_{uni} \times k_{1\text{-}2uni} / k_{2uni} / k_{1uni} \times Q_1 \qquad (10.10)$$

where

$Q_1$ = emission rate of the primary pollutant,
$f_{er2}$ = E-R function slope of the secondary pollutant,
$k_{2uni}$ = depletion velocity of the secondary pollutant, and
$k_{1\text{-}2uni}$ = average of transformation velocity, defined as

$$k_{1\text{-}2}(x) = F_{1\text{-}2}(x) / c_1(x) \qquad (10.11)$$

with

$F_{1\text{-}2}(x)$ = creation flux of secondary pollutant per $m^2$ of horizontal surface, and
$c_1(x)$ = primary concentration increase at x due to $Q_1$.

This can be written in the same form as Eq.10.7

$$D_{2uni} = f_{er2} \times \rho_{uni} / k_{2uni,eff} \times Q_1 \qquad (10.12)$$

if an "effective depletion velocity" is defined

$$k_{2uni,eff} = k_{2uni} \times k_{1uni} / k_{1\text{-}2uni} \qquad (10.13)$$

**Table 10.1.** Values of $k_{uni}$ for different pollutants, used for the "uniform world model". Except for CO, they have been derived by nonlinear regression to dispersion calculations with EcoSense.

| a) Primary pollutants, $D_{uni}$ of Eq.10.7. | $K_{uni}$ [m/s] |
|---|---|
| $PM_{10}$ (including heavy metals, dioxins) | 0.0067 |
| $SO_2$ | 0.0073 |
| $NO_2$ | 0.0147 |
| CO | 0.001 |
| **b) Secondary pollutants, $D_{2uni}$ of Eq.10.12.** | **$K_{2uni,eff}$ [m/s]** |
| $NO_2$ transformed to nitrates | 0.0071 |
| $SO_2$ transformed to sulfates | 0.0173 |

Numerical values were determined by fits to EcoSense data. Table 10.1 shows the values of $k_{uni}$ for different pollutants. For CO 0.001 m/s is assumed based on life time estimates of 36 to 110 days given by Manahan (1994), about a factor ten longer than for $PM_{10}$ and $SO_2$.

In the region bounded by Sicily to the South, Portugal to the West, Scotland to the North and Poland to the East, the average population density is 80 persons/km$^2$. This is about half the average EU15 population density of 158 persons/km$^2$ per land area because it includes much water. Examining the results of detailed site specific calculations for more than fifty installations in the EU15 countries (ExternE, 1998), Spadaro (1999) found that the UWM, with $k_{uni}$ of Table 10.1, does indeed provide representative results for typical industrial installations. Of the many comparisons provided by Spadaro (1999) as typical example the one for sulfate impacts is cited in Table 10.2. The ExternE calculations for the respective regions are based on a relatively small number of sites and it is not known how typical they really are of the region in question; since the variation between sites is larger than the difference between the 4th and 5th columns in Table 10.2, the agreement between UWM and the detailed calculations of ExternE is as close as is meaningful. More recent comparisons with ExternE-like calculations in the USA and in Asia confirm this conclusion.

As it turns out, the damage costs thus calculated are essentially the same as the damage costs for industrial sources in France. Therefore we refer to the results reported in Sect. 13.3.

**Table 10.2.** Comparison of ExternE (1998) damage costs and Simple UWM estimates for sulfate impacts in Europe. ExternE results are average values for the country or region in which the precursor emissions take place.

| Source location | Countries affected | $\rho_{REG}$ pers/km$^2$ | UWM Euro/ton | ExternE Euro/ton |
|---|---|---|---|---|
| Finland | Finland, Norway, Russia, Sweden | 10 – 20 [a] | 725-1450 | 1044 |
| Ireland | France, Ireland, UK | 20 –30 [a] | 1450-2146 | 2276.5 |
| Spain | France, Italy, Portugal, Spain | 50 – 60 [a] | 3596-4321 | 3755.5 |
| France | France | 105 | 7540 | 6380 |
| Europe | ALL Western Europe, except Scandinavia | 80 [b] | 5742 | 4495 |
| Central Europe | ALL Western Europe | 46 [c] | 3306 | 4089 |

[a] $\rho_{REG}$ is the receptor density inside a 1000 km circle about the source
[b] Impact area includes Mediterranean Sea
[c] Impact area includes Mediterranean, North and Baltic seas

# 11 Uncertainty

## 11.1 General Remarks

Since the uncertainties are large, an analysis of the uncertainties is crucial for the credibility of the results. To begin, one needs to identify and quantify the individual sources of uncertainty. It is appropriate to group them into different categories, even though there may be some overlap:

1. data uncertainty, e.g. slope of a dose-response function, cost of a day of restricted activity, and deposition velocity of a pollutant;
2. model uncertainty, e.g. assumptions about causal links between a pollutant and a health impact, assumptions about form of a dose-response function (e.g. with or without threshold), and choice of models for atmospheric dispersion and chemistry;
3. uncertainty about policy and ethical choices, e.g. discount rate for intergenerational costs, and value of statistical life;
4. uncertainty about the future, e.g. the potential for reducing crop losses by the development of more resistant species;
5. idiosyncrasies of the analyst, e.g. interpretation of ambiguous or incomplete information.

The first two categories (data and model uncertainties) are of a scientific nature. They are amenable to analysis by statistical methods, combining the component uncertainties over the steps of the impact pathway, to obtain formal confidence intervals around a mid estimate. For ethical choice and for uncertainty about the future a sensitivity analysis may be more appropriate, indicating how the results depend on these choices and on the scenarios for the future.

Quantifying the sources of uncertainty in this field is problematic because of a general lack of information. Usually one has to fall back on subjective judgment, preferably by the experts of the respective disciplines. That has been done, in an informal manner, for this project. We would have liked to carry out a formal survey of experts, but unfortunately within the very limited means of the present project it was not possible to solicit opinions from a representative sample of experts. We also used all the relevant information we could find in the literature.

To find the resulting uncertainty of the damage cost from the uncertainties of the components of the analysis, the approach of Rabl and Spadaro (1999) was followed which is based on lognormal distributions and multiplicative confidence intervals. The justification for this approach lies in the observation that typical damage estimates correspond to the simple "uniform world model" described in Sect. 10.2, as demonstrated by extensive numerical calculations for a large number of sites in Europe, Asia and America, including all the sites studied by ExternE (Spadaro and Rabl, 1999). Since the "uniform world model" is multiplicative, the uncertainty of the damage estimate is approximately lognormal thanks to the central limit theorem of statistics. Thus it suffices to specify geometric mean and geometric standard deviations, or equivalently, multiplicative confidence intervals about the geometric mean (which is usually quite close to the median). This can greatly simplify the expert judgment elicitation: multiplicative confidence intervals are much easier to specify than an entire probability distribution. Likewise it facilitates the interpretation and communication of the final result.

However, not all uncertainties are amenable to such a quantitative analysis. For the presentation of the ExternE results it is desirable to provide some indication of the full uncertainty which is larger than the uncertainties that have been quantified explicitly. This point has been addressed in Sect. 11.5 where heavy error bars are used to indicate what we call the "parametric uncertainties", i.e. the ones found by quantitative analysis, and light error bars ("comprehensive uncertainties") indicate our subjective judgement about reasonable upper and lower limits of the damage costs. Because of the limitations of present knowledge it is sometimes difficult to draw sharp distinctions between different types of uncertainty. For example in Sect. 11.4 a framework for quantifying the uncertainties due to the lack of knowledge about the health effects of specific components of particulate air pollution is presented; but while the framework is quantitative, the application rests on subjective judgement. Thus Sect. 11.4 is used to help estimate "comprehensive uncertainties" in Sect. 11.5.

## 11.2 Methodology for Estimating Uncertainties

Since the "uniform world model" states the estimate of typical damages as a simple product of several factors, it is appropriate to examine the uncertainty of a product

$$Y = Z_1 Z_2 \dots Z_n \tag{11.1}$$

of n independent random variables $Z_i$, each characterised by a probability distribution $p_i(Z_i)$. Even though there may be correlations between impacts calculated along parallel pathways (e.g. inhalation and food intake), for the multiplicative processes along a single pathway, independence of the factors is usually a valid assumption. For example, atmospheric dispersion and dose-response function are clearly independent. Taking the logarithm of Eq.11.1 gives

$$\ln(Y) = \ln(Z_1) + \ln(Z_2) + \dots + \ln(Z_n) \,. \tag{11.2}$$

Let us define the geometric mean $\mu gi$ for each variable as the expectation value of the logarithm

$$\ln(\mu_{gi}) = \int dZ_i\, p_i(Z_i) \ln(Z_i) \; ; \tag{11.3}$$

for the product in analogous manner it can be defined

$$\ln(\mu_g) = \int dY\, p(Y) \ln(Y) \, . \tag{11.4}$$

Assuming independence of the probability distributions $p_i(Z_i)$ one finds

$$\mu_g = \mu_{g1}\, \mu_{g2} \ldots \mu_{gn} \tag{11.5}$$

It is convenient to work with dimensionless quantities by dividing each variable by its geometric mean

$$z_i = Z_i / \mu_{gi} \text{ and } y = Y / \mu_g \tag{11.6}$$

Let us now define the geometric standard deviation $\sigma_{gi}$ as

$$[\ln(\sigma_{gi})]^2 = \int dZ_i\, p_i(Z_i)\, [\ln(Z_i) - \ln(\mu_{gi})]^2 \tag{11.7}$$

Invoking again independence of the distributions, we find that the geometric standard deviation $\sigma_g$ of the product Y is given by

$$[\ln(\sigma_g)]^2 = [\ln(\sigma_{g1})]^2 + [\ln(\sigma_{g2})]^2 + \ldots + [\ln(\sigma_{gn})]^2 \tag{11.8}$$

This is the key equation on which the uncertainty analysis will be based. It is valid for any set of independent distributions with finite geometric means and geometric standard deviations. The interpretation of the geometric standard deviation in terms of a confidence interval requires an assumption on the probability distribution. For this purpose is instructive to look at the lognormal distribution.

The lognormal distribution of a variable x is obtained by assuming that the logarithm of x has a normal distribution. The central limit theorem of statistics, applied to Eq.11.2, implies that the lognormal distribution is the "natural" distribution for multiplicative processes, the same way that the normal (gaussian) distribution is natural for additive processes. Although the lognormal distribution becomes rigorous only in the limit of infinitely many factors, in practice it can be a good approximation even for a few factors, provided the distributions with the largest spread are not too far from lognormal. As far as we were able to verify, this is indeed the case for damage costs of air pollution. For a lognormal variable the geometric mean $\mu_g$ and the geometric standard deviation $\sigma_g$, defined by Eqs.11.16

and 11.20, have a simple interpretation in terms of multiplicative confidence intervals such as

- $[\mu_g/\sigma_g, \mu_g \sigma_g]$ for approximately 68% and
- $[\mu_g/\sigma_g{}^2, \mu_g \sigma_g{}^2]$ for approximately 95%.

For the uncertainty of atmospheric dispersion models a geometric standard deviation in the range from two to five is sometimes cited, but without making a distinction between episodic values and averages over space or time. In fact, atmospheric models are far more accurate for averages than for episodic values. This is an important consideration since for policy applications, of interest here, one needs long term average values rather than episodic values. For example, the European tracer experiment (ETEX) (Dop et al., 1998) has provided validation for a variety of dispersion models, but on an episodic basis; the relatively large discrepancies between measured and calculated values are therefore no indication of the accuracy that can be expected for long term averages.

Among the many parameters and input data of an atmospheric model, most have only a relatively minor effect on the calculation of long term average concentrations. To see which parameters are the most important, it is instructive to look at the "uniform world model", because it yields the damage costs for typical conditions. In the "uniform world model" the key parameters of a dispersion model are those that affect the depletion velocity, defined by Eq.10.5. For an indication of the kind of distribution that can be expected, in Fig. 11.1 a histogram is shown for data for dry deposition velocity of $SO_2$, based on a review by Sehmel (1980). Visibly, a logarithmic scale is much more appropriate for these data than a linear one. The geometric standard deviation is approximately 2.5 for this sample. The variability of this sample reflects two distinct effects: variability due to different surface materials, and uncertainty for a given surface material. In practical applications the distinction gets blurred by uncertainties in the geographic distribution of surface materials. More recent data would probably have a smaller standard deviation, but a survey as comprehensive as that of Sehmel could not be found.

For particles we refer to Fig. 19.3 of Seinfeld & Pandis (1998), which likewise suggests a lognormal distribution. Dry deposition of small particles has been reviewed by Nicholson (1988), who points out the large variability of measured deposition velocities with the nature of the surface and the conditions of the observation. The spread of values seems to be comparable to Fig. 11.1 for $SO_2$.

Deposition velocities for reactive nitrogen compounds have been reviewed by Hanson and Lindberg (1991); here the variability with the conditions of the absorbing surface is further enhanced by the high chemical reactivity of nitrogen compounds.

The possibility of low values of dry deposition velocities could imply very large damages under dry conditions. However, for the wet climates typical of Europe, long distance dispersion will be limited by wet deposition. This has been verified for particulate matter by varying the dry deposition velocity in the EcoSense model (Krewitt et al., 1995). Thus the uncertainty of total deposition in Europe appears to be significantly smaller than suggested by dry deposition data.

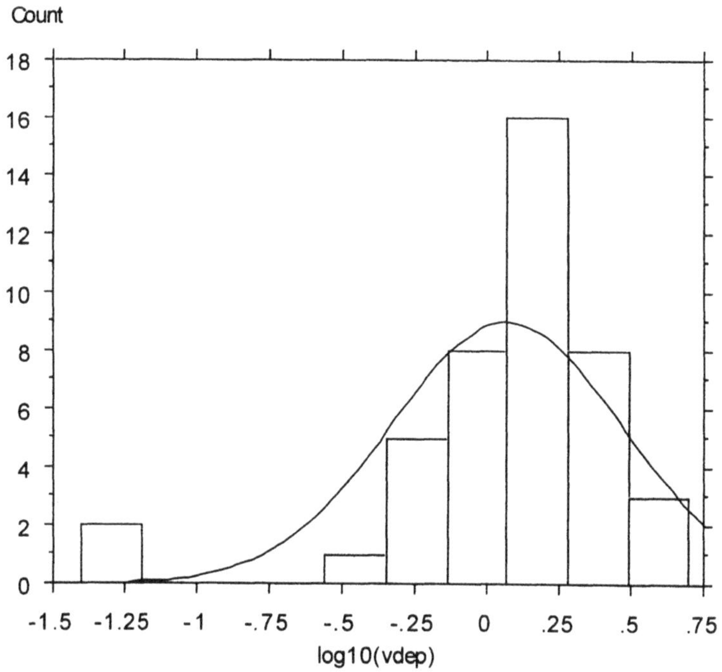


**Fig. 11.1.** Distribution and lognormal fit of maximum values, in the review of Sehmel (1980), for dry deposition velocity [in cm/s] of $SO_2$ over different surfaces.

In view of all these considerations we are led to suggest a $\sigma_g$ around 1.8 for dispersion modeling of primary air pollutants, as far as total damages are concerned. These numbers are consistent with estimates by McKone and Ryan (1989). For secondary pollutants there is additional uncertainty due to chemistry, especially in the case of ozone, but the uncertainty due to inaccuracies in the spatial distribution of concentration values relative to receptors is much smaller because secondary pollutants form only gradually at distances removed from the source.

Our estimates of the component uncertainties has been improved, relative to the values in ExternE (1998), on the basis of continuing advice from the experts of the respective disciplines involved in the impact calculations. The current assessment for key impact categories is indicated in Table 11.1 and Tablc 11.2. They show the uncertainties of the damage cost per unit of a product, such as a kWh, (per functional unit, in the terminology of life cycle assessment); thus they include the uncertainty in the emitted quantity Q per functional unit. Also included in Table 11.1 and Table 11.2 are the steps for determining how the confidence intervals are to be placed about the mean estimate, according to the method described in Sect. 11.3.

For $PM_{10}$, $PM_{2.5}$, $SO_2$ via sulfates and $NO_2$ via nitrates the same E-R functions are assumed, therefore the contributions to the uncertainty are the same for each of these pollutants, with the exception of

**Table 11.1.** Uncertainty of damage cost estimates per functional unit (e.g. per kWh) for chronic mortality. Sample calculations of geometric standard deviation Total $\sigma_g$, using Eq.11.8 and the component uncertainties, followed by determination of multiplicative confidence intervals about the mean.

| Step of calculation | Lognormality | PM | Sulfates | Nitrates |
|---|---|---|---|---|
| Emission | probably yes | 1.2 | 1 | 1 |
| Dispersion | yes | 1.8 | 2.2 | 2.5 |
| FCR regression | no | 1.3 | 1.3 | 1.3 |
| FCR (transfer composition) | ? | 1.5 | 1.5 | 1.5 |
| YOLL, given RelRisk | probably yes | 1.2 | 1.2 | 1.2 |
| VSL (value of statistical life) | yes | 2 | 2 | 2 |
| Value of YOLL, given VSL | ? | 1.3 | 1.3 | 1.3 |
| Discount rate&latency | probably yes | 1.3 | 1.3 | 1.3 |
| Total $\sigma_g$ | | 3.08 | 3.41 | 3.72 |
| mean/median, total uncorrected | | 1.88 | 2.12 | 2.37 |
| $\sigma_g$ dispersion | | 1.80 | 2.20 | 2.50 |
| mean/median, dispersion | | 1.19 | 1.36 | 1.52 |
| mean/median,total, corrected for dispersion | | 1.33 | 1.14 | 1.02 |
| median/mean | | 0.75 | 0.88 | 0.98 |
| lower/mean | | 0.24 | 0.26 | 0.26 |
| upper/mean | | 2.31 | 2.99 | 3.63 |

**Table 11.2.** Uncertainty of damage cost estimates per functional unit (e.g. per kWh) for cancers. Same format as Table 11.1.

| Radionuclides | Lognormality | As, Cd, Cr, Ni | Radionuclides |
|---|---|---|---|
| Emission data | approximately | 1.2 | 1.5 |
| Dispersion | yes | 1.8 | 1.8 |
| factor for noninhalation pathways | ? | 2 | 2 |
| dose-response function (unit risk) | ? | 1.5 | 2 |
| dose and dose-rate effectiveness factor | | | 1.5 |
| fraction that is fatal | | | 1.3 |
| YOLL (years of life lost) probably | yes | 1.2 | 1.5 |
| VSL (value of statistical life) | yes | 2 | 2 |
| Value of YOLL, given VSL | ? | 1.3 | 1.3 |
| Latency (including discount rate) probably | yes | 2 | 1.3 |
| Total $\sigma_g$ | | 4.24 | 4.84 |
| mean/median,total uncorrected | | 2.84 | 3.47 |
| $\sigma_g$ dispersion | | 1.80 | 1.80 |
| mean/median, dispersion | | 1.19 | 1.19 |
| mean/median,total corrected | | 2.01 | 2.45 |
| median/mean | | 0.50 | 0.41 |
| lower/mean | | 0.12 | 0.08 |
| upper/mean | | 2.11 | 1.97 |

- atmospheric dispersion and chemistry (we assume different geometric standard deviations: $\sigma_g = 1.8$ for $PM_{10}$ and $PM_{2.5}$, $\sigma_g = 2.2$ for sulfates, and $\sigma_g = 2.5$ for nitrates).
- the relative scale of the E-R functions, allowing for different toxicities of sulfates and nitrates relative to $PM_{10}$, as analyzed in Sect. 11.4.

## 11.3 Placement of the Confidence Intervals

In ExternE (1998) insufficient attention had been paid to the question how the confidence intervals should be placed relative to the damage costs estimated by ExternE. In fact, one needs to consider whether the key parameters of the calculations have been estimated as means, medians or something else, for instance modes (= point where the probability distribution of possible parameter values has is maximum). An inquiry of the practice of researchers in this field leads us to the conclusion that the typical choice is the mean. For example, the value VSL = 3.1 MEuro has been derived as a mean of the value of life studies carried out in Europe.

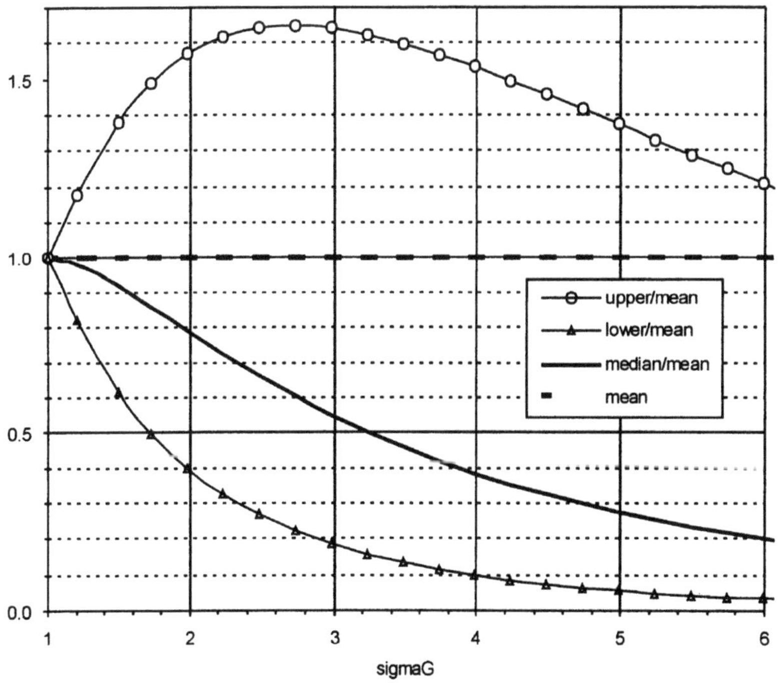


**Fig. 11.2.** Median, upper bound and lower bound, relative to mean, as function of the geometric standard deviation $\sigma_g$ for a lognormal distribution.

Since the confidence intervals that have been estimated for the damage costs are symmetric around the median on a logarithmic scale, their placement relative to the quoted damage costs has to be modified. For a lognormal distribution the ratio of mean <x> and median $\mu_g$ is given by

$$<x>/\mu_g = \exp(0.5\ \Phi^2) \quad \text{with } \Phi = \ln(\sigma_g) \tag{11.9}$$

There is a sizeable difference between median and mean, as can be seen in Fig. 11.2 where the ratios median/mean, upper/mean and lower/mean are plotted as function of the geometric standard deviation $\sigma_g$ for a lognormal distribution.

To see what this implies for the placement of the upper and lower bounds, let us consider the numbers for PM in Table 11.1. If each factor in the formula had a ratio mean/median according to its geometric standard deviation $\sigma_g$ (as per Fig. 11.2), the mean/median ratio of the damage cost would be 1.88 for a geometric standard deviation of 3.08. However, the depletion velocity k is in the denominator rather than the numerator. The mean/median ratio of 1/k is in fact the inverse of the mean/median ratio of k. Since $\sigma_g$ for the depletion velocity k is 1.8, the mean/median ratio of 1/k is 1/1.19. To correct for this feature 1.88 is to be divided by the square of 1.19, which yields a mean/median ratio of 1.33 for the damage or median/mean = 0.75. Thus the 1 $\sigma_g$ lower bound is 0.75/3.08 = 0.24 and the upper bound is 0.75*3.08 = 2.31.

## 11.4 Alternative Hypotheses about Causes

### 11.4.1 The Nature of Particulate Matter

One of the main sources of uncertainty lies in the difficulty of establishing causal relationships, in particular for health effects of air pollution, the dominant term among the damage costs. Epidemiologists are increasingly reaching the conclusion that the association between air pollution and various health end points can be considered causal, but it is not clear what exactly the contribution of specific pollutants is. How much is due to $SO_2$ and how much is due to particles (and which characteristics of particles: chemical composition, solubility, size, mass, number of particles, ...)?

In view of this situation the best one can do is to evaluate and compare the consequences of different assumptions about causal relationships. The focus is put on the E-R (exposure-response) functions for particulate matter since they account for most of the damage costs of ExternE. The starting point is the relation between the concentrations that have been used in epidemiological studies and the concentrations calculated by ExternE. The epidemiological studies are based on whatever mixture of primary and secondary particles happens to be in the air in the cities under consideration, as measured by $PM_{10}$ or $PM_{2.5}$ (or in some case other indices such as TSP or black smoke).

**Table 11.3.** Typical results for composition of ambient particulate matter.

| Region | $\mu g/m^3$ | C(elem) | C(org) | $NH_4$+ | $NO_3$- | $SO_4$-- | Other |
|---|---|---|---|---|---|---|---|
| Remote | 4.8 | 0.3% | 11% | 7% | 3% | 22% | 57% |
| Nonurban continental | 15 | 5% | 24% | 11% | 4% | 37% | 19% |
| Urban | 32 | 9% | 31% | 8% | 6% | 28% | 18% |
| Rubidoux, CA | 87.4 | 3% | 18% | 6% | 20% | 6% | 47% |

Unfortunately, data on the composition of these particles are rarely if ever available for these studies because that is usually not measured by monitoring stations. In fact, one should note that PM designates anything that collects on the filters of the detectors, including droplets of sulfuric acid. Detailed data on composition or size of ambient particles are only known from a few spot measurements. The main components are:

- primary combustion particles (soot, fly ash, ...)
- sulfates ($H_2SO_4$, $(NH_4)HSO_4$, $(NH_4)_2SO_4$, ...)
- nitrates ($NH_4NO_3$, $HNO_3$ absorbed onto particles, ...)
- other (soil, road dust, ...).

Table 11.3 indicates some typical results that have been reported in the literature, as cited in Table 2.18 of Seinfeld and Pandis (1998).

### 11.4.2 A Model for Assessing the Uncertainty Due to Composition of PM

ExternE calculates the concentrations of primary particles (i.e. emitted by the source of pollution), and of nitrates and of sulfates, resulting from the respective emissions of $NO_x$ and of $SO_2$. Then the damages are calculated by making the following assumptions:

- the E-R functions are linear;
- the damage is proportional to the mass of particulate matter in the air, as measured by $PM_{10}$ or $PM_{2.5}$;

  in addition, for a given end point
- the E-R function for primary particles from power plants is equal to that for $PM_{10}$;
- the E-R function for primary particles from cars is equal to that for $PM_{2.5}$;
- the E-R function for sulfates is 1.67 times that for $PM_{10}$;
- the E-R function for nitrates is equal to that for $PM_{10}$.

Due to the differences between the PM data of the epidemiological studies and the particle concentrations calculated by ExternE there are major uncertainties, unavoidable at the present state of knowledge. The best one can do is a sensitivity study to evaluate how the results can change with different assumptions that can plausibly be made about the toxicity of the components of particles in ambient air.

For that purpose let us consider four main components:

- Primary combustion particles;
- Nitrates;
- Sulfates; and
- Other,

designated by the subscripts P, N, S and O, respectively. For each of these components a linear E-R function with slope $f_{er,i}$ is assumed where i = P, N, S and O, the effects being independent of each other (i.e. without positive or negative interference). Then the slope of the E-R function for $PM_{10}$ is

$$f_{er,PM10} = p_P\, f_{er,P} + p_S\, f_{er,S} + p_N\, f_{er,N} + p_O\, f_{er,O} \qquad (11.10)$$

where $p_i$ designates the percentage of the respective component in the ambient air at the sites where the epidemiological study was carried out. Of course the $p_i$ must satisfy the constraint

$$p_P + p_S + p_N + p_O = 1\ . \qquad (11.11)$$

It is convenient to rewrite Eq.11.10 as

$$p_P\, t_P + p_S\, t_S + p_N\, t_N + p_O\, t_O = 1 \qquad (11.12)$$

by defining the relative toxicity $t_i$ of component i relative to $PM_{10}$ as

$$t_i = f_{er,i} / f_{er,PM10} \qquad (11.13)$$

Now alternative hypotheses can be evaluated by varying $p_i$ and $t_i$, within reasonable limits subject to the constraints of Eq.11.11 and Eq.11.12. Here we take

$$p_P = 33\%\ ,\ p_S = 33\%\ ,\ p_N = 7\%\ ,\ p_O = 27\% \qquad (11.14)$$

for the compositions, in view of the data in Table 11.3. As an example, let us consider the total damage per kWh for a coal fired power plant with emissions per kWh in Table 11.4.

The assumptions of ExternE (1998) correspond to

$$t_P = 1.0\ ,\ t_S = 1.7\ ,\ t_N = 1.0\ ,\ t_O = 0.2 \qquad (11.15)$$

**Table 11.4.** Total damage per kWh for a new coal fired power plant, for the damage cost per tonne of pollutant as calculated by ExternE (1998) for typical conditions in Europe (Spadaro and Rabl, 1998).

| | Euro/kg | g/kWh | mEuro/kWh |
|---|---|---|---|
| Primary particles | 13.53 | 0.2 | 2.7 |
| $NO_2$ | 16.92 | 1.0 | 16.9 |
| $SO_2$ | 11.87 | 2.0 | 23.7 |

the latter being implied by Eq.11.12. For the corresponding damage cost per kg of pollutant the values calculated by the "uniform world model" of Sect. 10.3 are taken because they have been found to represent typical conditions for power plants in Europe (Spadaro and Rabl 1999); this model has the simple form

$$D_i = f_{er;i}\,\rho\,Q_i / k_i = t_i\,f_{er;PM10}\,\rho\,Q_i / k_i \tag{11.16}$$

where

| | | |
|---|---|---|
| $Q$ | = | emission [kg/yr] |
| $k$ | = | depletion velocity [ms/s] of pollutant |
| $\rho$ | = | population density |
| $f_{er}$ | = | slope of E-R function |

The generalisation to secondary pollutants is given by Eq.10.12. The total damage per kWh is

$$D_{tot} = \sum D_i \tag{11.17}$$

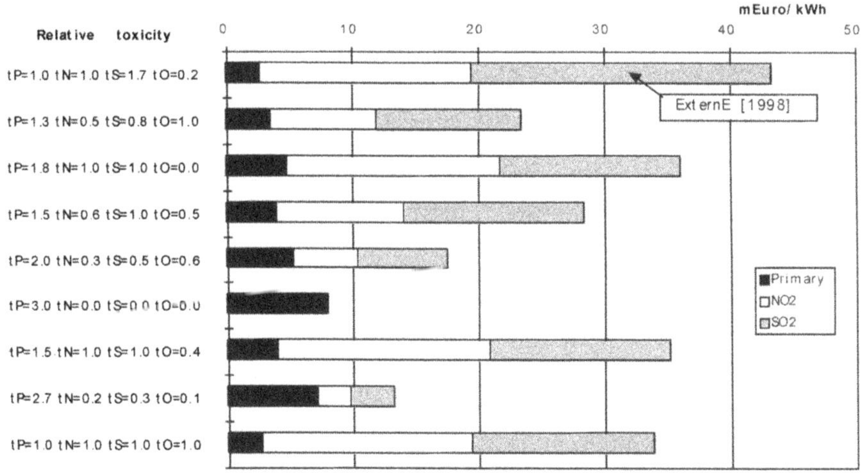


**Fig. 11.3.** Total damage per kWh for a new coal fired power plant, for different hypotheses about relative toxicity t of components (P, N, S, and O) of particulate matter.

is shown in Fig. 11.3 for a wide range of possible hypotheses about the relative toxicities. For the alternatives shown in this figure, the choice of ExternE is an upper bound. If nitrate and sulfate aerosols are less toxic than assumed by ExternE, the damage per kWh could be a factor of two to three smaller. But note that this variation of total cost depends very much on individual emissions.

Unfortunately there is no good answer to an important question of environmental policy: what is the benefit of reducing the emissions of each pollutant separately? That question is particularly troubling in situations where one has to make tradeoffs between reductions of different pollutants. For instance, certain control technologies for diesel engines reduce the emissions of particles while increasing those of Nox; for this case the large uncertainty of nitrate damages precludes any firm conclusion.

### 11.4.3 Analogies with Smoking

Since the lion's share of the damage costs arises from chronic mortality, it was examined what can be learned from analogous studies of another air pollutant from combustion: tobacco smoke. Parallels and differences between exposure to tobacco smoke and air pollution are listed in Table 11.5. Many of the primary air pollutants are common to both. Probably the most important differences are nicotine, $SO_2$, sulfates and nitrates.

The parallels go even further for passive smoking because the concentrations are comparable. According to measurements by Dockery and Spengler (1981) the presence of a smoker increases the concentration of $PM_{2.5}$ in a home by about 20 $\mu g/m^3$, on average. Among the cities of the air pollution study of Pope et al. (1995), the difference between highest and lowest $PM_{2.5}$ levels is 24.5 $\mu g/m^3$.

**Table 11.5.** Parallels and differences between exposure to tobacco smoke and air pollution.

| | |
|---|---|
| Common in both | soot, tar, volatile organic compounds, CO, $NO_x$, fly ash |
| Only in tobacco smoke | nicotine |
| Only in ambient air | $SO_2$, sulfates, nitrates, toxic metals, crustal particles |

**Table 11.6.** Comparison of relative risks between smoking and air pollution.

| Cause of death | Smoking, active | Smoking, passive | Sulfates (19.9 $\mu g/m^3$) | $PM_{2.5}$ (24.5 $\mu g/m^3$) |
|---|---|---|---|---|
| All | 2.1 [a] | | 1.15 [a] | 1.17 [a] |
| Lung cancer | 9.7 [a] | 1.35 [b] | 1.36 [a] | 1.03 [a] |
| Cardiopulmonary | 2.3 [a] | | 1.26 [a] | 1.31 [a] |
| Ischemic heart disease | 1.7 [c] | 1.2 [c] | | |

[a] Pope et al. (1995)
[b] "Tabagisme passif" Report of Académie Nationale de Médecine (1997)
[c] Wells AJ 1994. "Passive smoking as a cause of heart disease". J American College of Cardiology, 24, pp.546-554.

It is therefore instructive to compare, in Table 11.6, the relative risks between smoking and air pollution. The risks for passive smoking are roughly of the same order of magnitude as for air pollution, confirming that exposure to low concentration of air pollutants can indeed have the sort of impacts found in epidemiological studies.

## 11.5 Comprehensive Uncertainty

Figure 11.4 shows the uncertainties of the damage costs, in € per kg of pollutant. Two confidence intervals are indicated: "parametric uncertainty" (heavy error bars) and "comprehensive uncertainty" (light error bars). The former covers those uncertainties for which a quantitative error analysis appears meaningful. The latter is an attempt to estimate overall upper and lower bounds; this involves considerable subjective judgement.

The uncertainty of atmospheric dispersion and chemistry can be considered part of the "parametric uncertainty", with lognormal probability distribution. The uncertainty due to different relative toxicities is more problematic because there is almost no epidemiological or toxicological evidence one way or the other; here simple upper and lower bounds seem more appropriate without any interpretation in terms of probabilities. For the comprehensive upper and lower bounds the following assumptions are made (somewhat subjective):

- $PM_{10}$, $PM_{2.5}$, $NO_2$, $SO_2$ and $O_3$: In view of results by Krupnick et al. (2000) it is estimated that the value of a life year lost due to air pollution could be about a factor 5 smaller than ExternE has assumed, and it is unlikely to be higher.
- $PM_{10}$, $PM_{2.5}$, nitrates and sulfates: For the transfer of E-R functions from the USA to Europe the decision had been made by the epidemiologists of ExternE that the slope of an E-R function from the USA should be reduced by a factor of 2, in view of the observation that acute E-R functions for PM in Europe tend to have about half the slope of the USA functions. It is considered that this reduction factor could be anywhere between 1 and 2.
- $SO_2$: In view of the limited epidemiological studies for health impacts of sulfates, it is assumed that the toxicity of sulfates relative to that of $PM_{10}$ could be as low as 0.3 but is unlikely to be higher than 1.0 (the ExternE assumption).
- $NO_2$: In view of the lack of epidemiological studies for health impacts of nitrates and the fact that ammonium nitrate is an extremely soluble and non-aggressive substance, the assumption is that the toxicity of nitrates relative to that of $PM_{10}$ could be as low as 0.1 but is unlikely to be higher than 1.0 (the ExternE assumption).
- Primary $PM_{10}$ and $PM_{2.5}$: PM in ambient air is a mixture of combustion particles (primary pollutant), nitrates and sulfates (secondary pollutants), and other particles (crustal, ...). The toxicity of primary particles relative to PM in ambient air is constrained by assumptions about the relative toxicity of these other components. With the above assumptions for nitrates and sulfates it turns

out that the relative toxicity of primary particles could be as high as 2.2 but is unlikely to be lower than 1.0 (see Fig. 11.3).

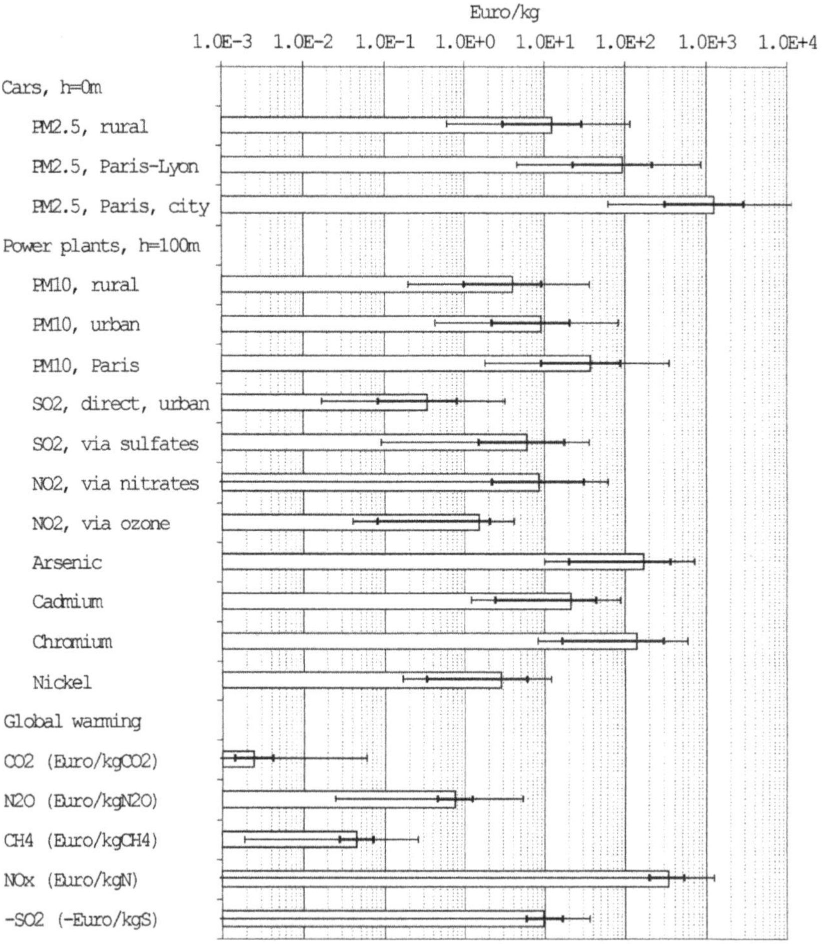


**Fig. 11.4.** Presentation of uncertainty for the damage cost, in € per kg of pollutant emitted in France. The heavy error bars indicate the "parametric uncertainty", the light error bars the "comprehensive uncertainty" (i.e. the overall upper and lower bounds).

# 12 Quantification of Total and Average Externalities (Aggregation)

The previous chapters described the approach for calculating the marginal external costs of transport (the costs caused by an additional vehicle). This chapter describes the issues and methodology associated with aggregating this externality data to levels required by policy instruments and for the analysis of energy balances.

## 12.1 Aggregation Issues

The previous phase of the ExternE Project (European Commission, 1999) developed and demonstrated an aggregation method for assessing the externalities of a country's electricity generation sector. Two approaches were adopted. The first was a simplified methodology applied by each team in each country, which assessed damages by multiplying total sector emissions by country-specific pollutant damage values (Euro per tonne of pollutant). The second was a detailed methodology using the multi-source EcoSense model, applied to Germany and the UK only.

These approaches provide an important starting point for transport sector aggregation. However, there are new theoretical and practical issues for transport, which mean a different approach is needed.

The first issue is with respect to speed. Transport emissions vary significantly with speed and this variation is non-linear. Moreover, for road transport, the speed on particular road types (highway, main road, etc.) will also vary with the numbers of other vehicles on the road. As a result, even emission aggregation becomes an issue for transport (in contrast to the assessment of the power sector).

The second issue centres on the assessment of mobile emission sources. The emissions from vehicles vary over time as well as by location. As we are dealing with linear dose-response functions, the issue of time dependency is less important (all emissions from any one location for any time period can effectively be assessed as an annual emission). In contrast, the issue of location is of major concern.

The previous aggregation analysis concluded that the geographical transferability of global impacts and regional impacts (within one country) were acceptable but the transferability of local impacts was poor. For power stations this was not a major issue; emissions from the electricity generation sector are generally released from high stacks, away from major urban areas, and local impacts are a small part of overall damages. In contrast, local impacts are often dominant with transport, because emissions are released at ground level, often in

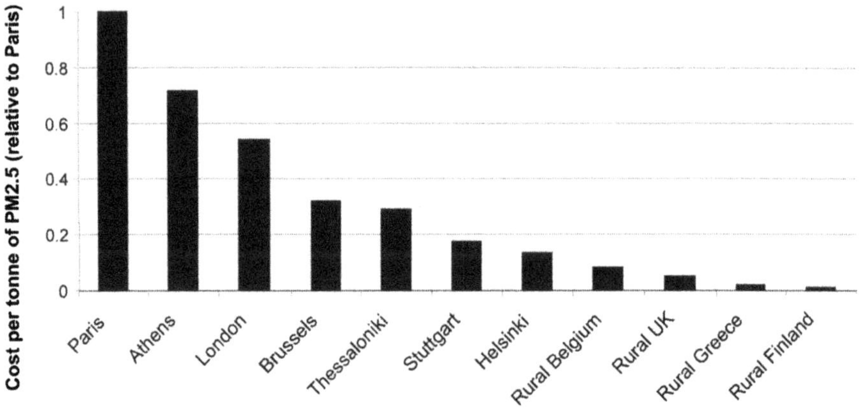


**Fig. 12.1.** Relative damage per tonne of $PM_{2.5}$ in different urban and rural locations

areas of high population density. The importance of these local impacts can be seen in Fig. 12.1, which shows the difference in damage costs in major urban and rural areas in a number of European countries.

The difference between the rural and urban areas is attributable to local impacts. As Fig. 12.1 shows, these local impacts dominate damages (increasingly so as urban size/population density rises). The implications of this are clear. A much greater level of geographical resolution is required for the local level assessment for transport sector aggregation. At the very least, an aggregation analysis should differentiate between urban and non-urban areas. Moreover, the greater the level of data resolution within urban areas, the better the accuracy of the aggregation results.

The dominance of local impacts also has another implication – the multi-source EcoSense model is not as powerful a tool for the transport sector, as it does not have the resolution to capture all the local impact variations. However, it still provides the most accurate way to deal with the regional damage impacts of transport emissions.

Finally, there is one additional issue that warrants discussion. In contrast to the power sector, there are a large number of possible aggregation levels required for policy level analysis for transport. These include:

- Aggregation for a single road, or for a transport scheme;
- Aggregation for a network or at the urban area level;
- Aggregation at a regional or national level.

Obviously as the size of the area under investigation increases, the level of detail with respect to speed, traffic flow, meteorological conditions, etc. must be reduced to keep the assessment manageable.

The first two aggregation levels can provide useful information to local transport operators or local policy makers (local government) and are particularly useful for looking at ways of improving local air quality. Aggregation for a single road, or for example a public transport scheme along a specific transport corridor can best be undertaken using the ExternE computer tool, taking into account detailed characteristics for speed, local meteorological data, etc. A similar method can be used for analysis at the network or urban area level, using marginal costs from the computer tool for a specific area. Examples of these types of studies are included in the marginal cost analysis and the policy case studies. Aggregation at a regional or national level requires a very different approach. This chapter therefore concentrates on the issues and approaches that can be used for such assessments.

## 12.2 Aggregation of Impacts

Transferability of dose-response functions is generally not problematic, except for the complex ecological systems that are still not monetised. There are, however, some issues regarding the linearity of air pollution modelling at a regional level for secondary pollutants, especially for ozone and for secondary pollutants formed from $NO_x$ and $SO_2$ emissions (sulphates and nitrates).

A discussion of the non-linearity of ozone impacts (and a discussion of marginal and average costs) were presented in depth in an earlier chapter and are not repeated here. For other secondary pollutants (such as sulphate and nitrate aerosols) there are similar but less dramatic issues of non-linearity. The issue has been explored through the use of the multi-source EcoSense model, both in previous ExternE phases and in the current phase. By looking at the impacts of the entire transport sector from any one country, and comparing against the sum of marginal cost estimates, it is possible to look at non-linearities in the modelling of secondary pollutants. These studies have shown that differences between marginal and average costs do exist, though they are small – and less than the errors in other parts of the analysis. For the national aggregation studies in Chap. 14, therefore linearity was assumed and geographical transferability for regional impacts within countries.

## 12.3 Aggregation Methodology

For aggregation at the national level, there are two possible approaches. The first aggregates emissions and then applies area specific costs per tonne, whilst the second undertakes marginal cost assessment in detail and then combines with national vehicle km data. These are described below.

### 12.3.1 Aggregating by Emissions

This approach estimates emissions from transport and applies relevant unit pollution costs per tonne from marginal cost runs.

For global and regional impacts, there is no problem transferring damage costs per tonne (for regional impacts the costs within one country are transferable,

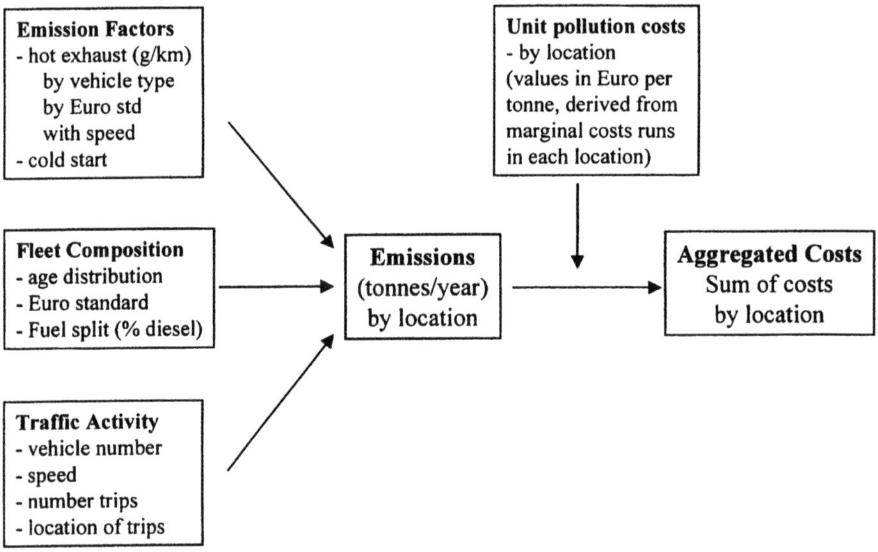


**Fig. 12.2.** Methodology for Road Transport Aggregation by Emissions

though of course accuracy can be increased by undertaking assessments at several locations). For local impacts, the previous sections have shown transferability is an issue.

The exact approach used for assessing local impacts depends on the level of data availability within any country. Unlike the electricity sector, data is not always available in a readily useable form, i.e. total emissions of pollutant from the transport sector dis-aggregated by location. It is therefore worth discussing how usable estimates can be generated. This approach has been used for the UK road sector aggregation presented in Chap. 14. Figure 12.2 shows the outline methodology for this approach.

Data are first needed on the National vehicle fleet, specifically the number of vehicles, the fuel split (gasoline and diesel) and the age distribution/technology standard (pre-Euro, Euro I, Euro II, Euro III). In cases where the breakdown by EURO standard is not available, a good estimate can be derived from basing on the year of first registration (e.g. EURO I passenger cars were introduced in January 1993, EURO II passenger cars in January 1997).

Data is also needed on the trip km and speed characteristics. The level of data with respect to trip type varies considerable across Europe. Ideally, information is needed on the number of kilometres travelled on different roads (e.g. motorway/highway, main road, minor road) in different areas (e.g. on each road type in rural areas, in towns, in cities). This information allows emission factors to adjust for speed characteristics (for example using characteristic average speeds on different road types) and for location (using different unit costs for each area type).

In some cases a further level of detail can be introduced, for example if data is available on journey lengths then cold start emissions can be taken into account. Similarly, if the mileage distribution of different vehicles is available then the final aggregation can take into account that newer vehicles (especially diesels) tend to have higher annual mileage.

Once the data is assembled, the unit pollution values can be applied to estimate damage costs, where possible using as greater dis-aggregation for local impacts as possible. These estimates are derived from marginal cost runs using the EcoSense Transport model. The results from such an approach are presented with the UK national road sector aggregation in Chap. 14.

### 12.3.2 Aggregating by Cost per Kilometre

The alternative approach is to estimate marginal costs per km for all vehicle types for each road or location type. This approach has been used for the road sector aggregation for Belgium presented in Chap. 14. Marginal costs runs (using the EcoSense Transport model) have been undertaken for each vehicle type (pre-EURO, EURO I, EURO II) for each vehicle type (diesel car, gasoline car, diesel LDV, etc) for six drive types. The six drive types (urban peak, urban non-peak, highway peak, highway non-peak, rural peak, and rural non-peak each assume representative average speeds. The analysis then uses these costs per km with the national data on fleet distribution and national vehicle km to estimate total costs. The results from such an approach are presented for the Belgian national aggregation in Chap. 14.

### 12.3.3 Discussion

The question arises of which is the better methodology to use? Both are valid and for the studies presented in Chap. 14 for the UK (aggregation by emissions) and for Belgium (aggregation by cost per km), there is little to choose between them. However, the second approach requires a large number of marginal costs runs, especially where a greater dis-aggregation by speed exists and thus aggregating by emissions (used by the UK and German teams in Chap. 14) is quicker to undertake and is therefore recommended for wider application.

Finally, for both approaches there an additional issue is the differences in data categorisation between countries. This means there is a problem of comparability between studies. National level transport statistics varies enormously across Europe, from countries with little speed or location dis-aggregation, through countries that separate between urban and non-urban kilometres, and finally to countries which have vehicle km data separated by road type, speed and location at a very detailed level.

Within this study, the approach has been for each country to take the best data readily available and use it for aggregation. The alternative was for all teams to use the same level of data aggregation. The latter approach has an important downside as it constrains the whole analysis to the lowest data classification. As a result, the three national aggregation examples (for the UK, Belgium and the Netherlands) have all used different levels of data categorisation (though the level of resolution for the UK and Belgium are very similar). Of course this means that some care should be taken in comparing the aggregation results for the different countries presented in Chap. 14. Indeed, strictly speaking they are not directly comparable! For this phase of the work, which has concentrated on developing methodologies (rather than producing European wide results for policy analysis) this is not so important. However, for future policy analysis it is a major issue and one that is highlighted as a priority for future research.

## 12.4 Economic Aspects

It is anticipated that a main user group of the research undertaken in the ExternE project will be policy makers, at the EU, national and even local level. To make the project results relevant to policy analysis, it is important that the marginal damage cost valuations derived are aggregated to a policy-relevant level. The previous section outlined the aggregation method for the assessment of the physical environmental impacts. This section outlines issues that are relevant in the aggregation of economic assessment. We also discuss the issues that arise in the aggregation of data where monetary valuation is not possible.

The assumption of linearity has been used in the aggregation of the marginal damage costs within ExternE, with the Euro monetary unit the common denominator for aggregation of marginal damage costs. However, there are two contexts in which the assumption of linearity needs to be modified in order to ensure accuracy of results.

First, it may be the case that the degree of change in the environmental impact is greater than has been measured in the marginal valuation. In the instance of the types of policy measures considered in the case studies, the environmental benefits such as the decreased risk to health are, in fact, likely to correspond with the changes in risk measured in the monetary valuation. However, it is conceivable that the EcoSense model may be applied in the analysis of more radical policy measures where, for example, the health risk changes are non-marginal and where the actual risk valuation assumes a non-linear function. As a result, the true aggregate value may differ significantly from the value derived from the modelling work. This problem of potential non-linearity in the monetary values is common to all impact categories. In such a case, it would be necessary for new empirical research to be carried out in order to account fully for these non-linearities. The ExternE team suggests that this research is most needed in the case of health risk valuation since this impact category tends to dominate environmental damage estimates.

The second context in which non-linearity is a potential problem is where the impact is non-marginal and is not confined to a single market. As described in detail in Chap. 8, a straight aggregation of the marginal damages estimated by multiplying price by the change in quantity may no longer be likely to give a precise enough estimate of the welfare loss. In the case studies undertaken in this project we judge that the policy measures at a European or National level are most likely to result in non-marginal effects (though even here, the effects in other connected markets are not likely to be significant). In the instance of potentially significant non-marginal effects, it is recommended that either extended partial equilibrium or general equilibrium modelling be undertaken – depending on the expected size of the effect in the wider economy. Once this method has been followed it is straightforward to sum the resulting monetary valuation of the welfare loss that can be included in a cost-benefit analysis of the specific policy.

An additional aggregation problem will arise when one or more impacts cannot be measured in monetary terms. This is likely to be the case, for example, when certain ecosystem damages are considered or where the complexity of the impact is likely to negate the formation of rational, perfectly informed, consumer preferences. Two possibilities exist to deal with this problem. If, in fact, consumers are able to establish preferences between monetised damages and non-monetised damages, a weighting system can be derived that will produce proxy monetary values for the original non-monetised damages – as outlined in Chap. 8. These values can be aggregated in the normal way. Where this is not possible, there appears to be no alternative but to present the impacts in a qualitative or quantitative way alongside the monetarily aggregated data. The policy maker will then be obliged to consult both the monetised and non-monetised data and make an informed judgement as to how the optimal policy can best be reached. Whilst this situation has not arisen in the case studies considered, it is envisaged that future research would explore this issue in an applied context to demonstrate how policy analysis can best be aided.

In general, the monetary values adopted in the EcoSense model are transferred, by assumption, across European regions. This means that there is no sensitivity for regional variation. However, this is in effect accounted for in the derivation of the values where an averaging of valuation estimates produces a figure acceptable for aggregation purposes across disparate regions within the EU. We judge this to be a justifiable method avoiding, as it does, the issue of differences in valuations reflecting income distribution differences across Europe. Chapter 9 on global warming does, however, directly address the issue of income differences since it attempts to aggregate impact values for all world regions, where it may be less justifiable to transfer monetary values (conversely, it may be objectionable to aggregate values that are estimated on a national basis whereby a life lost in Bangladesh counts less than a life lost in Germany if the differences in values simply reflect income differences). In this instance, the results are presented in four variations:

1. EU impacts only, with EU values ('EU only');
2. 1 plus impacts in other regions with local values ('regional values');
3. 1 plus impacts in other regions with globally averaged values ('world average');
4. plus impacts in other regions with EU values ('EU values').

The alternatives are ordered in terms of the estimated severity of the marginal cost estimates. Alternative 1 ignores impacts outside the EU, closest to realpolitik. Alternative 2 is based on the expressed willingness-to-pay of people outside of the EU, but suffers from the welfare-economic inconsistencies induced by aggregating money measures over people with very different incomes. Alternative 3 is based on the perspective of a benevolent world leader. Alternative 4 follows the moral perspective of a European decision-maker inclined to follow Kantian ethics. The approach for global warming is to reject alternative 2 as internally inconsistent, though the outcomes are reported, and use alternative 3 – which represents a compromise between alternatives 1 and 4 – for the central estimates given.

As noted in the previous ExternE study, it has been shown that there may be further problems involved in aggregating across different sectors when the services provided by the different sectors are complements. This is the case, for example, when the consumer would buy less of one service if the price of the other went up, such as boating and leisure, which is dependent on good health. The damage costs will therefore be under-estimated. Conversely, for substitutes, damages will be over-estimated. As before, the conclusion remains that there appears to be no easy way to avoid the errors of aggregation associated with the interrelationships between the different impacts. Though these errors therefore have to be accepted it is likely that they will be dwarfed by other more significant measurement problems.

Our conclusion is that in the policy case studies considered in this study, aggregation is likely to be straightforward – from a summation of relevant marginal values. However, it is noted that the issues discussed in this section are likely to arise in other foreseeable EU transport policy contexts, hence an awareness of these issues is therefore required by policy analysts.

## 12.5 Conclusions

This chapter discusses the new issues facing the extension of the aggregation methodology to the transport sector. It shows that the issue of geographical transferability becomes extremely important for transport, because of the importance of local impacts on damages. The chapter discusses how these issues can be addressed. The chapter also presents the updated debate concerning the issue of economic valuation in aggregation and discusses how these should be dealt with. Finally, the issue of data compatibility (by location) between countries for national level aggregation is raised as an issue warranting further research. This is particularly important for producing consistent European wide results for future policy analysis.

# 13 Marginal Costs

## 13.1 Belgium

### 13.1.1 Road Transport

#### *13.1.1.1 Routes*

The selected trajectories should allow to evaluate the relative importance of factors that determine the external costs (like population density), to cover the variety of traffic situations and be representative as a basis for aggregation. As in this book the wide variety of situations, e.g. small and bigger cities, cannot be presented data for three trajectories that can be categorised as urban, rural or highway were selected. It has to be noted that – as population density is high in Belgium (326 inhabitants/km²), and especially in Flanders (425 inh./km²) – the rural and highway trajectories are both situated in a relatively high populated area, compared to the European average.

To select three "typical" trajectories, one for each category about 30 trajectories had been screened (indicated in dark grey in Fig. 13.1), located throughout the country. In each of the three categories, the trajectory was selected that yielded a central value for external costs per vehicle km. This does not necessarily mean that these trajectories are especially important in terms of the total annual mileage. Although the results for the selected trajectories are typical, there remains some geographical variation in each category. Urban, rural and highway trajectories can be found for which the externalities are between half and double the value that was calculated for the chosen trajectory.

It was aimed at a detailed modelling exercise, with high precision for roads and population density. The spatial resolution at which trajectories were digitised roughly equals the resolution of the 1/10.000 NGI topographical maps which is much higher than the resolution of the population or meteo datasets (see below).

***Urban trajectory***. This trajectory was chosen to represent a typical commuting trip between the centre of a large city and its outskirts. To ensure that the result could be compared with urban trajectories in other European countries, only trajectories in Antwerp and Brussels were screened. Finally a trajectory was

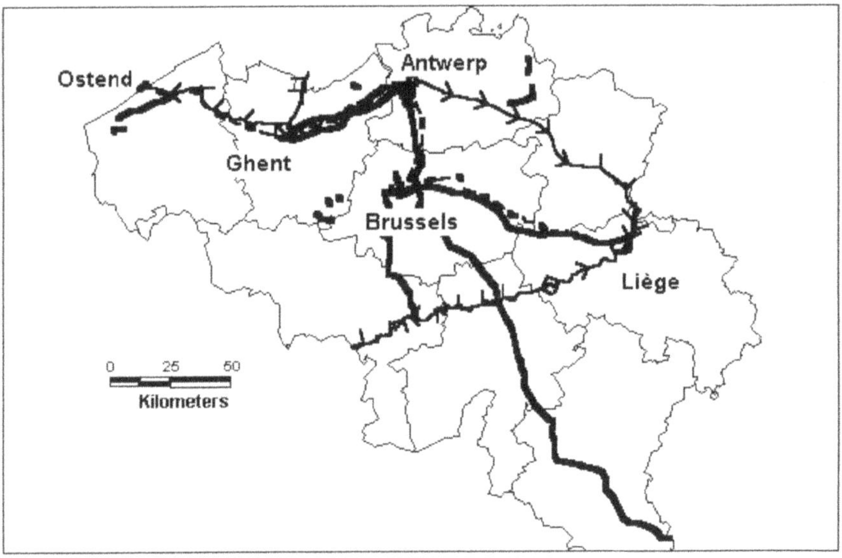


**Fig. 13.1.** Trajectories of roads (===), railways (- - -) and waterways (->-->-)

chosen in Brussels between the Berlaymont building of the European Community and the city ring to the north of the city. This trajectory is 10 kilometres long and crosses some of the most densely populated areas as well as business and industrial districts with lower population densities.

Population densities are highest between 2 and 5 kilometres from the trajectory, highlighting the lower population in Brussels inner centre. Densities remain higher than 3000 inh/km$^2$ (a typical value for built-up areas in Belgium) over several kilometres (Fig. 13.2).

***Rural trajectory.*** Rural areas in Belgium can hardly be compared with rural locations elsewhere. Because of the high average population density and the poor land-use planning, typical rural sites still have relatively high population densities. To look for places with an extremely low population density (e.g. in the Ardennes) would not yield a 'typical' trajectory. In addition it appeared necessary to include only trajectories within or between villages because these are thought to be more relevant for aggregation. Of course this increases even more the average population density around the chosen roads. Typical population densities within (rural) villages or (urban) cities in Belgium are nearly always around 3000 inh/km$^2$.

Finally a trajectory was chosen in Sint-Gillis-Waas a small village in the north of Flanders between the cities of Ghent and Antwerp. The trajectory is 2705 meters long and connects the centre of the village to the nearby national road.

Population densities drop of to low numbers even at a few 100 meters from the trajectory because houses are concentrated along the main roads (ribbon building). Densities rise again to the Flemish average at a distance of only 3 or 4 kilometres, the typical distance between villages in this region (Fig. 13.2.).

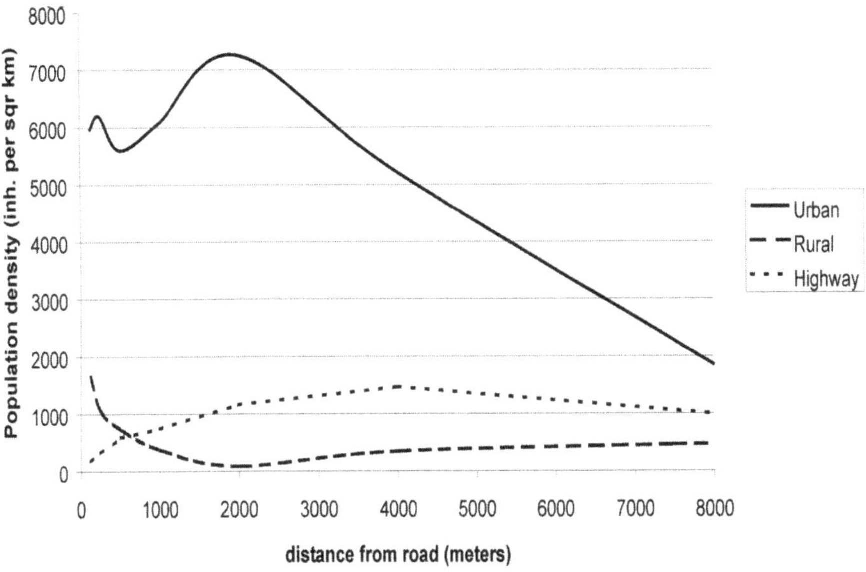


**Fig. 13.2.** Population density around selected trajectories.

***Highway trajectory.*** The selection of a typical highway trajectory proved to be just as difficult as the selection of a rural trajectory. Clearly most highway traffic can be found on the highways towards Antwerp and Brussels. The highways between Antwerp and Brussels cross some of the most densely populated regions of the country. The highways from the south towards Brussels cross some regions with much lower population densities. Therefore it was decided to take the E17 highway between Ghent and Antwerp, in the north of Flanders, as the reference trajectory in this category. Because of its proximity to the rural reference trajectory, this choice makes comparisons easier. Compared to the rural trajectory, densities are lowest close to the road, but these rise quickly because of the proximity of several small cities.

***Population maps for local modelling.*** Vito has constructed detailed population maps that allocate population to the housing districts within each community. To that purpose, a map with the administrative borders of all Belgian communities was updated with census data from different (official) sources for each of the three regions. Most data are for the year 1995 (Flanders and Brussels) or 1994 (Wallonia). Missing data for a few communities were taken for different years.

To match the resolution of the EcoSense grid and the receptor data, the population of each community was assigned to the housing districts within that community. For this calculation, a number of applicable categories were selected from the official land-use planning maps. In this way the houses that lie outside of the designated dwelling areas are neglected, but this error is obviously much smaller than assigning a uniform population distribution to the entire surface area of each municipality. Finally, as the fringes of the local grid often cross the

Belgian border, some data for The Netherlands, Luxemburg, Rheinland-Pfalz and Nordrhein-Westfalen were added.

***Meteorological maps and data for local modelling.*** For this type of analysis, the major differences in yearly windspeeds in Belgium are those between urban and rural areas. Therefore two different meteorological datafiles were used for the local modelling. Data on windspeed profiles and directions for a typical rural location were obtained from the national weather bureau KMI. This file was adapted, on theoretical grounds, to resemble the conditions in urban settings (essentially scaling down the windspeeds at low altitudes). The criterion to choose either urban or rural profile was based on a combination of population density and size of the community.

#### 13.1.1.2 Vehicle Technologies, Emission Factors and Average Speeds

Vehicle technologies that were selected for the case studies on the Belgian trajectories are mostly identical to EcoSense standard. Only for LPG (liquefied petroleum gas) vehicles, the database was modified. Because the EcoSense standard emissions tables do no include emission factors for particulates and carcinogenic VOC's, the results for LPG would come out very much lower compared to petrol fuelled cars. To avoid spurious results, additional emission data from other sources was included (Table 13.1).

Emissions however are strongly dependent not only on vehicle type and emission control technology, but also on the (average) speed. This speed depends on the type of trajectory and may be different for similar trajectories in different countries.

Values of average speeds used in Belgium and Flanders are based on measurements done by KMS (Royal Military School) several years ago. Furthermore, these values are not available for every vehicle type and road type. So, Vito made some assessments based on its knowledge of on-the-road measurements and national maximum speed limits for different types of vehicles and roads.

**Table 13.1.** Additional emission factors for LPG vehicles (source: (CBS, 1995)).

| | | Conventional (EURO 0) LPG | Controlled (EURO I) LPG |
|---|---|---|---|
| PM 2.5 | Urban | 0.018 g/km | 0.003 g/km |
| | Rural | 0.011 g/km | 0.002 g/km |
| | Highway | 0.014 g/km | 0.002 g/km |
| Benzene | Urban | 0.07% of VOC | 0.31% of VOC |
| | Rural | 0.05% of VOC | 0.37% of VOC |
| | Highway | 0.12% of VOC | 0.26% of VOC |
| Butadiene | Urban | 0.76% of VOC | 0.39% of VOC |
| | Rural | 0.76% of VOC | 0.34% of VOC |
| | Highway | 0.75% of VOC | 0.43% of VOC |

**Table 13.2.** Average speeds (km/h) for different vehicle types and road types under normal traffic conditions (source: (MIRA-2, 1996))

| Vehicle type | Average speed (km/h) urban | rural | highway |
|---|---|---|---|
| Passenger car | 22 | 51 | 110 |
| Public bus | 15 | 45 | 80 |
| Coach | 20 | 45 | 85 |
| Light duty | 22 | 51 | 110 |
| Heavy Duty (>32 t) | 30 | 45 | 85 |

Also normal traffic conditions were distinguished from dense traffic conditions. Average measured speeds on passenger cars in Brussels and Antwerp during rush hours, form the basis for the average speeds for all vehicle categories under dense traffic conditions. Table 13.2 gives an overview of the average speeds under normal traffic conditions for the vehicle types that were modelled in this study. It shows a pronounced difference between the average driven speed in urban traffic by different vehicles. Busses have a lower, and motorcycles a higher average speed than passenger cars because of the different number of stops (e.g. at bus stops or traffic lights). For heavy-duty vehicles, the lightest weight class has the lowest average speed; it is assumed that these vehicles drive also in the city centres (inclusive small streets), whereas the heavier ones generally take main streets.

## 13.2 Finland

### 13.2.1 Road Transport

#### *13.2.1.1 Urban Case Study*

The street, Mannerheimintie in the urban case study starts 4 km distance north-west of the Helsinki-city centre and ends in the city-centre. The route is presented in Fig. 13.3.

Data for the urban case study was provided by the Helsinki Metropolitan Area Council (YTV). The average speed and traffic volumes for an average day and night as well as for morning and evening traffic peak hours were available separately for 12 sections of the street. The speed data covers the average speeds in the sections and delays in the nodes. The data describes the traffic on average weekdays. The traffic volumes for the trajectory were formed by weighting the traffic volumes per section with the length of the section. Basic data on the urban case route "Mannerheimintie" is presented in Table 13.3.

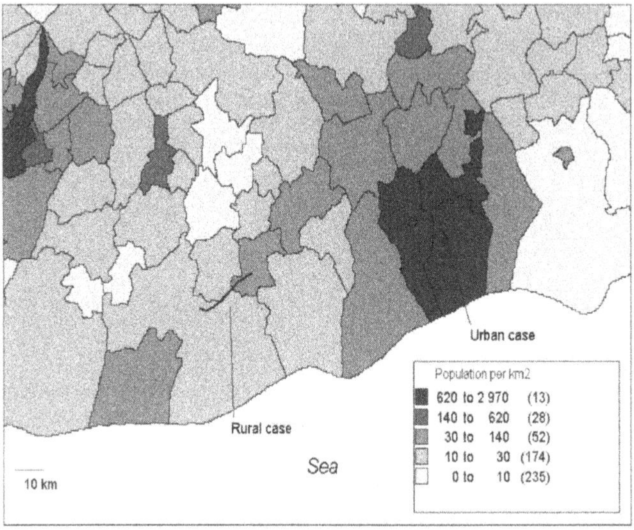


**Fig. 13.3.** The Finnish case study routes. Urban Case (Helsinki), Rural Case (Tammisaari-Karjaa)

The population data was available in the 100 x 100 m² cells for a distance of 1 km from the road and outside that area in the 1x1 km² cells, covering the Helsinki Metropolitan Area. The area extends just over 20 km to the north and close to 20 km to the east and west from the road. About 2 km south of the road, there is the sea. The maximum population density in 100 x 100 m² cells was 552 and the mean 64. The maximum population density in 1 x 1 km² cells was 15 396 and the mean 1 058. The average population densities (per km²) in the area of southern Finland can be seen in Fig. 13.3.

**Table 13.3.** Basic data on the urban case route "Mannerheimintie" in Helsinki.

| | | |
|---|---|---|
| Length | 4.1 | km |
| Sections | 12 | |
| Average day | | |
| Vehicles/day | 14739 | |
| Velocities (per section) | 31 - 39 | km/h |
| Morning rush hour | | |
| Vehicles/hour | 1265 | |
| Velocities (per section) | 23 - 42 | km/h |
| Evening rush hour | | |
| Vehicles/hour | 769 | |
| Velocities (per section) | 25 - 42 | km/h |

In the Helsinki area there are on average 1.3 people travelling in a car. In the whole of Finland the average number of people per car is 1.4. The loading of heavy duty trucks in Finland is on average 50% of the maximum tonnage, when empty driving is also included. The tonnage size of heavy duty vehicles in Finland is bigger than on average in Europe. In Finland the 53 – 60- tonne size class is the most common, which reduces the external costs per tonne-km compared with the average European heavy duty vehicles.

#### *13.2.1.2*
#### *Rural Case Study*

The rural case road is the road section between Tammisaari and Karjaa, located in southern Finland (see Fig. 13.3). Tammisaari is 90 km and Karjaa is 70 km west of Helsinki. The length of the road section is 20 km and the rate of the vehicles varies from 1 500 to 6 000 per day. According to the data measured from the road the velocities on the road vary from 70 to 100 km/h. A figure of 90 km/h was used in the calculations.

The data on population was available in 5 x 5 $km^2$ cells. In Tammisaari, within the area of 750 $km^2$, the average population density is 19.7 inh/$km^2$, in Karjaa (215 $km^2$) 41.4 inh/$km^2$. The average population density in Finland was 15,06 inh/$km^2$ in 1993.

#### *13.2.1.3*
#### *Meteorological Data*

The local meteorological data for Helsinki is from the reference meteorological year 1979 measured at the Helsinki-Vantaa meteorological station. The reference year has been statistically selected using a long-term time series of 1968 - 1983 in selection criteria definition. The same data was applied in the rural case, as it located on the southern coast of Finland as well. The most common direction of the wind was between south and west, about 40 % of all cases. The remaining 60% is quite evenly divided among all other directions.

### 13.2.2
### Rail Transport

#### *13.2.2.1*
#### *Route Data*

The tram transport in the city of Helsinki (Helsinki city transport HCT) was studied. The average population density in Helsinki is 2 961 inh/$km^2$ (see also Fig. 13.3). The basic data on trams by the Helsinki City Transport is presented in Table 13.4.

**Table 13.4.** Basic data on the tram transport in Helsinki

| | | |
|---|---|---|
| Number of lines | 12 | |
| Length of lines | 76 | km |
| Transport performance (1997) | | |
| Line kilometre | 5.4 | mill. km |
| Seat kilometre | 636 | mill. km |
| Load rate | 18,9 % | |
| Energy consumption 1997 | | |
| per vehicle kilometre | 3.8 | kWh/km |
| per seat kilometre | 0.17 | kWh/km |
| Electricity supply (Helsinki Energy) | | |
| Coal | 68 % | (CHP) |
| Natural gas | 32 % | (CHP) |

### 13.2.2.2 Environmental Costs of the Energy Supply for Trams

The external costs of tram traffic (ECU/passenger km) were extimated on the basis of the electricity consumption of the tram traffic. The trams of the HCT use the electricity supplied by Helsinki Energy. Helsinki Energy is the biggest energy company in Finland producing and distributing electricity and district heat.

The energy for use in the Helsinki area is produced in the company's own co-generation plants located in this area. Two of the plants use coal and one natural gas. The calculated proportion of electricity production from gas and coal according to the 1997 production data are 68% and 32%. The 1997 data obtained from Ekono-Energy's energy production database was applied in this case study. After the year 1997 a new gas power unit has been brought into use, after which the proportion of gas in the total fuel used has grown by over 50%. The emission factors per MJ fuel used in the power plants according to the 1997 data are presented in Table 13.5.

The allocation of fuel use to electricity and heat in the combined heat and power production (CHP) is based on the Finnish fuel taxation in 1997. In CHP production 95% efficiency is assumed for heat production (in 1997). The remaining fuel is assumed to be consumed for electricity production.

**Table 13.5.** Emission factors of the fuels

| Emission | Coal plants mg/MJFU | Gas plants mg/MJFU |
|---|---|---|
| $SO_2$ | 120 – 180 | 0 |
| $NO_x$ | 140 – 210 | 50 |
| TSP | 3 – 26 | 0 |
| $CO_2$ | 100000 | 59000 |

The dispersion calculations, impact assessment and valuation for the power plants of Helsinki Energy were performed with the EcoSense model, point source calculation procedure. The stack height applied to the coal-fired plants were 150 m and to the gas-fired plant 65 m. The results of the environmental costs (Eurocent/kWh) for the coal-fired plants and gas-fired plant were aggregated on the basis of the calculated proportions of coal and gas to reach the average result for the trams in Helsinki.

## 13.3 France

### 13.3.1 Road Transport

#### *13.3.1.1 Tailpipe Emissions*

Table 13.6a) shows the tailpipe emissions assumed. For existing cars they are based on emissions measured by Joumard et al. (1995), as interpreted by Spadaro et al. (1998) for a driving cycle of urban trajectories, representative of the likely usage of electric vehicles. For new cars they are taken as the regulatory limits imposed in France for all new vehicles as of January 1, 1997 (with exceptions noted). The real emissions may turn out to be different, but they are unlikely to be much lower since urban driving always involves the driving modes with the highest emissions. Greenhouse gas emissions are not regulated but they can be determined from the fuel consumption.

There are no regulations for particle emissions from gasoline cars, an item not considered significant in the past. However, in view of the high damage cost per kg of particles even small emissions can make an appreciable contribution to the total cost. They have been measured by CONCAWE (1998) for two gasoline cars of the kind being sold now with catalytic converter, and $PM_{2.5}$ calculations are based on this book. The uncertainties of this item are high, not only because it is based solely on two cars but because the emissions are close to the detection limits of the instrumentation. The numbers in the CONCAWE report are interpreted to imply a $PM_{2.5}$ emission of 2 mg/km for urban driving (with uncertainty range from 1 to 4 mg/km).

#### *13.3.1.2 Emissions from Fuel Production*

Emissions from the production of fuel for vehicles are not negligible. Lewis and Gover (Lewis and Gover, 1995) have assembled data from five life-cycle assessment (LCA) studies (two in the UK, and one each in the Netherlands, Sweden and the USA) for the production of gasoline and diesel, acknowledging that the disparity among them is considerable although they all claim to take the entire fuel cycle into account. Table 13.6b) presents the average of these studies.

**Table 13.6.** Emissions and damage costs, per km. "<97" = pre 1997, ">97" = post 1997.

a) Tailpipe emissions

| | Emissions, g/km | | | | Damage costs, Euro/km | | | |
|---|---|---|---|---|---|---|---|---|
| | Diesel >97 | Gasoline >97 | Diesel <97 | Gasoline <97 | Diesel >97 | Gasoline >97 | Diesel <97 | Gasoline <97 |
| $PM_{2.5}$ | 0.08[a] | 0.002[c] | 0.21[d] | 0.0214[d] | 0.1030 | 0.0026 | 0.2750 | 0.0273 |
| $SO_2$ | 0.026[b] | 0.026[b] | 0.05[b] | 0.07[b] | 0.0006 | 0.0006 | 0.0010 | 0.0016 |
| $NO_2$ | 0.511[a] | 0.125[a] | 1.24[d] | 0.47[d] | 0.0051 | 0.0012 | 0.0123 | 0.0047 |
| NMVOC | 0.189[a] | 0.375[a] | 0.45[d] | 1.37[d] | 0.0002 | 0.0003 | 0.0004 | 0.0013 |
| $CO_{2equiv}$ | 192 | 224 | 361 | 616 | 0.0005 | 0.0005 | 0.0009 | 0.0015 |
| CO | 1[a] | 2.2[a] | 2.4[d] | 12.8[d] | 0.0000 | 0.0000 | 0.0000 | 0.0003 |
| Total | | | | | 0.1093 | 0.0052 | 0.2896 | 0.0366 |

b) Refinery emissions, based on Lewis and Gover (Lewis and Gover, 1995) with fuel consumption of Table 15.5.

| | Emissions, g/km | | | | Damage costs, Euro/km | | | |
|---|---|---|---|---|---|---|---|---|
| | Diesel >97 | Gasoline >97 | Diesel <97 | Gasoline <97 | Diesel >97 | Gasoline >97 | Diesel <97 | Gasoline <97 |
| $PM_{2.5}$ | 0.0108 | 0.00946 | 0.020 | 0.026 | 0.0001 | 0.0001 | 0.0002 | 0.0002 |
| $SO_2$ | 0.112 | 0.145 | 0.211 | 0.399 | 0.0007 | 0.0009 | 0.0013 | 0.0024 |
| $NO_2$ | 0.113 | 0.134 | 0.213 | 0.369 | 0.0011 | 0.0013 | 0.0021 | 0.0037 |
| NMVOC | 0.179 | 0.306 | 0.337 | 0.841 | 0.0002 | 0.0003 | 0.0003 | 0.0008 |
| $CO_{2equiv}$ | 22.7 | 35.4 | 42.7 | 97.3 | 0.0001 | 0.0001 | 0.0001 | 0.0002 |
| CO | | | | | | | | |
| Total | | | | | 0.002 | 0.003 | 0.004 | 0.007 |

[a] Emissions for >97 vehicles are limits imposed by directive 94/12/CEE, March 23, 1994 applicable to >97 vehicles as of 1/1/97. The limits are specified for CO, HC + $NO_x$, and fine particles. HC and $NO_x$ emissions are calculated with a split factor equal to the ratio of HC:$NO_x$ in the emission factors from Joumard et al. (1995), calculated for our representative driving cycle.

[b] Calculated by multiplying fuel consumption by 0.05%, the admissible sulfur content as of 10/1/96 (Certu, 1997).

[c] Based on CONCAWE (1998).

[d] Based on Joumard et al. (1995), as interpreted by Spadaro et al. (1998).

### *13.3.1.3 Emissions from Electricity Production*

The electricity for charging the batteries is assumed to come from nuclear power plants because they provide approximately 80% of the total kWh produced in France, and 100% of the base load. Fossil fuels are used only during cold weather, providing 4 to 8% of the total, and the remainder comes from hydro. The damage cost of nuclear electricity in France has been calculated by the ExternE Project. This includes all stages of the fuel chain, even waste disposal and major accidents (although any estimate of the latter items is controversial). The value for zero discount rate, i.e. the upper bound, is 2.5 mEuro/kWh.

**Table 13.7.** Typical damage costs per kg of pollutant. Values for large combustion installations correspond to uniform world model Duni of Eq.10.7 and Eq.10.12 with population density 80 persons/km$^2$ (Europe).

| Pollutant | Impact | Cost, €/kg$_{poll}$ |
|---|---|---|
| Ground level emissions | | |
| $PM_{2.5}$ (primary) cars, Paris | mortality and morbidity | 1.30E+03 |
| $PM_{2.5}$ (primary) cars, Paris-Lyon | mortality and morbidity | 9.20E+01 |
| $PM_{2.5}$ (primary) cars, rural | mortality and morbidity | 1.20E+01 |
| $SO_2$ direct, Paris | mortality and morbidity | 28 |
| $SO_2$ direct, highway Paris-Lyon | mortality and morbidity | 2.2 |
| $SO_2$ direct, rural France | mortality and morbidity | 0.3 |
| CO, Paris | morbidity | 0.02 |
| CO, highway Paris-Lyon | morbidity | 0.002 |
| Emissions from 100 m stack | | |
| $PM_{10}$ (primary) | mortality and morbidity | 8.90E+00 |
| $SO_2$ (primary) | mortality and morbidity | 3.50E-01 |
| As (primary) | cancer | 1.70E+02 |
| Cd (primary) | cancer | 2.10E+01 |
| Cr (primary) | cancer | 1.40E+02 |
| Ni (primary) | cancer | 2.90E+00 |
| Emissions from any height | | |
| CO | morbidity | 2.00E-03 |
| $SO_2$ (via sulfates) | mortality and morbidity | 5.90E+00 |
| $NO_2$ (via nitrates) | mortality and morbidity | 8.40E+00 |
| $NO_2$ (via $O_3$) | Crops, mortality and morbidity | 0.6 |
| VOC (via $O_3$) | Crops, mortality and morbidity | 1.2 |
| Dioxins, TEQ | cancer | 1.90E+07 |
| Global warming | | |
| $CO_2$ (€/kg$CO_2$) | | 2.40E-03 |
| $N_2O$ (€/kg$N_2O$) | | 7.50E-01 |
| $CH_4$ (€/kg$CH_4$) | | 4.50E-02 |
| $NO_x$ from air craft (€/kgN) | | 3.40E+02 |
| $SO_2$ (€/kgS) | | -9.80E+00 |

Notes:
- For a given relation of source and receptors $PM_{2.5}$ is 1.67 times as toxic as $PM_{10}$.
- Multipliers for variation with site (proximity of big city, local climatic conditions) and stack conditions (stack height h, temperature T, exhaust velocity v):

  No variation for $CO_2$;

  Weak variation for dioxin because non-inhalation pathways dominate: ≈0.7 to 1.5;

  Weak variation for secondary pollutants: ≈0.5 to 2.0;

  Strong variation for primary pollutants: ≈0.5 to 5 for site, ≈0.6 to 3 for stack conditions (up to 15 for ground level emissions in big city).

#### 13.3.1.4
#### *Costs per kg of Pollutant*

Damage costs per kg of pollutant for typical conditions in France are shown in Table 13.7.

#### 13.3.1.5
#### *The Canyon Effect*

The ExternE calculations carried out so far are essentially "flat terrain". Effects of nonflat terrain are taken into account only indirectly and very incompletely, by means of the roughness parameter of the Roadpol gaussian plume model, or in so far as they affect the wind data used for the model. Spadaro and Rabl (1999) address a potentially significant contribution to the total damage, not yet taken into account: the increased concentration of pollutants inside urban street canyons. This canyon effect is appreciable only for primary pollutants, the time constants for the formation of secondary pollutants being long compared to the residence time in the canyon. The fluid dynamics software package FLUENT is used to model the dispersion inside a street canyon for a wide range of rectangular geometries and wind velocities. The results suggest that the canyon effect can increase the total damage estimate for ground level emissions in large cities by roughly 10 to 20% of the total. The importance of the canyon effect is, of course, highly variable with local conditions; it could be much smaller but it is unlikely to add more than 100% to the flat terrain estimate. Because of the large uncertainty of the magnitude of the canyon effect, the numbers in the present study have not been increased according to this effect.

### 13.3.2
### Rail Transport

In France the rail system runs almost exclusively on electricity. This electricity can be allocated to the nuclear power sector because in France all of the base load and intermediate load is provided by nuclear power plants, and because the consumption due to trains is base load or intermediate load.

The starting point for quantifying the upstream impacts of the is the technical data for this type of train, in particular the power consumption (6 MW) and the cruising speed (300 km/h). This implies an electricity use of 50 kWh per kilometer with an external cost of 0.25 Eurocents per kWh of electricity. Hence, the damage cost per trainload is equal to 12.5 Eurocents/vkm, and assuming an average capacity of 500 people, the damage cost per person comes out to 0.025 Eurocents/pkm. This is very small compared to the damage cost for private cars. It is important to note that damage costs are highly sensitive to the fuel mix used to produce the electricity, and, therefore, can vary significantly amongst different countries.

# 13.4 Germany

The German case studies cover the 3 most important transport systems: road, rail and air transport. For road and rail transport, both, passenger and goods transport are examined. Due to the variety of vehicle types in the road vehicle fleet, a large number of vehicle categories and technologies are covered, including passenger cars, LDVs, HDVs, coaches, buses and motorcycles. This allows not only to compare different modes e.g. private car and public transport but also the evolution over time in vehicle technology e.g. due to emission regulations. Diverse technologies, e.g. diesel and gasoline cars with their respective emission characteristics are taken into account.

For rail transport, both, diesel and electric traction are covered. To allow an exact comparison with road transport, the two diesel train case studies are situated close to the respective roads.

## 13.4.1 Road Transport

Four routes, located in distinct regions of Germany, were studied for the differences of marginal external costs due to various parameters. The regions were selected based on typical meteorology and population density. Other parameters are vehicle speed depending on the road category and location of drives: outside and inside built-up areas. Especially the location of the road largely influences the impacts on the local range, while the location within Europe influences impacts from secondary pollutants by courtesy of synoptical meteorology and background emissions. Also the potential of affected receptors is important for the magnitude of damages, as e.g. pollutants emitted in the north-east of Germany are in general transported to the Baltic sea and to the low populated Scandinavian countries, while pollutants emitted in the south affect the densely populated areas in Germany.

Two of the routes are located in the south-west of Germany: the Stuttgart city drive (S-CD) and the Heimsheim-Karlsbad highway drive (HD). The Berlin city drive (B-CD) and the Güstrow-Neustrelitz rural drive (RD) are located in the east of Germany. Stuttgart is characterised by a very densely populated city centre due to it's location in a steep basin surrounded by a large and heavily populated commuter belt. Moreover, the Stuttgart area has a very low yearly average wind speed, which has considerable influence on the level of local air pollution exposure of the population. The second city drive is situated in Berlin, the capital and with 3.4 Mio people the largest city of Germany. Unlike Stuttgart, Berlin's yearly average wind speed is one of the highest in Germany apart from the coastal line of the North Sea.

In addition to the low-speed city-drives, highway and inter-urban drives are studied. Fast going vehicles on highway drives cause increased emissions and therefore potentially cause considerable external costs. The cities Heimsheim and Karlsbad are located near the federal highway A8, which links Stuttgart and

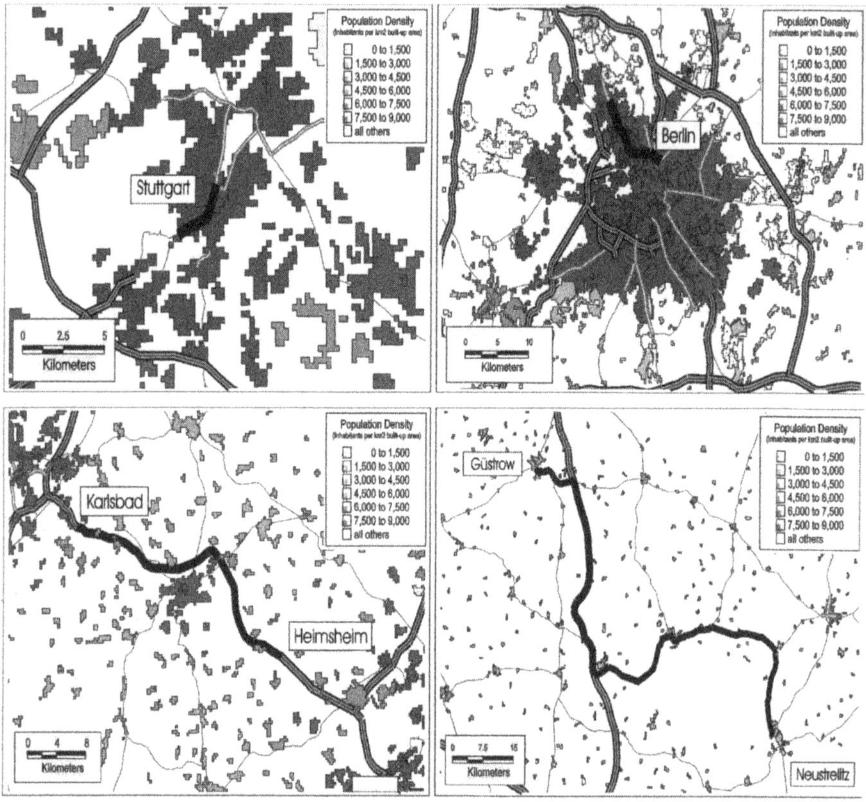


**Fig. 13.4.** Location of German case-study routes, population distribution, population density and major roads.

Karlsruhe. The population density, 328 people per km² in this area, is little below the average of the federal state of Baden-Württemberg. Finally, the rural drive (or inter-urban drive) is located in the federal state of Mecklenburg-Vorpommern (MV) in the north-east of Germany. With 78 inhabitants per km², MV is the least densely populated federal state in Germany. Table 13.8 presents the population, the area of the administrative district, the area-related and the population density on the built-up ground, showing the twofold effect of differences in settlement patterns.

Effort was made for the accurate modelling of the population distribution. The average population density on administrative areas as often given in official statistics, was not considered as adequate. Using average figures, impact assessment on the local scale would not distinguish sufficiently between urban and extra-urban drives. Exposure estimates would result in evenly distributed low values, whereas in reality nobody lives along extra-urban roads and all people live within built-up areas. Moreover, most of the people spend most of their time within built-up areas. The respective GIS maps were modelled using land-use data

**Table 13.8.** Population densities in case-study areas of Germany

| | Population | Area in $km^2$ | Pop. dens. per $km^2$ | Pop. dens. per $km^2$ built-up area |
|---|---|---|---|---|
| Berlin city area | 3,425,800 | 890.77 | 3846 | 7928[a] |
| Stuttgart city area | 707,100 | 307.21 | 2302 | 8368[a] |
| Baden-Württemberg | 10,396,600 | 35,752 | 291 | 4506[a] (avg.) |
| Mecklenburg-Vorpommern | 1,807,800 | 23,170 | 78 | 2048[a] (avg.) |

All numbers taken from Statistical yearbook 1999 (Federal Statistial Office, 1999), reference date 31.12.1997, except [a]: own calculations.

with a resolution of 250 x 250 $m^2$. The number of inhabitants on community level (ArcData Publishing Program, 1995) was attributed to the areas of the land-use map which were classified as urban built-up areas. Other land-use classes as industrial areas, parks, or transport infrastructure were not taken into account.

Figure 13.4 shows the road sections of the case studies within the total major road network, as well as the population maps. The differences in population distribution and population density catches the eye.

Care was also taken for using appropriate meteorological statistical data. The data sets used consisted of 10 year's average of 3-hourly measured data by the German meteorological service. It turned out, that differences in yearly average wind speed had great influence on the exposure estimates. Figure 13.5 shows the high share of low wind speed for the Stuttgart area and the distinct distribution for Berlin and Mecklenburg-Vorpommern. For the Heimsheim-Karlsbad highway drive, the same meteorology was taken as for Stuttgart.

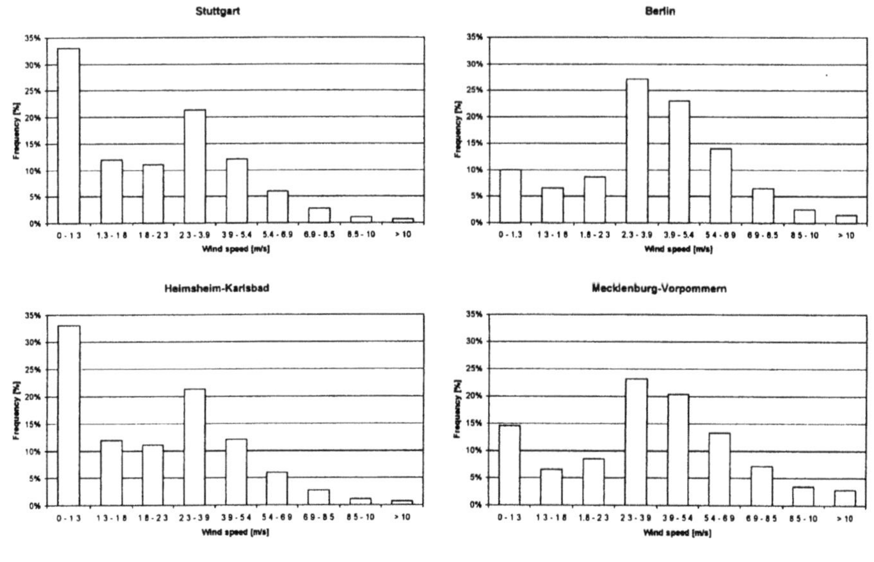


**Fig. 13.5.** Frequency distribution of wind speed in the case-study areas (1981-1990 average from measurements (Deutscher Wetterdienst, 1999).

**Table 13.9.** Vehicle speeds on different roads for the road transport case studies. (Derived from (Umweltbundesamt and INFRAS, 1997).

| Km/h | Car | LDV | HDV | Coach | Urban bus | Motor-cycle |
|---|---|---|---|---|---|---|
| Urban drive | 39 | 36 | 36 | 36 | 39 | 41 |
| Extra-urban drive | 66 | 65 | 65 | 65 | - | 74 |
| Highway drive | 115 | 86 | 86 | 103 | - | 107 |

The vehicle technologies used were principally those listed under 'reference technologies' in Chap. 3, with in addition motorcycle, LDV gasoline EURO I, LDV diesel EURO I and HDV 3.5-7.5 t. The vehicle speeds used for the various vehicle categories on the respective roads are given in Table 13.9.

## 13.4.2 Rail Transport

About half of the German railway network is electrified, and 80% of the transport volume is performed by electric trains. Diesel traction is only of importance for side tracks which are not electrified.

For electric traction the damages due to airborne pollutants are not related to the location where a train is running, but to the locations of the power plants generating the electricity used. Externalities per kWh are calculated for the power plant mix of the "Deutsche Bahn AG" for trains and applicable to all electric trains in Germany. For light rail and tram the electricity mix of the "Stuttgarter Straßenbahnen AG" is applicable to the local light rail and tram network of Stuttgart. Table 13.10 shows the power plant mix of the electricity for trains and for light rail and tram in Stuttgart. Table 13.12 and Table 13.13 present the technical data of the trains considered.

For diesel trains the methodology for assessing marginal costs is the same as for road transport. Emission factors were based on information given by Deutsche Bahn (1997) and combined with vehicle data from Bialonski et al. (1990). Table 13.11 shows the emission factors for diesel trains, Table 13.12 and Table 13.13 give the technical data.

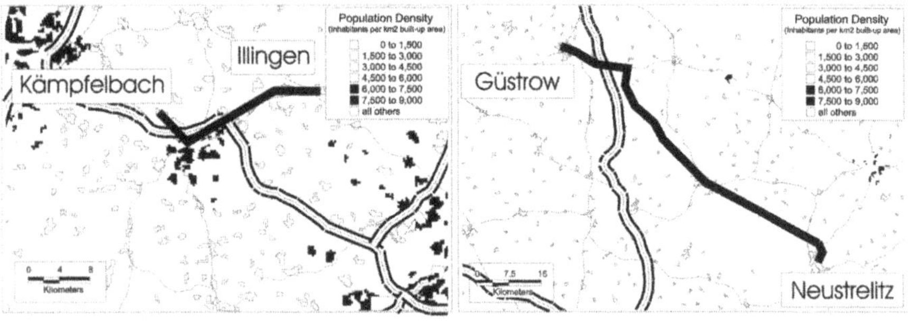


**Fig. 13.6.** Location of case-studies for rail.

**Table 13.10.** Power plant mix of the electricity generation for Rail (Deutsche Bahn, 1997) and Light Rail and Tram in Stuttgart (Fahl et al., 1994)

| | Rail | Light Rail and Tram (Stuttgart) |
|---|---|---|
| Coal | 34.4% | 5.0% |
| Nuclear | 22.1% | 88.0% |
| Oil and Gas | 13.2% | 5.0% |
| Hydro | 10.1% | 2.0% |
| Transformation from the public grid | 20.2% | - |

**Table 13.11.** Emission factors for diesel trains in g/kWh based on Deutsche Bahn (1997)

| Emission | g/kWh |
|---|---|
| CO | 0.89 |
| VOC | 0.40 |
| $NO_x$ | 4.30 |
| $PM_{10}$ | 0.17 |
| $SO_2$ | 0.12 |
| $CO_2$ | 284.70 |

While the costs for electric train operation are independent of the route, the damages due to diesel train operation depend on the location of the route. Two trajectories were selected for analysis, both in extra-urban areas, as in urban areas the majority of trains are operated electrically. For reasons of comparability, the routes were selected close to the respective road transport trajectories. The train line running near the motorway between Heimsheim and Karlsbad leads from Illingen to Kämpfelbach (see Fig. 13.6). The train line from Neustrelitz to Güstrow is more direct and therefore shorter than the road trajectory. The meteorology data used are the same as for the road trajectories.

**Table 13.12.** Vehicle data for passenger trains (taken from Bialonski et al. (1990))

| Traction | Train type | Capacity [Persons] | Average load [%] | [Persons] | Mean speed [km/h] | Energy use [kWh] |
|---|---|---|---|---|---|---|
| Electric | ICE[a] | 634 | 49.4 | 313 | n.a. | 21.9 |
| Electric | Intercity | 600 | 41.4 | 248 | 113.6 | 16.6 |
| Electric | Local Train | 449 | 29.1 | 130 | 84.0 | 12.2 |
| Electric | Suburban Train | 685 | 23.0 | 157 | 53.7 | 11.1 |
| Electric | Light Rail | 190 | 18.0 | 34 | 31.0 | 5.1 |
| Electric | Tram | 105 | 18.0 | 19 | 19.0 | 2.1 |
| Diesel | Intercity | 399 | 37.6 | 150 | 82.2 | 36.0 |
| Diesel | Local Train | 226 | 20.3 | 46 | 68.0 | 33.1 |

[a] Data taken from Prognos (1995); n.a. = not available

**Table 13.13.** Vehicle data for goods trains (taken from Bialonski et al., 1990)

| Traction | Train type | Capacity [t] | Average load [%] | [t] | Mean speed [km/h] | Energy use [kWh] |
|---|---|---|---|---|---|---|
| Electric | Goods Train | 835 | 65.0 | 543 | 65.3 | 20.9 |
| Diesel | Goods Train | 782 | 71.3 | 558 | 49.6 | 50.1 |

## 13.4.3 Aircraft Transport

The newly developed approach assessing aircraft emissions with the impact pathway approach was applied to the international airport Tegel, Berlin. Three airports are located in Berlin: Tegel (TXL), Schönefeld (SXF) and Tempelhof (THF). 1997, 11.6 Million passengers were serviced in total, 8.7 Million of them in Tegel. Tegel airport ranks on position 96 of the most busiest airports in the world. 122,022 aircraft movements were registered 1997 on Tegel airport.

One of the main reasons to select Tegel airport was, that Tegel is located near the city boundaries of Berlin, which implies that local impacts are much more relevant than for other airports, which are located far outside the city.As one of the most often built and most frequently used aircraft, a Boeing 737-300/400 was taken for this case study. The main characteristics are as follows:

- Maximum take-off weight: 67900 kg
- Start speed: ca. 300 km/h
- Runway: 2800 m
- Power plant: 2 x CFM56-3C
- Capacity: 146-168
- First flight: 19.02.1988
- Units built: 1400

**Table 13.14.** LTO-emissions from a Boeing 737-300/400 at Tegel International Airport, Berlin (g per LTO).

| | approach | idle/taxi | take-off | climb | Total |
|---|---|---|---|---|---|
| $NO_x$ | 1467.6 | 1407.6 | 2104.3 | 2132.8 | 7112.4 |
| CO | 500.0 | 8773.2 | 91.5 | 107.8 | 9472.5 |
| VOC | 11.3 | 464.9 | 3.0 | 4.8 | 484.0 |
| PM | 17.3 | 2.2 | 34.3 | 20.6 | 74.4 |
| $SO_2$ | 161.3 | 327.4 | 215.2 | 119.8 | 823.7 |
| Benzene | 0.2 | 8.8 | 0.1 | 0.1 | 9.2 |
| 1,3 Butadiene | 0.2 | 8.4 | 0.1 | 0.1 | 8.7 |
| $CO_2$ | 660619.0 | 62206.5 | 320219.1 | 711760.4 | 1754805.0 |

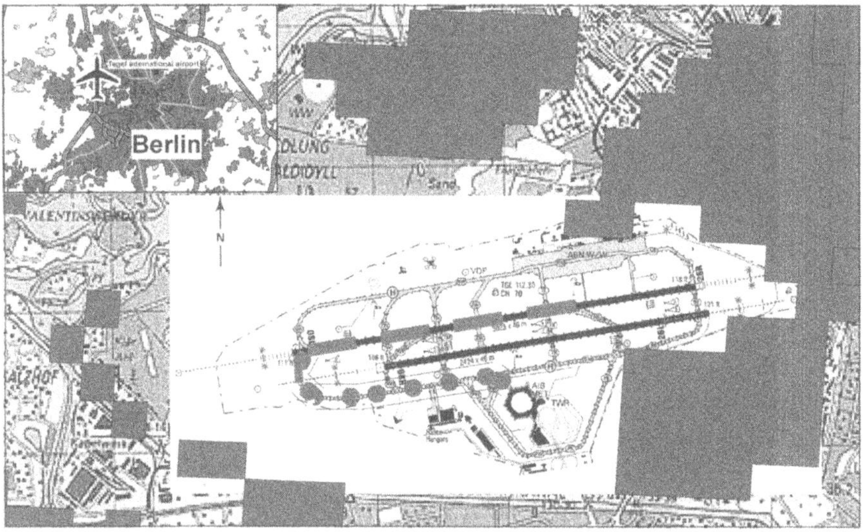


**Fig. 13.7.** Map of Tegel airport with surrounding population map. The lines on the air field indicate the operation mode: idle/taxi (dotted line) and take-off (dashed-line).

Emissions were calculated according to the method described in Chap. 3. The totals, distinguished by operation modes (approach, idle/taxi, take-off and climb) are given in Table 13.14.

## 13.5 Greece

### 13.5.1 Road Transport

Road transport externalities and especially the local impacts are very site-specific, and depend heavily on the population density of the area around the location of the transport task. Six different locations in Greece were selected in order to examine the effects of population density on the marginal costs of road transport. These areas included two large urban agglomerations (Athens and Thessaloniki), two towns (Lamia and Larissa) and two rural areas in central Greece.

The greater Athens area covers an area of about 1450 km², of which 544 km² is built-up and the rest consists of semi-rural areas, agricultural land, forest preserves and grassland. The estimated population of Athens in 1997 was about 3,700,000 residents. The average population density is about 5,100 residents per km² of urban area, but in densely populated areas it often exceeds 20,000 residents per km². Because of the large area and the variation of the population density within the Athens basin two different locations were selected for transport tasks, one in the urban core of the city and one in the suburbs (Fig. 13.8). Thessaloniki is the second largest city of Greece, located in its northern part. Its total population in

1991 was about 750,000 residents. The total built-up area is about 45 km², being significantly smaller than Athens. However, its urban core population density is similar to that of Athens and reaches 30,000 residents per km². The location of the transport task was in the city centre. Lamia and Larissa are two smaller towns

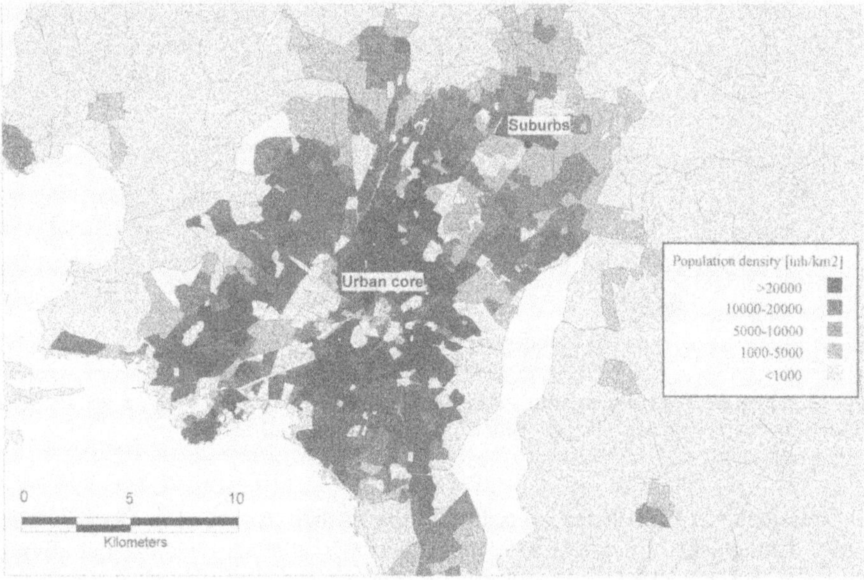


**Fig. 13.8.** Population density in Athens and location of transport tasks

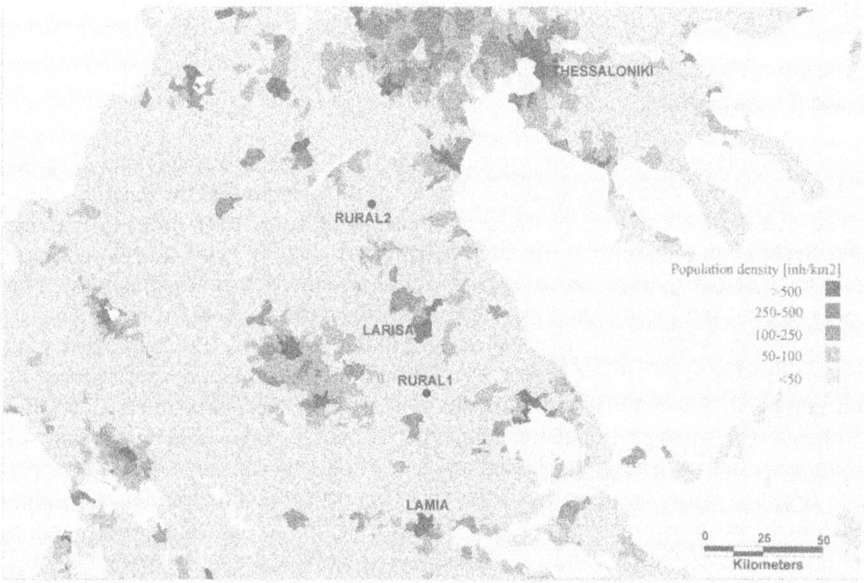


**Fig. 13.9.** Population density and locations of transport tasks

having populations of 44,000 and 113,000 respectively. The population density for Lamia and Larissa greater areas does not exceed 1,500 residents per $km^2$. The locations of the transport tasks for both towns were within each city's centre. The locations of the two rural sites are shown in Fig. 13.9. The population densities for these areas were 60 and 30 residents per $km^2$ respectively. The wind roses for all the locations examined are shown in Fig. 13.10.

In all locations the roads selected had an overall length of 1 km. A range of velocities between 15 and 90 km/h were applied to all trajectories in order to examine the effect of traffic speed on the prevailing externalities. The vehicle types and technologies examined are listed in Table 13.15. The emission factors included in EcoSense were used for all vehicle technologies, except for the CNG buses. The emission factors used for CNG buses are listed in Table 13.16.

**Table 13.15.** Vehicle types and technologies examined

| Mode | Fuel | Technology | Mode | Fuel | Technology |
|---|---|---|---|---|---|
| Passenger Cars | Petrol | ECE 15-04, <1.4 l | Heavy Duty Vehicles | Diesel | Bus, Early 90s |
| | | ECE 15-04,1.4-2.0 l | | | Bus, EURO I |
| | | EURO I, <1.4 l | | | Bus, EURO II |
| | | EURO I, 1.4-2.0 l | | | Coach, Early 90s |
| | | EURO II, <1.4 l | | | Coach EURO I |
| | | EURO II, 1.4-2.0 l | | | Coach EURO II |
| | | EURO III, <1.4 l | | | Truck 3.5-7.5 t |
| | | EURO III, 1.4-2.0 l | | | Truck 3.5-7.5 t, EURO I |
| | Diesel | Uncontrolled, <2.0 l | | | Truck 7.5-16 t |
| | | EURO I | | | Truck 7.5-16 t, EURO I |
| | | EURO II | | | Truck 16-32 t |
| | | EURO III | | | Truck 16-32 t, EURO I |
| | LPG | Uncontrolled | | | Truck 32-40 t |
| | | EURO I | | | Truck 32-40 t, EURO I |
| | | | | | Truck 32-40 t, EURO II |
| | | | | CNG | Urban bus |
| Light Duty Vehicles | Petrol | Uncontrolled | Motorcycles | Petrol | Uncontrolled, <250 cc |
| | | EURO I | | | Uncontrolled, 250-750 cc |
| | Diesel | Uncontrolled | | | Controlled |
| | | EURO I | | | |
| | | EURO II | | | |

**Table 13.16.** Emission factors for the CNG buses (Corbo et al., 1999), (Hitchcock et al., 1999), (Craps and D. De Keukeleere, 1999)

| Pollutant | CO | $NO_x$ | PM | $CH_4$ | NMVOC | $CO_2$ | CNG Consumption |
|---|---|---|---|---|---|---|---|
| Emissions [g/kWh] | 1.12 | 0.94 | 0.05 | 0.20 | 0.01 | 660 | 26.4 MJ/km |

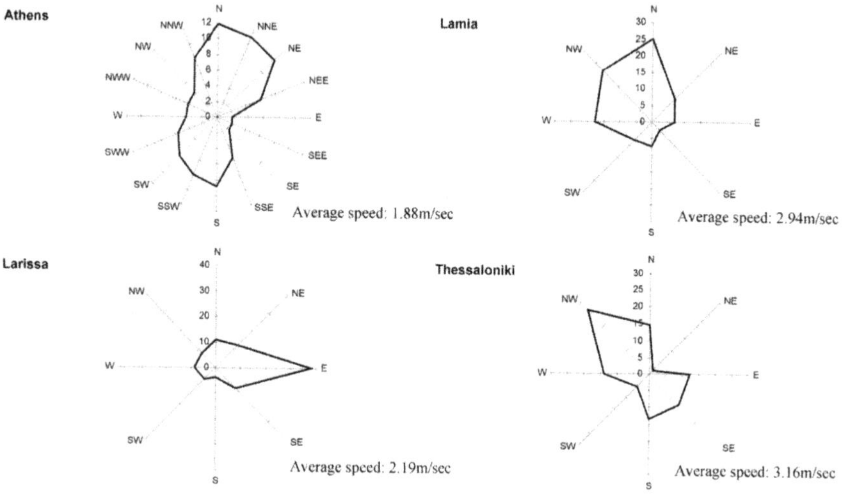


**Fig. 13.10.** Wind roses

**Table 13.17.** Ship types and emission factors (European Commission, 1998b); Hotelling and manoeuvring emission factors are per hour of operation, cruising emission factors per km)

| Ship type | Operating mode | CO [g] | $NO_x$ [kg] | VOC [g] | PM [g] | $SO_2$ [kg] | $CO_2$ [kg] | Fuel [kg] |
|---|---|---|---|---|---|---|---|---|
| Solid bulk | Hotelling | 0 | 1.281 | 112.67 | 352.08 | 11.267 | 901.3 | 281.67 |
| | Manoeuvring | 107.03 | 3.442 | 478.83 | 1408.33 | 22.533 | 1802.7 | 563.33 |
| | Cruising | 18.30 | 0.296 | 3.61 | 106.15 | 1.698 | 135.9 | 42.46 |
| Liquid bulk | Hotelling | 0 | 1.560 | 137.16 | 428.64 | 13.716 | 1097.3 | 342.92 |
| | Manoeuvring | 130.31 | 4.190 | 582.96 | 1714.58 | 27.433 | 2194.7 | 685.83 |
| | Cruising | 22.47 | 0.363 | 4.43 | 130.32 | 2.085 | 166.8 | 52.13 |
| General cargo | Hotelling | 0 | 0.806 | 70.9 | 221.56 | 7.090 | 567.2 | 177.25 |
| | Manoeuvring | 67.36 | 2.165 | 301.32 | 886.25 | 14.180 | 1134.4 | 354.5 |
| | Cruising | 11.54 | 0.187 | 2.28 | 66.94 | 1.071 | 85.7 | 26.8 |
| Container | Hotelling | 0 | 2.498 | 219.6 | 686.25 | 21.960 | 1756.8 | 549 |
| | Manoeuvring | 208.62 | 6.709 | 933.3 | 2745 | 43.920 | 3513.6 | 1098 |
| | Cruising | 26.76 | 0.433 | 5.28 | 155.2 | 2.483 | 198.7 | 62.1 |
| Passenger | Hotelling | 351.15 | 1.820 | 292.62 | 1234.87 | 17.557 | 1872.8 | 585.3 |
| | Manoeuvring | 316.03 | 6.402 | 5852.5 | 2434.64 | 35.115 | 3745.6 | 1170.5 |
| | Cruising | 42.56 | 0.443 | 35.47 | 147.55 | 2.128 | 227.0 | 70.9 |
| High-speed ferry | Hotelling | 402.1 | 2.084 | 335.08 | 1414.05 | 20.105 | 2144.5 | 670.2 |
| | Manoeuvring | 361.89 | 7.331 | 6701.6 | 2787.89 | 40.210 | 4289.1 | 1340.3 |
| | Cruising | 23.69 | 0.246 | 19.74 | 82.13 | 1.184 | 126.3 | 39.5 |

### 13.5.2
### Marine Transport

The trajectory selected for the examination of sea transport impacts is one of the most busy sea routes in the Aegean and connects two of the most important ports of Greece, Piraeus and Iraklio in Crete. Piraeus is considered to be part of the greater Athens area and its population characteristics are mentioned in paragraph 13.5.1. Iraklio is the main port of Crete and has a total population of 116,000 residents. The population density exceeds 2,000 residents per $km^2$.

The sea route was split into three sections, one close to Piraeus, one in the open sea, and one close to Iraklio, in order to examine the effects of location to the resulting externalities. Three operating modes - cruising, hotelling and manoeuvring - were examined. The ship types considered and the emission factors used are listed in Table 13.17.

## 13.6
## The Netherlands

The marginal cost analysis for the Netherlands considers road, rail, water and air transport. In the road, rail and water transport analysis, transport tasks (transport from location A to location B) are analysed while for air transport only a take-off at Schiphol airport is considered. The transport tasks are presented in Fig. 13.11. The tasks are discussed in more detail in the following sub-paragraphs. The technologies that are analysed are fossil fuel and electricity driven vehicles, ships and aircraft.

### 13.6.1
### Road Transport

#### *13.6.1.1*
#### *Routes*

***Amsterdam Central Station to Groningen Central Station.*** In this transport task passengers and goods are transported from Amsterdam Central Station to Groningen Central Station by a 186 km road tripe. The short title for this transport task is AG. For road transport this task is divided into 4 building blocks. The first trajectory, AG1 (4.6 km), is an urban drive from the city centre of Amsterdam to the main road N10 with an average speed of 28 km/h. The second trajectory is AG2 (6.9 km) on main road N10 to the highway A10 with an average speed of 68 km/h. Then AG3 (171.5 km), a highway drive with an average speed of 105 km/h on highway A10, A1, A6 and A7 to the city border of Groningen. Last trajectory AG4 (3.0 km) is again an urban drive, from the highway exit to the Central Station in the city centre of Groningen with an average speed of 28 km/h.

***Amsterdam Central Station to Schiphol Airport.*** Passengers and goods are transported from Amsterdam Central Station to Schiphol Airport Railway Station by road over a distance of 16.3 km. Short title for this transport task is AS, and the

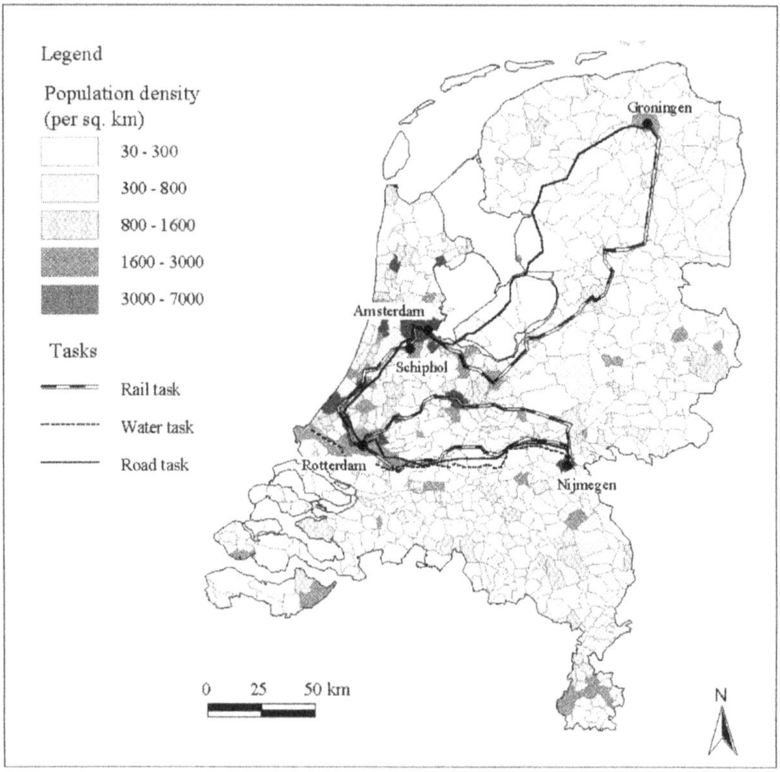


**Fig. 13.11.** Road, rail, water transport tasks analysed in the Netherlands

building blocks of this task are AS1, AS2 and AS3. The first trajectory is an urban drive (4.3 km) from the centre of Amsterdam to the highway A10. The second trajectory is a highway drive (11.0 km) on highway A10, A4 and A9 to the main road $N_2 32$. The last trajectory is a drive with an urban pattern (28 km/h) from main road $N_2 32$ to the railway station of Schiphol Airport.

***Schiphol Airport to Rotterdam Central Station.*** Transport task SR from Schiphol Airport Railway Station to Rotterdam Central station (67.1) is divided into 3 building blocks, named SR1, SR2 and SR3. SR1 is a 0.9 kilometre drive from the Railway station at Schiphol Airport to the highway with an average speed of 28 km/h. SR2 is a highway drive via A9 and A6 (61.9 km) to Rotterdam city border. The last trajectory, SR3 (4.3 km), is an urban drive from the exit to the Central Station in the city centre of Rotterdam.

***Rotterdam Central Station to Nijmegen Central Station.*** The transport task RN by road (111.0 km) is divided into 3 building blocks. The first trajectory is RN1, an urban drive from the Central Station in the centre of Rotterdam to the

**Table 13.18.** Technologies for road transport tasks

| Transport type | Fuel type | Technology | Category | Occupancy rate/ Load capacity |
|---|---|---|---|---|
| Passenger transport technologies | | | | (Passengers/vkm) |
| Passenger car | Gasoline | ECE 15/4 | 1.4 l<cc<2.0 l | 1.43 |
| Passenger car | Gasoline | Euro I | 1.4 l<cc<2.0 l | 1.43 |
| Passenger car | Gasoline | Euro II | 1.4 l<cc<2.0 l | 1.43 |
| Passenger car | Gasoline | Euro III | 1.4 l<cc<2.0 l | 1.43 |
| Passenger car | Diesel | Uncontrolled | cc<2.0 l | 1.43 |
| Passenger car | Diesel | Euro I | all cat. | 1.43 |
| Passenger car | Diesel | Euro II | all cat. | 1.43 |
| Passenger car | Diesel | Euro III | all cat. | 1.43 |
| Passenger car | LPG | Conventional | all cat. | 1.43 |
| Passenger car | LPG | Euro I | all cat. | 1.43 |
| Passenger car | Ethanol (E85) | Euro I | 1.4 l<cc<2.0 l | 1.43 |
| Passenger car | Methanol (M85) | Euro I | 1.4 l<cc<2.0 l | 1.43 |
| Passenger car | CNG | Euro I | 1.4 l<cc<2.0 l | 1.43 |
| Passenger car | Hybrid | | | 1.43 |
| Passenger car | Electric | | | 1.43 |
| Urban Bus | Diesel | Conventional | all cat. | 13.10 |
| Urban Bus | Diesel | Euro I | all cat. | 13.10 |
| Urban Bus | Diesel | Euro II | all cat. | 13.10 |
| Urban Bus | Methanol (M100) | Euro I | all cat. | 13.10 |
| Urban Bus | Ethanol (E95) | Euro I | all cat. | 13.10 |
| Coach | Diesel | Conventional | all cat. | 13.70 |
| Coach | Diesel | Euro I | all cat. | 13.70 |
| Coach | Diesel | Euro II | all cat. | 13.70 |
| Goods transport technologies | | | | (tonne/ vkm) |
| LDV | Diesel | Uncontrolled | all cat. | 0.275 |
| LDV | Diesel | Euro I | all cat. | 0.275 |
| LDV | Diesel | Euro II | all cat. | 0.275 |
| LDV | Hybrid | | | 0.275 |
| HDV1 | Diesel | Conventional | 7.5-16 t | 2.830 |
| HDV2 | Diesel | Conventional | 16-32 t | 6.690 |
| HDV1 | Diesel | Euro I | 7.5-16 t | 2.830 |
| HDV2 | Diesel | Euro I | 16-32 t | 6.690 |
| HDV1 | Ethanol (E100) | Conventional | 7.5-16 t | 2.830 |
| HDV2 | Ethanol (E100) | Conventional | 16-32 t | 6.690 |

highway (5.1 km). The second trajectory (RN2) is a 104.9 km highway drive on A15 to Nijmegen. The third trajectory is RN3, 1.0 km from the highway exit to the Central Station of Nijmegen with an average speed of 28 km/h.

### *13.6.1.2*
### *Vehicle Technologies*

For all transport tasks described above, calculations for marginal costs are made with the 33 technologies presented in Table 13.18. For the electric car an electricity consumption of 0.25 kWh/km is assumed (Funk and Rabl, 1999).

## 13.6.2
## Rail Transport

### *13.6.2.1*
### *Routes*

Rail transport can be divided in passenger train, goods train, tram and metro transport. For location of the transport tasks see Fig. 13.11. The routes are described separately below.

***Amsterdam Central Station to Groningen Central Station.*** This transport task is 217 km long for both passenger and goods trains. There is no tram or metro on this trajectory.

***Amsterdam Central Station to Schiphol Airport Railway Station.*** This task is 19 km for passenger and goods trains. There will be tram and/or metro for this task in the future. The task will be 16 km for tram and 14 km for metro.

***Schiphol Airport Railway Station to Rotterdam Central Station.*** This transport task is 67 km long for both passenger and goods train. There is no tram or metro on this trajectory.

**Table 13.19.** Technologies for rail transport tasks

| Transport type | Fuel type | Technology | Occupancy rate/ Load capacity |
|---|---|---|---|
| Passenger transport technologies | | | (Passengers/vkm) |
| Train | Electric | Intercity | 84.7 |
| Train | Electric | Local | 62.4 |
| Train | Electric | High speed | 269.9 |
| Tram | Electric | | 19.4 |
| Metro | Electric | | 28.5 |
| Goods transport technologies | | | (tonne/vkm) |
| Train | Diesel | DE6400 | 522.2 |
| Train | Diesel | DE1024 | 522.2 |
| Train | Electric | 1.5 kV | 522.2 |
| Train | Electric | 25 kV | 522.2 |

***Rotterdam Central Station to Nijmegen Central Station.*** This transport task is 133 km for passenger and 122 km for goods trains. The connection for passengers is longer (via the city Utrecht) but faster according to the timetable of de Nederlandse Spoorwegen (Dutch Railway Company) (www.NS.nl). There is no tram or metro on this trajectory.

#### *13.6.2.2 Vehicle Technologies*

For all transport tasks described above, calculations for marginal costs are made with the following technologies mentioned in Table 13.19.

### 13.6.3 Inland Shipping

#### *13.6.3.1 Routes*

Only one transport task for shipping is analysed: Rotterdam docks to Nijmegen docks. For location see Fig. 13.11. The transport task Rotterdam docks Nijmegen docks (RN) is 101 km on the river Waal.

#### *13.6.3.2 Vehicle Technologies*

A review of inland shipping emission factors was performed by Dorland et al. (2000). They derive an equation for estimating the emissions based on data from Lloyd's Register (1990b, 1990a, 1993, 1995) and Dings et al. (1997). In this case study the average emission factors for medium inland ships (river Rhine barges) and push tugs in the year 2000 are used.

Data on the non-operation life cycle stage emissions are available for three different classes of ships: bulk ships, tankers and containers. It is assumed that the emission factors for a medium ship are representative for the emissions for a medium size bulk ship and a tanker and that the emission factors for a push tug are representative for a container ship. Hence, in this study externalities for bulk ships, tankers and containers are presented, see Table 13.20.

**Table 13.20.** Technologies for ship transport tasks

| Transport type | Fuel type | Technology | Load capacity |
|---|---|---|---|
| Goods transport technologies | | | (tonne/vkm) |
| Ship | Diesel | Bulk | 1550 |
| Ship | Diesel | Container | 5400 |
| Ship | Diesel | Tanker | 1550 |

## 13.6.4
## Aircraft Transport

Air transport is analysed at runway 1 at Schiphol airport, see Fig. 13.12. Only one take-off, including the taxi, start and climb phases, is analysed (the EcoSense run was performed by S. Schmid at IER). The average duration of the different phases of this task are presented in Table 13.21.

**Table 13.21.** Duration of take-off phases at Schiphol airport

| Phase | Duration in seconds |
|---|---|
| Taxi | 1200 |
| Start | 42 |
| Climb | 78 |

Source: Huygen (1990)

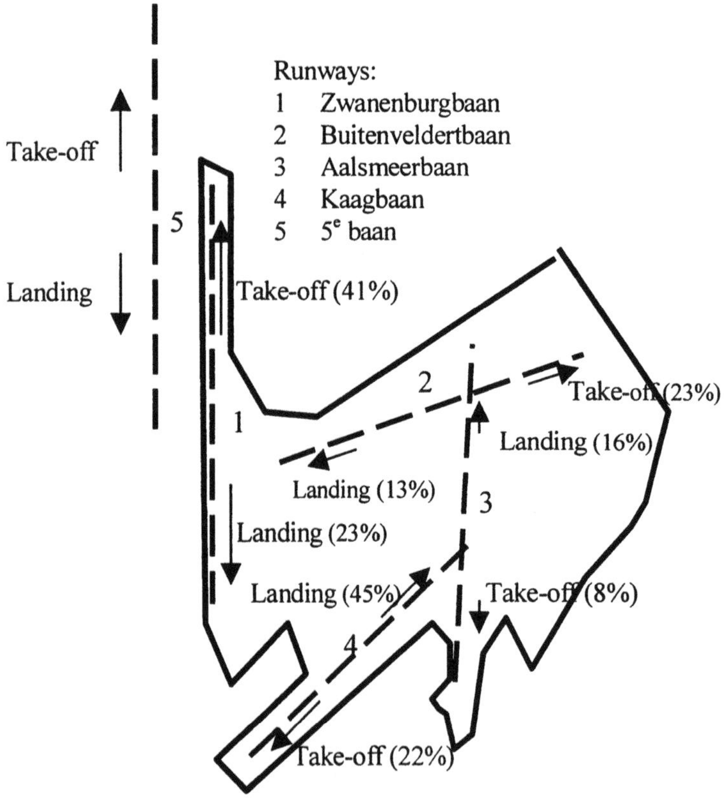


**Fig. 13.12.** Airport Schiphol including landing and take-off directions and distributions in 1998 (CGS, 1999) (CGS, 1999).

The technology analysed for this transport task was a Boeing 737-300/400 type aircraft. This type of aircraft is very common (12% of all aeroplane movements) at this airport.

## 13.7 United Kingdom

The UK marginal cost analysis assessed the external costs of road, rail, air and shipping, looking at both passenger and freight transport. Details by mode are presented below.

### 13.7.1 Road Transport

For the road sector, a large number of different locations and routes were assessed, shown in Fig. 13.13.

The locations were selected so as to include all the area/road types needed for the aggregation and policy case study tasks. They also provide data on the relationship between population density/urban size and damages. For the summary

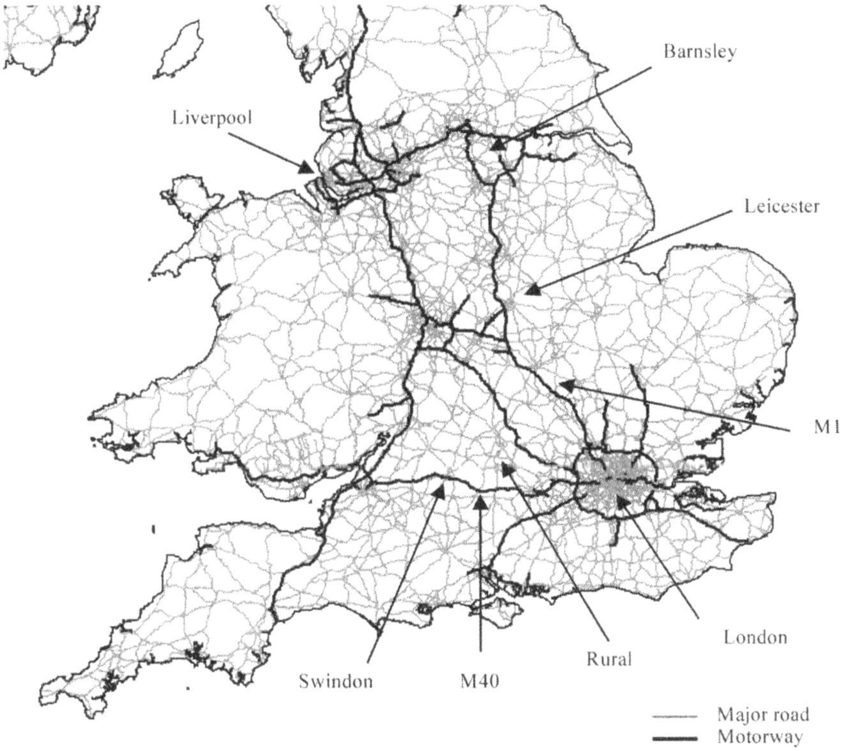


**Fig. 13.13.** The UK Major Road Network and Study Routes and Locations

**Table 13.22.** Details of the UK Locations and Trajectories

| Location/trajectory | Type | Route (km) | Average kph |
|---|---|---|---|
| A5 Edgware Road London | Central London | 8.1 | 28 |
| A4 London | Outer London | 8.1 | 30 |
| A57 Liverpool | Large City | 10.7 | 50 |
| A6 Leicester | City | 7.7 | 50 |
| A635 Barnsley | Large Town | 2.4 | 50 |
| A4311/A4259 Swindon | Town | 9.0 | 50 |
| M40 J5 to J10 | Motorway | 40.4 | 110 |
| M1 J12 to J17 | Motorway | 60.0 | 110 |
| A329 Oxfordshire | Rural | 12.7 | 75 |
| A338 Oxfordshire | Rural | 30.2 | 75 |

Average speed data was taken from National road speed statistics (DETR, 1998a-e).

of results later in this Chapter, results from only five of the locations are presented. The km of each trip and the average speed are shown in Table 13.22.

Emissions data was taken from the ExternE Transport computer tool for all the reference technologies. Additional data on alternative fuel vehicles was taken from ETSU (1996:1998) and MEET deliverable 26 (EC, 1998), also included in the computer tool.

The geographical data for the UK road network was based on DETR data, and extracted from the National Atmospheric Emissions Inventory (NAEI). Population data for the UK was based on the UK 1991 census data (and the 1995 update) at enumeration district level, summed to provide population on 1 $km^2$ resolution. Meteorological boundary layer depth analysis data (5-10 year average) for each area was used (supplied by the UK Meteorological Office). The data actually used were the frequencies of Pasquill stability category, wind speed and wind direction and the average boundary layer depth for each Pasquill stability category.

To calculate damages per passenger km, an average of 1.58 passengers per car was assumed (TSGB, 1998), and an average of 12 passengers per urban bus and 10 passengers per coach. For goods vehicles, a maximum load of 2 tonnes for LGVs and 25 tonnes for HGVs was assumed (the latter based on a 32 tonne vehicle). Average freight loads were calculated based on estimates that the UK freight sector only uses around 62% of the available capacity, falling to 44% when empty running is added (McKinnon, 1996).

### 13.7.2 Rail Transport

For rail transport, two similar regional trips in the South of the UK were assessed, with journeys from London to Leicester and London to Peterborough (journey distances 162 and 123 km respectively). These routes are shown in Fig. 13.14.

Both passenger and freight transport trips were assessed (diesel locomotives only). Emission factors were taken from the UK emission factors database (London Research Centre, 2000) and are presented in Table 13.23. Note the factors are for generic trip types and do not differentiate by speed.

**Fig. 13.14.** The UK Rail Network (Southern England/Wales) and Rail Routes (Source: LRC for NAEI)

The values for $PM_{10}$ are based on total suspended particulate (TSP) emission factors and assume all TSP is $PM_{10}$. To derive benzene and 1,3-butadiene emissions, the HC emissions were scaled, consistent with the method used within the UK NAEI for assessing emissions of these pollutants from trains.

**Table 13.23.** UK Rail Emissions (g/km)

| Type | $CO_2$ | CO | $NO_x$ | $PM_{10}$ | $SO_2$ | HC |
|---|---|---|---|---|---|---|
| Pacer Diesel Multiple Unit per car | 1,349 | 10.10 | 12.80 | 0.10 | 1.80 | 0.50 |
| Sprinter DMU per car | 1,690 | 1.36 | 19.15 | 0.10 | 1.30 | 0.70 |
| Turbo DMU per car | 1,199 | 1.20 | 30.60 | 0.10 | 1.60 | 0.50 |
| Inter-City 125 high speed train per power car* | 12,170 | 28.10 | 97.40 | 8.50 | 16.30 | 15.10 |
| Class 47 locomotive plus 7 passenger coaches | 9,764 | 39.90 | 127.60 | 5.10 | 13.10 | 11.60 |
| Class 37 freight per locomotive (typical load) | 11,268 | 24.50 | 51.80 | 5.10 | 15.10 | 12.60 |
| Class 37 freight per pair of loco (typical load) | 22,536 | 62.40 | 132.80 | 10.20 | 30.20 | 32.00 |
| Class 47 freight per locomotive (typical load) | 16,720 | 26.10 | 80.10 | 5.10 | 22.40 | 5.40 |
| Class 56 freight per locomotive (typical load) | 21,439 | 43.20 | 129.60 | 5.10 | 28.70 | 22.40 |
| Class 58 freight per locomotive (typical load) | 21,439 | 22.50 | 103.50 | 5.10 | 28.70 | 12.00 |
| Class 60 freight per locomotive (typical load) | 20,152 | 21.60 | 129.60 | 4.70 | 26.90 | 10.80 |
| Class 66 freight per locomotive (typical load) | - | 6.80 | 120.00 | 2.86 | 5.95 | 3.23 |

Source LRC, 2000. * IC 125 trains consists of two power cars plus 8 passenger coaches.

To calculate damages per passenger km, a maximum of 74 passengers per dedicated passenger carriage was assumed. For the passenger/locomotive units, the passenger capacity varies considerable within each category. For this study, maximum passenger capacities of 50, 80 and 94 were assumed for the pacer, sprinter and turbo unit respectively. Detailed data on average passenger numbers by train type was not available and so typical load factors were used – these factors were 0.5 for sprinters and turbos, 0.6 for pacers and class 47 locomotives, and 0.7 for IC125s. It was also not possible to calculate the detailed damages per tonne km, due to a lack of detailed data on freight tonnage for different locomotive types. To calculate values per tonne km, average UK freight data (average tonnes per freight train journey) was applied to all locomotives. This is likely to lead to errors in the relative impact of different freight locomotive types and the team is currently investigating the split by train type in more detail to improve these results at a later stage.

### 13.7.3 Aircraft Transport

The UK study also assessed the marginal costs of regional aircraft trips in the UK. The trip assessed was an internal flight from Heathrow (London). A number of aircraft were assessed. Landing and Take-off cycle (LTO) data was taken from the

**Table 13.24.** Total LTO cycle Emissions (g/lto)

| Aircraft type | Engine type and No | CO | $NO_x$ | PM | VOC | $SO_2$ | $CO_2$ | Benzene | Butad. |
|---|---|---|---|---|---|---|---|---|---|
| Boeing 737-200 | PW JT8D-15A (2) | 6354 | 7297 | 287 | 995 | 817 | 2850793 | 19.3 | 17.9 |
| Boeing 737-400 | CFMI CFM56-3C1 (2) | 11182 | 9621 | 50 | 574 | 807 | 2816429 | 11.1 | 10.3 |
| Boeing 747-100 | PW JT9D-7A (4) | 114587 | 49166 | 352 | 48433 | 2887 | 10073447 | 939.6 | 871.8 |
| Boeing 747-200 | RR RB211-524D4 (4) | 16892 | 80494 | 831 | 3219 | 3283 | 11454720 | 62.4 | 57.9 |
| Boeing 747-400 | RR RB211-524H2 (4) | 20572 | 98005 | 1622 | 1980 | 3518 | 12272376 | 38.4 | 35.6 |
| Boeing 757-200 | RR RB211-535C (2) | 12060 | 18554 | 202 | 1105 | 1280 | 4466587 | 21.4 | 19.9 |
| Boeing 757-200 | RR RB211-535E4 (2) | 8079 | 23428 | 82 | 222 | 1229 | 4289491 | 4.3 | 4.0 |
| Boeing 767-300 | RR RB211-524H (2) | 10286 | 49003 | 811 | 990 | 1759 | 6136188 | 19.2 | 17.8 |
| Boeing 777-200 | GE GE90-92B (2) | 12284 | 60344 | 122 | 424 | 2191 | 7645008 | 8.2 | 7.6 |
| Boeing 777-200 | GE GE90-76B (2) | 14096 | 39361 | 105 | 558 | 1886 | 6580812 | 10.8 | 10.0 |
| Airbus A320 | CFMI CFM56-5A1 (2) | 6185 | 9011 | 119 | 570 | 694 | 2420827 | 11.1 | 10.3 |

ICAO (ICAO, 1999) - i.e. for one approach/landing, taxi-in, taxi-out and take-off/climb out (shown in the Table below). The times in mode (take-off, climb-out, approach and idle) are ICAO standard times (0.7, 2.2, 4 and 26 minutes respectively). In practice these times will be airport, aircraft and airline dependant.

The method of assessment used in the case study was not as detailed as for the German case study. The ICAO data has a further split into ground level LTO emissions and total LTO emissions - for the analysis it was assumed that only ground level LTO emissions contributed to local air pollution and that all LTO emissions (below 1000 m) contributed to regional air pollution. Approximate emissions for the rest of the journey (aircraft cruise $CO_2$ emissions per km) was taken from MEET and from industry data.

Data on maximum plane passenger capacity and maximum freight capacity for dedicated freighter planes were taken from the Boeing web site (Boeing, 2000). To calculate impacts per passenger and freight km, load factors of 0.7 for passenger transport and 0.6 for dedicated freight transport were assumed.

### 13.7.4 Marine Transport

The final transport task considered in the UK was North Sea shipping. The study assessed a typical freight trip from Felixstowe to Rotterdam (197 km) for a number of different ship types (solid and liquid bulk, general cargo, and container). Emissions were taken from the MEET project (EC, 1997) assuming a fuel sulphur content of 0.04%. Emissions were calculated using a simplified methodology, based on average speeds for the different freight ships (assuming medium speed diesel engines). The values presented are for cruising only and do not include additional emissions and damages from loading, hoteling and manoeuvring at ports. No data was available on specific emissions for ships of specific deadweight (i.e cargo carrying capacity). Instead typical ship cargos were used, together with a load factor of 0.6 to estimate values per tonne km.

## 13.8 Results of Marginal Cost Case Studies

Table 13.25 gives an overview of the locations considered.

### 13.8.1 Site-Specific Results

In the following section damages are presented per unit of pollutant emitted. By this way of presentation, the impacts at different locations can be compared directly to each other. On the local scale, the population density is directly reflected in the damage costs per tonne, because air chemistry is only of minor importance and because health impacts are dominating the costs. For illustration Fig. 13.15 shows the damage costs per tonne of $PM_{2.5}$ emitted in different locations. For each country two locations are included: one representing an urban environment and one representing a site with extra-urban or rural character. As can be seen in Fig. 13.15, the local scale damage differs with the size of a city or agglomeration. And the damages for locations outside urban areas in the different countries vary considerably.

On the regional scale the geographic position in Europe determines the size of impact, as illustrated in Fig. 13.16 and Fig. 13.17. The damage costs for the locations in one country differ only little, except for Germany, where one location

**Table 13.25.** Overview of locations considered

| Country | Location | Character |
|---|---|---|
| Belgium | Brussels | City centre |
| Belgium | Rural | Rural village |
| Belgium | Highway | Highway drive in rural location |
| Finland | Helsinki | City centre |
| Finland | Tammisaari – Karjaa | Rural |
| France | Paris | City centre |
| France | Rural | Extra-urban in rural France |
| Germany | Berlin | City centre |
| Germany | Stuttgart | City centre |
| Germany | Heimsheim – Karlsbad | Rural location in Baden-Württemberg (highway) |
| Germany | Illingen – Kämpfelbach | Rural location in Baden-Württemberg (train) |
| Germany | Güstrow – Neustrelitz | Rural location in Mecklenburg-Vorpommern |
| Germany | Berlin Tegel | Airport close to Berlin city |
| Greece | Athens | City centre |
| Greece | Athens suburbs | Suburbs |
| Greece | Thessaloniki | City centre |
| Greece | Larissa | Town centre |
| Greece | Lamia | Town centre |
| Greece | Rural1 | Rural location between Lamia and Larissa |
| Greece | Rural2 | Rural location between Larissa and Thessaloniki |
| Greece | Piraeus | Harbour close to Athens |
| Greece | Iraklio | Harbour of capital of Crete |
| Greece | Aegean sea | Open sea far from populated areas |
| Netherlands | Amsterdam – Groningen | Inter-urban trajectory, parts of different character |
| Netherlands | Amsterdam – Schiphol | Inter-urban trajectory, parts of different character |
| Netherlands | Schiphol – Rotterdam | Inter-urban trajectory, parts of different character |
| Netherlands | Rotterdam – Nijmegen | Inter-urban trajectory, parts of different character |
| Netherlands | Schiphol | Airport near Amsterdam |
| United Kingdom | London | City centre |
| United Kingdom | Liverpool | Large city |
| United Kingdom | Leicester | City |
| United Kingdom | Barnsley | Large Town |
| United Kingdom | Swindon | Town |
| United Kingdom | Motorway | Motorway in rural location |
| United Kingdom | Rural | Oxfordshire |
| United Kingdom | London – Leicester | Rail trip |
| United Kingdom | London – Peterborough | Rail trip |
| United Kingdom | Heathrow | Airport close to London |
| United Kingdom | Felixstowe – Rotterdam | Ship trip across the Channel |

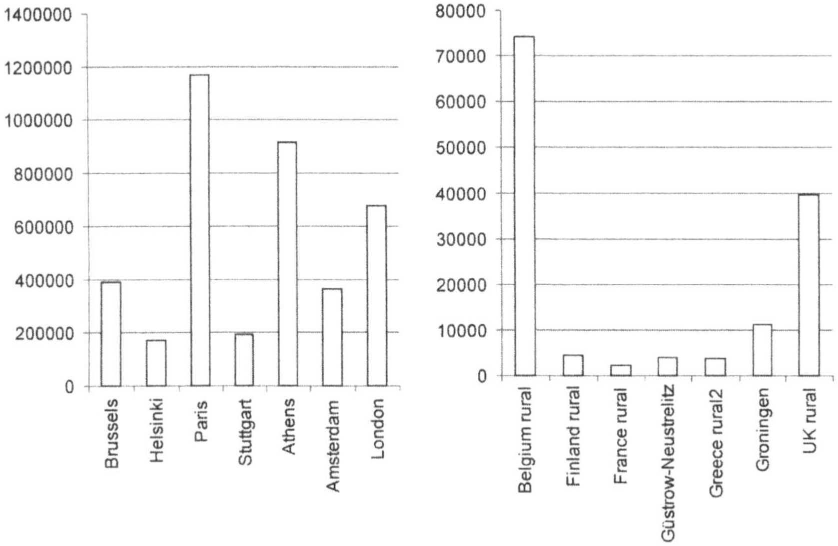


**Fig. 13.15.** Local scale damages in Euro per tonne of $PM_{2.5}$

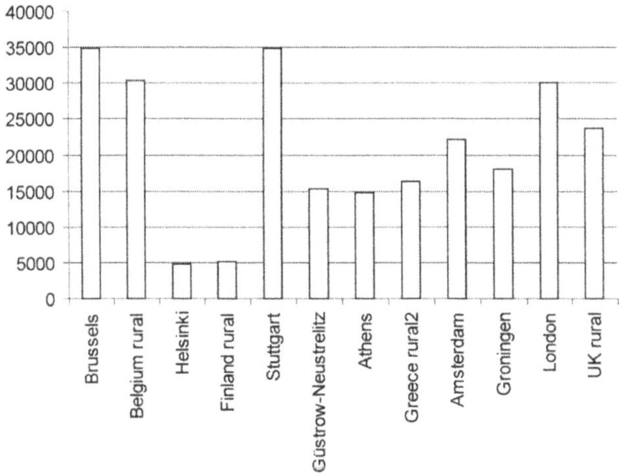


**Fig. 13.16.** Regional scale damages in Euro per tonne of $PM_{2.5}$

is situated in the southwest (Stuttgart), the other in the northeast of Germany (Güstrow-Neustrelitz). The prevailing west wind transports emissions in the northeast of Germany towards the baltic sea and Scandinavia. Thus little population is affected compared to the densely populated areas in Eastern Europe, where emissions from the Southwest are carried. The variation in damage cost is higher in Fig. 13.17, because air chemistry is involved in the formation of secondary particles and thus geographic differences may cause higher differences.

The total costs given in Table 13.26 are the combination of local and regional effects, where double counting was avoided.

The formation of sulphates and nitrates is influenced by the background concentrations of the reactive species involved and the ratio of $SO_2$ and $NO_x$ emitted. Therefore, the damages due to nitrates are attributable to both $NO_x$ and $SO_2$ emissions as well as sulphate damages. For this reason, the estimates presented are based on model runs with either $SO_2$ or $NO_x$ emissions set to zero, so that all changes in secondary particles are caused by one precursor only. The same method had to be applied for ozone, where runs with either $NO_x$ or NMVOC emission increments were carried out.

But for ozone the situation is still more complicated. As can be seen in Table 13.26, in Belgium, Germany, the Netherlands, and the United Kingdom the emission of $NO_x$ via ozone leads to negative costs, i.e. benefits. This means, that an additional unit of $NO_x$ leads to a reduction in ozone and thus a decrease in ozone damages. Due to the complex and non-linear ozone formation processes, for certain $NO_x$ / NMVOC ratios an increase of $NO_x$ emissions at first leads to increasing ozone concentrations and after passing a "hill" to decreasing ozone concentrations. Therefore for the effect of $NO_x$ changes it is crucial, on which side of the hill one is starting. For the countries mentioned above, the start lies beyond the hill, therefore an increase of $NO_x$ leads to benefits. It is desirable, to reach a situation "before the hill", but that requires massive changes in $NO_x$ and NMVOC emissions, which is not the subject of the marginal analysis carried out here. Furthermore, besides via ozone, $NO_x$ causes effects via nitrates and sulphates. Adding up the costs of both effects leads to overall damage costs when emitting $NO_x$. In contrast to $NO_x$, NMVOC emissions cause damages due to ozone in all countries considered.

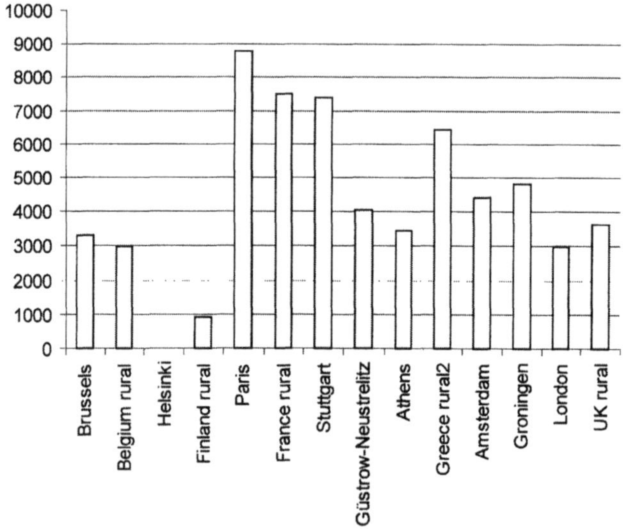


**Fig. 13.17.** Regional scale damages due to secondary particles in Euro per tonne of $NO_x$ (note: value for Helsinki is too small to be visible, see Table 13.26)

It has to be noted that these monetary estimates include only ozone formation on the regional scale. Local and global scale effects could not be taken into account quantitatively.

**Table 13.26.** Damage costs per tonne of pollutant emitted in Euro per tonne (Loc = local, Reg = regional, Tot = total)

| Location of emission | Scale | $PM_{2.5}$ | $SO_2$ - direct | $SO_2$ – via secondary particles | $NO_x$ – via secondary particles | $NO_x$ – via ozone | NMVOC – via ozone | CO | DME | Benzene | 1,3-Butadiene |
|---|---|---|---|---|---|---|---|---|---|---|---|
| Belgium | Loc | 74172 | 1894 | | | | | 0.55 | 2050 | 445 | 16674 |
| rural | Reg | 30434 | 452 | 3945 | 2982 | -1829 | 1717 | 0.29 | 841 | 216 | 9127 |
| | Tot | 103490 | 2322 | 3945 | 2982 | -1829 | 1717 | 0.83 | 2860 | 654 | 25554 |
| Belgium | Loc | 388460 | 9923 | | | | | 2.86 | 10739 | 2330 | 87361 |
| Brussels | Reg | 34861 | 533 | 4440 | 3312 | -1829 | 1717 | 0.33 | 964 | 246 | 10317 |
| centre | Tot | 418608 | 10348 | 4440 | 3312 | -1829 | 1717 | 3.15 | 11572 | 2546 | 96543 |
| Finland | Loc | 170523 | 5219 | | | | | 1.25 | 4714 | 1018 | 38180 |
| Helsinki | Reg | 4732 | 58 | 830 | 816 | 354 | 261 | 0.05 | 131 | 36 | 1736 |
| centre | Tot | 174763 | 5264 | 830 | 816 | 354 | 261 | 1.30 | 4831 | 1052 | 39796 |
| Finland | Loc | 4376 | 112 | | | | | 0.03 | 121 | 26 | 984 |
| Tammisaari | Reg | 5083 | 53 | 915 | 932 | 354 | 261 | 0.06 | 141 | 39 | 1876 |
| -Karjaa | Tot | 9339 | 162 | 915 | 932 | 354 | 261 | 0.09 | 258 | 65 | 2834 |
| France | Loc | 1170000 | 25200 | | | | | n.a. | n.a. | 7140 | 260000 |
| Paris | Reg | n.a. | n.a. | 6300 | 8800 | 630 | 1152 | n.a. | n.a. | n.a. | n.a. |
| | Tot | n.a. | n.a. | 6300 | 8800 | 630 | 1152 | 20.0 | 960 | n.a. | n.a. |
| France | Loc | 2360 | 270 | | | | | n.a. | n.a. | 18 | 680 |
| rural | Reg | n.a. | n.a. | 5300 | 7500 | 630 | 1152 | n.a. | n.a. | n.a. | n.a. |
| | Tot | n.a. | n.a. | 5300 | 7500 | 630 | 1152 | 0.20 | 310 | n.a. | n.a. |
| Germany | Loc | 192965 | 4927 | | | | | 1.42 | 5335 | 1157 | 43384 |
| Stuttgart | Reg | 34828 | 2819 | 6870 | 7400 | -1136 | 1651 | 0.35 | 963 | 252 | 10769 |
| city centre | Tot | 222746 | 7636 | 6870 | 7400 | -1136 | 1651 | 1.73 | 6158 | 1379 | 53004 |
| Germany | Loc | 4005 | 102 | | | | | 0.03 | 111 | 32 | 1869 |
| Güstrow– | Reg | 15338 | 1281 | 3140 | 4050 | -1136 | 1651 | 0.17 | 424 | 118 | 5336 |
| Neustrelitz | Tot | 18909 | 1383 | 3140 | 4050 | -1136 | 1651 | 0.20 | 523 | 146 | 7086 |
| Greece | Loc | 916802 | 23402 | | | | | 6.74 | 24673 | 5497 | 206116 |
| Athens | Reg | 14820 | 166 | 3123 | 3455 | 1016 | 684 | 0.15 | 399 | 101 | 4751 |
| centre | Tot | 926778 | 23470 | 3123 | 3455 | 1016 | 684 | 6.86 | 24942 | 5569 | 209781 |
| Greece | Loc | 3764 | 96 | | | | | 0.03 | 101 | 23 | 846 |
| Rural2 | Reg | 16313 | 152 | 4281 | 6437 | 1016 | 684 | 0.17 | 439 | 116 | 5428 |
| | Tot | 19637 | 238 | 4281 | 6437 | 1016 | 684 | 0.20 | 528 | 136 | 6171 |

| Location of emission | Scale | $PM_{2.5}$ | $SO_2$ - direct | $SO_2$ – via secondary particles | $NO_x$ – via secondary particles | $NO_x$ – via ozone | NMVOC – via ozone | CO | DME | Benzene | 1,3-Butadiene |
|---|---|---|---|---|---|---|---|---|---|---|---|
| Netherlands | Loc | 11276 | 264 | | | | | 0.08 | 324 | 62 | 2322 |
| Groningen | Reg | 18017 | 1156 | 2582 | 4803 | -2020 | 1376 | 0.19 | 518 | 136 | 5994 |
| (ag4) | Tot | 29293 | 1420 | 2582 | 4803 | -2020 | 1376 | 0.27 | 842 | 198 | 8316 |
| Netherlands | Loc | 361915 | 8464 | | | | | 2.44 | 11522 | 1987 | 74526 |
| Amsterdam | Reg | 22129 | 1278 | 3011 | 4402 | -2020 | 1376 | 0.23 | 705 | 162 | 7015 |
| (as2) | Tot | 384044 | 9743 | 3011 | 4402 | -2020 | 1376 | 2.66 | 12226 | 2149 | 81541 |
| United | Loc | 675041 | n.a. | | | | | 4.97 | 18950 | 4047 | 151751 |
| Kingdom | Reg | 30064 | n.a. | 3018 | 2991 | -1394 | 997 | 0.28 | 844 | 204 | 8537 |
| London | Tot | 705105 | 17789 | 3018 | 2991 | -1394 | 997 | 5.24 | 19794 | 4252 | 160288 |
| United | Loc | 39521 | n.a. | | | | | 0.29 | 1110 | 237 | 8899 |
| Kingdom | Reg | 23766 | n.a. | 3018 | 3633 | -1394 | 997 | 0.23 | 668 | 165 | 7036 |
| Rural | Tot | 63287 | 1392 | 3018 | 3633 | -1394 | 997 | 0.52 | 1778 | 403 | 15935 |

n.a. = not available

## 13.8.2 Road Transport

In the following section the costs quantified are presented per vehicle kilometre (vkm). This allows comparison of different fuels and emission standards for a vehicle category. Comparisons of certain vehicle technologies between locations include the aspects location and emission factor. The former was presented in the previous section, the latter is determined by local conditions – here mainly vehicle speed –, which varies between countries.

The composition of damage costs and the variation between an urban and a rural location is illustrated for Greece. The picture is the same for other countries. The damages due to a petrol and a diesel car, both complying with the EURO II emission standard, are presented for a trip in Athens city centre and on a rural motorway. Figure 13.18 shows the shares of different pollutants in the damage costs on the local and regional scale and the total. The amount in Euro per 100 vehicle kilometres is given on top of every bar. Besides local regional damages, the total includes damages due to greenhouse gas emissions.

The damage costs are dominated by primary ($PM_{2.5}$) and secondary (Sulphates, Nitrates) particles. The shares of benzene and 1,3-butadiene are too small to be visible. sulphates and nitrates do not appear on the local scale, as air chemistry is not relevant on this scale. Diesel motor emissions (DME) are only emitted by diesel cars.

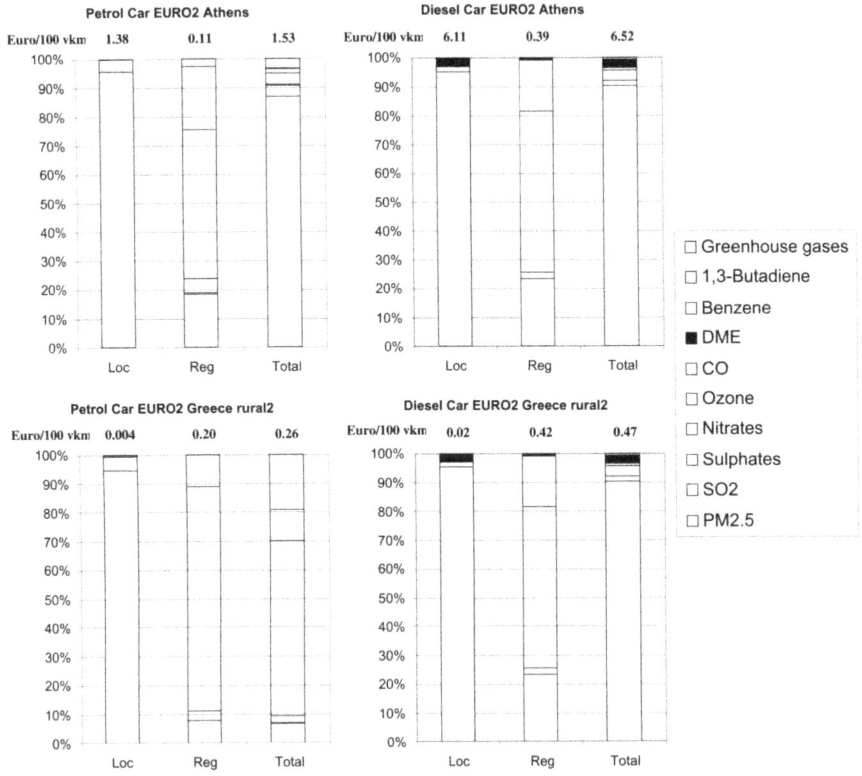


**Fig. 13.18.** Share of different pollutants in damage costs for different fuels and locations

The importance of local and regional scale is very different for the different locations: while the damages for Athens are dominated by the local scale, the regional scale is determining the size of damages for the rural location. The structure of damages is similar for petrol and diesel car, but the absolute value is very different, as the diesel car emits more primary particles. Shares of $SO_2$ and Sulphates are smaller for the diesel car than for the petrol car. This seems to contradict the fact that diesel cars emit more $SO_2$ than petrol cars, but is correct: the absolute value of the damage due to $SO_2$ and sulphates is higher for diesel cars (see Table 13.27 and Table 13.28). But as total damages for the diesel car are much higher and dominated by $PM_{2.5}$, the relative share of $SO_2$ and sulphates is lower.

The shares of different damage categories are presented in Fig. 13.19. Costs are dominated by health effects. For Athens health effects make 90% of the damage costs for the petrol car and 98% for the diesel car. Due to the importance of local scale health damages, global warming, crop losses and material damages play only a minor role. For the rural location the picture changes a little, even though health effects still have a share of 71% (petrol car) and 82% (diesel car).

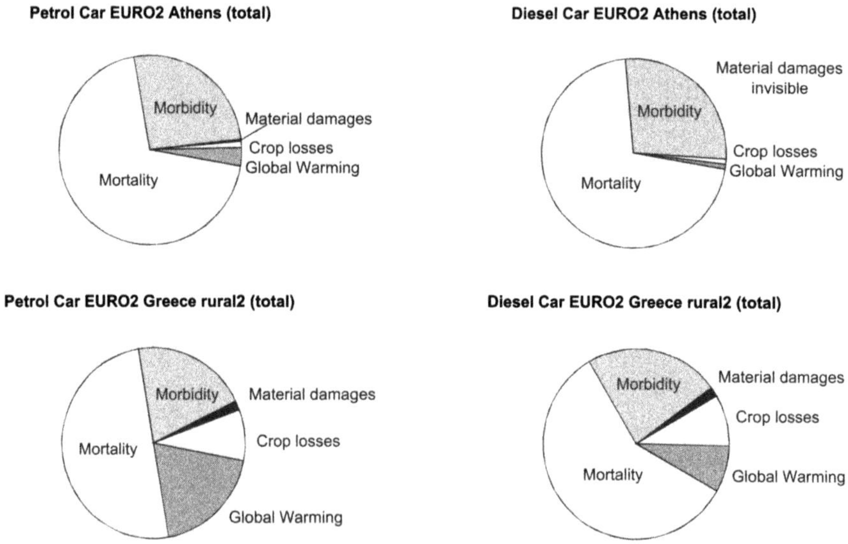


**Fig. 13.19.** Share of damage categories in damage costs for different fuels and locations

Figure 13.20 presents a comparison of damage costs for different car technologies in Athens. Diesel cars cause considerably higher damages than petrol cars due to their high particle emissions.

The introduction of EURO I, EURO II, and EURO III emission standards reduces the damage costs per vehicle kilometre by 65%, 75% and 90% respectively compared to an uncontrolled diesel car This is almost only due to the reduction of particle emissions. The reduction of damage costs due to petrol cars is considerable from ECE 15/04 to EURO I with 65%. The reductions due to the

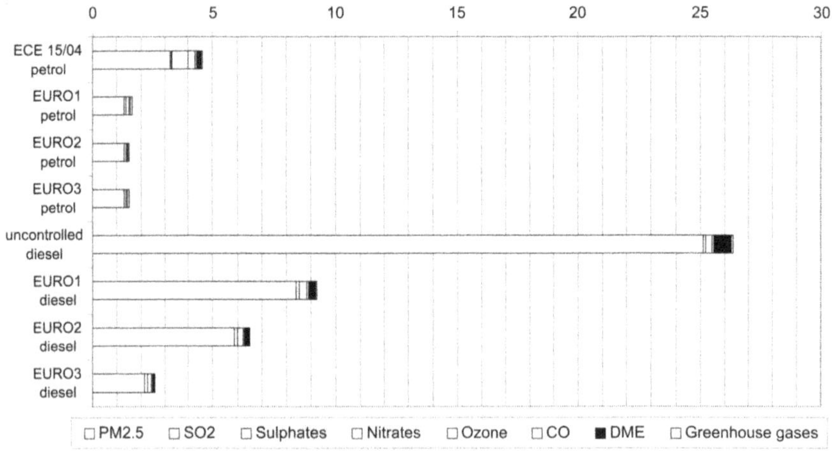


**Fig. 13.20.** Damage costs in Euro per 100 vkm for different car technologies in Athens

**Table 13.27.** Damage costs in Euro per 100 vkm for petrol cars EURO II, 1.4 - 2 l (Loc = local, Reg = regional, Tot = total)

| Location | Scale | $PM_{2.5}$ | $SO_2$ | Sulphates | Nitrates | Ozone | CO | Benzene | 1,3-Butadiene | Greenhouse gases | Total |
|---|---|---|---|---|---|---|---|---|---|---|---|
| Belgium | Loc | 0.09 | 0.003 | | | | 8.4e-5 | 7.3e-5 | 1.7e-5 | | 0.09 |
| Rural | Reg | 0.04 | 0.001 | 0.007 | 0.06 | -0.02 | 4.4e-5 | 3.4e-5 | 9.1e-6 | | 0.08 |
| | Tot | 0.13 | 0.004 | 0.007 | 0.06 | -0.02 | 1.3e-4 | 1.1e-4 | 2.6e-5 | 0.042 | 0.22 |
| Belgium | Loc | 0.62 | 0.027 | | | | 1.4e-3 | 8.1e-4 | 1.9e-4 | | 0.65 |
| Brussels | Reg | 0.05 | 0.001 | 0.014 | 0.08 | -0.02 | 1.4e-4 | 7.5e-5 | 2.0e-5 | | 0.12 |
| | Tot | 0.66 | 0.028 | 0.014 | 0.08 | -0.02 | 1.5e-3 | 8.8e-4 | 2.1e-4 | 0.069 | 0.84 |
| Finland | Loc | 0.21 | 0.009 | | | | 3.8e-4 | 2.6e-4 | 6.1e-5 | | 0.22 |
| Helsinki | Reg | 0.01 | 0.005 | 0.002 | 0.01 | 0.01 | 1.7e-5 | 9.3e-6 | 2.8e-6 | | 0.03 |
| | Tot | 0.22 | 0.014 | 0.002 | 0.01 | 0.01 | 4.0e-4 | 2.7e-4 | 6.4e-5 | 0.055 | 0.31 |
| Germany | Loc | 0.23 | 0.010 | | | | 3.6e-4 | 2.6e-4 | 6.2e-5 | | 0.24 |
| Stuttgart | Reg | 0.04 | 0.003 | 0.011 | 0.11 | -0.01 | 8.9e-5 | 5.7e-5 | 1.5e-5 | | 0.15 |
| | Tot | 0.27 | 0.013 | 0.011 | 0.11 | -0.01 | 4.4e-4 | 3.1e-4 | 7.5e-5 | 0.050 | 0.44 |
| Germany | Loc | 0.00 | 0.000 | | | | 6.1e-6 | 6.9e-6 | 2.8e-6 | | 0.00 |
| Güstrow- | Reg | 0.02 | 0.002 | 0.003 | 0.08 | -0.02 | 3.7e-5 | 4.3e-5 | 2.7e-5 | | 0.08 |
| Neustrelitz | Tot | 0.02 | 0.002 | 0.003 | 0.08 | -0.02 | 4.3e-5 | 4.9e-5 | 3.0e-5 | 0.048 | 0.14 |
| Greece | Loc | 1.32 | 0.060 | | | | 6.6e-5 | 1.8e-3 | 4.2e-4 | | 1.38 |
| Athens | Reg | 0.02 | 0.000 | 0.006 | 0.06 | 0.02 | 2.9e-3 | 3.3e-5 | 9.6e-6 | | 0.11 |
| centre | Tot | 1.33 | 0.060 | 0.006 | 0.06 | 0.02 | 3.0e-3 | 1.8e-3 | 4.3e-4 | 0.049 | 1.53 |
| Greece | Loc | 0.00 | 0.000 | | | | 4.5e-6 | 7.6e-6 | 5.2e-6 | | 0.00 |
| Rural2 | Reg | 0.02 | 0.000 | 0.007 | 0.16 | 0.02 | 5.5e-5 | 3.9e-5 | 3.3e-5 | | 0.20 |
| | Tot | 0.02 | 0.000 | 0.006 | 0.16 | 0.03 | 4.7e-5 | 3.9e-5 | 3.1e-5 | 0.049 | 0.26 |
| Netherlands | Loc | 0.02 | 0.001 | | | | 3.0e-5 | 1.9e-5 | 4.4e-6 | | 0.02 |
| Groningen | Reg | 0.02 | 0.003 | 0.006 | 0.08 | -0.02 | 7.5e-5 | 4.1e-5 | 1.1e-5 | | 0.09 |
| | Tot | 0.04 | 0.003 | 0.006 | 0.08 | -0.02 | 1.0e-4 | 5.9e-5 | 1.6e-5 | 0.061 | 0.17 |
| Netherlands | Loc | 0.37 | 0.018 | | | | 6.8e-4 | 7.4e-4 | 5.1e-4 | | 0.39 |
| Amsterdam | Reg | 0.02 | 0.003 | 0.006 | 0.11 | -0.05 | 6.3e-5 | 6.0e-5 | 4.8e-5 | | 0.10 |
| | Tot | 0.39 | 0.020 | 0.006 | 0.11 | -0.05 | 7.5e-4 | 8.0e-4 | 5.5e-4 | 0.053 | 0.54 |
| United | Loc | 0.90 | 0.042 | | | | 0.002 | 0.001 | 2.9e-4 | | 0.95 |
| Kingdom | Reg | 0.04 | 0.001 | 0.007 | 0.05 | -0.01 | 1.1e-4 | 6.1e-5 | 1.6e-5 | | 0.09 |
| London | Tot | 0.94 | 0.044 | 0.007 | 0.05 | -0.01 | 0.002 | 0.001 | 3.0e-4 | 0.061 | 1.10 |
| United | Loc | 0.03 | 0.002 | | | | 2.7e-5 | 6.9e-5 | 2.9e-5 | | 0.03 |
| Kingdom | Reg | 0.02 | 6.9e-4 | 0.004 | 0.06 | -0.02 | 2.1e-5 | 4.8e-5 | 2.3e-5 | | 0.07 |
| Rural | Tot | 0.05 | 0.003 | 0.004 | 0.06 | -0.02 | 4.8e-5 | 1.2e-4 | 5.3e-5 | 0.037 | 0.14 |

**Table 13.28.** Damage costs in Euro / 100 vkm for diesel cars EURO II, <2 l (Loc = local, Reg = regional, Tot = total)

| Location | Scale | $PM_{2.5}$ | $SO_2$ | Sulphates | Nitrates | Ozone | CO | DME | Benzene | 1,3-Butadiene | Greenhouse gases | Total |
|---|---|---|---|---|---|---|---|---|---|---|---|---|
| Belgium | Loc | 0.22 | 0.006 | | | | 1.5e-5 | 0.006 | 8.2e-5 | 6.7e-4 | | 0.23 |
| Rural | Reg | 0.09 | 0.001 | 0.013 | 0.17 | -0.09 | 7.7e-6 | 0.002 | 3.8e-5 | 3.6e-4 | | 0.19 |
| | Tot | 0.30 | 0.008 | 0.013 | 0.17 | -0.09 | 2.2e-5 | 0.009 | 1.2e-4 | 1.0e-3 | 0.036 | 0.45 |
| Belgium | Loc | 2.70 | 0.046 | | | | 2.3e-4 | 0.087 | 9.4e-4 | 7.7e-3 | | 2.85 |
| Brussels | Reg | 0.21 | 0.002 | 0.019 | 0.32 | -0.14 | 2.4e-5 | 0.007 | 8.7e-5 | 8.1e-4 | | 0.42 |
| | Tot | 2.91 | 0.048 | 0.019 | 0.32 | -0.14 | 2.5e-4 | 0.094 | 1.0e-3 | 8.5e-3 | 0.050 | 3.31 |
| Finland | Loc | 0.84 | 0.017 | | | | 7.0e-5 | n.a. | 3.1e-4 | 2.6e-3 | | 0.86 |
| Helsinki | Reg | 0.02 | 0.008 | 0.003 | 0.04 | 0.02 | 2.9e-6 | n.a. | 1.1e-5 | 1.1e-4 | | 0.10 |
| | Tot | 0.87 | 0.026 | 0.003 | 0.04 | 0.02 | 7.3e-5 | n.a. | 3.2e-4 | 2.7e-3 | 0.043 | 1.00 |
| Germany | Loc | 0.80 | 0.018 | | | | 6.4e-5 | 0.022 | 3.0e-4 | 2.5e-3 | | 0.84 |
| Stuttgart | Reg | 0.14 | 0.009 | 0.015 | 0.35 | -0.06 | 1.6e-5 | 0.004 | 6.5e-5 | 6.1e-4 | | 0.47 |
| | Tot | 0.92 | 0.027 | 0.015 | 0.35 | -0.06 | 7.8e-5 | 0.025 | 3.6e-4 | 3.0e-3 | 0.040 | 1.33 |
| Germany | Loc | 0.02 | 0.000 | | | | 9.1e-7 | 0.000 | 4.5e-6 | 2.9e-5 | | 0.02 |
| Güstrow- | Reg | 0.07 | 0.004 | 0.004 | 0.18 | -0.05 | 4.1e-6 | 0.002 | 1.7e-5 | 7.9e-5 | | 0.20 |
| Neustrelitz | Tot | 0.08 | 0.004 | 0.004 | 0.18 | -0.05 | 4.9e-6 | 0.002 | 2.1e-5 | 1.1e-4 | 0.037 | 0.25 |
| Greece | Loc | 5.83 | 0.104 | | | | 1.1e-5 | 0.161 | 2.1e-3 | 1.7e-2 | | 6.11 |
| Athens | Reg | 0.09 | 0.001 | 0.009 | 0.22 | 0.07 | 5.0e-4 | 0.003 | 3.8e-5 | 3.9e-4 | | 0.39 |
| centre | Tot | 5.89 | 0.104 | 0.009 | 0.22 | 0.07 | 5.0e-4 | 0.163 | 2.1e-3 | 1.7e-2 | 0.048 | 6.52 |
| Greece | Loc | 0.02 | 0.000 | | | | 3.3e-7 | 0.001 | 2.1e-6 | 2.3e-6 | | 0.02 |
| Rural2 | Reg | 0.08 | 0.001 | 0.011 | 0.28 | 0.04 | 4.0e-6 | 0.002 | 1.1e-5 | 1.4e-5 | | 0.42 |
| | Tot | 0.09 | 0.001 | 0.011 | 0.28 | 0.04 | 3.4e-6 | 0.002 | 1.1e-5 | 1.4e-5 | 0.036 | 0.47 |
| Netherlands | Loc | 0.07 | 0.001 | | | | 5.1e-6 | 0.000 | 2.1e-5 | 1.8e-4 | | 0.07 |
| Groningen | Reg | 0.10 | 0.005 | 0.011 | 0.28 | -0.11 | 1.3e-5 | 0.000 | 4.7e-5 | 4.6e-4 | | 0.30 |
| | Tot | 0.17 | 0.006 | 0.011 | 0.28 | -0.11 | 1.8e-5 | 0.000 | 6.9e-5 | 6.3e-4 | 0.046 | 0.41 |
| Netherlands | Loc | 2.13 | 0.029 | | | | 5.1e-5 | 0.001 | 2.0e-4 | 2.2e-4 | | 2.16 |
| Amsterdam | Reg | 0.13 | 0.004 | 0.010 | 0.21 | -0.09 | 4.8e-6 | 0.000 | 1.6e-5 | 2.1e-5 | | 0.26 |
| | Tot | 2.26 | 0.034 | 0.010 | 0.21 | -0.09 | 5.6e-5 | 0.001 | 2.2e-4 | 2.4e-4 | 0.038 | 2.46 |
| United | Loc | 3.92 | 0.075 | | | | 3.3e-4 | 0.109 | 0.001 | 0.012 | | 4.11 |
| Kingdom | Reg | 0.18 | 0.002 | 0.013 | 0.18 | -0.08 | 1.9e-5 | 0.005 | 7.1e-5 | 6.5e-4 | | 0.30 |
| London | Tot | 4.09 | 0.077 | 0.013 | 0.18 | -0.08 | 3.5e-4 | 0.114 | 0.001 | 0.012 | 0.046 | 4.46 |
| United | Loc | 0.10 | 0.004 | | | | 2.7e-6 | 0.003 | 2.3e-5 | 7.5e-5 | | 0.11 |
| Kingdom | Reg | 0.06 | 0.001 | 0.009 | 0.12 | -0.05 | 2.1e-6 | 0.002 | 1.6e-5 | 5.9e-5 | | 0.15 |
| Rural | Tot | 0.16 | 0.005 | 0.009 | 0.12 | -0.05 | 4.8e-6 | 0.005 | 3.9e-5 | 1.3e-4 | 0.032 | 0.29 |

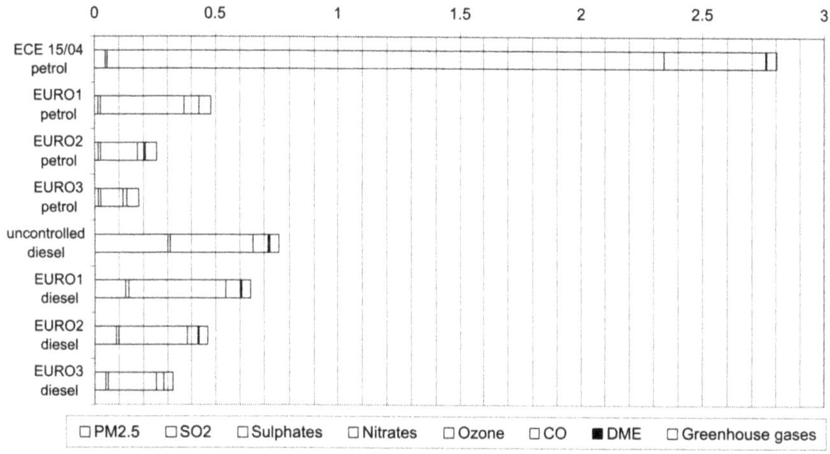


**Fig. 13.21.** Damage costs in Euro per 100 vkm for different passenger car technologies in rural environment in Greece (rural2)

introduction of EURO II and EURO III are only small, because particle emissions are only reduced marginally, but dominate the damages in the urban environment.

The picture is different when looking at the rural location. Damages are dominated by regional scale damages due to nitrates and therefore, the reduction of $NO_x$ emissions leads to considerable reduction of costs per kilometre. But of course, the absolute value of the costs is to be taken into account, which is an order of magnitude lower than for Athens.

Table 13.27 gives the damage costs per 100 vkm for petrol cars complying with EURO II and a cylinder capacity between 1.4 and 2 litres. Table 13.28 presents the costs for diesel cars EURO II with a cylinder capacity up to 2 litres.

### 13.8.3 Rail Transport

For electric traction the damages due to airborne pollutants are not related to the location where a train is running, but to the locations of the power plants generating the electricity used. Figure 13.22 presents costs due to electricity production per vehicle kilometre for electric trains. The differences between the train types are due to the electricity mixes and thus damage costs per kWh used and different energy consumption factors.

Damages due to diesel trains are site specific and depend on location and emission factors. The composition of the cost is similar to diesel road vehicles. Figure 13.23 shows the results for selected train types.

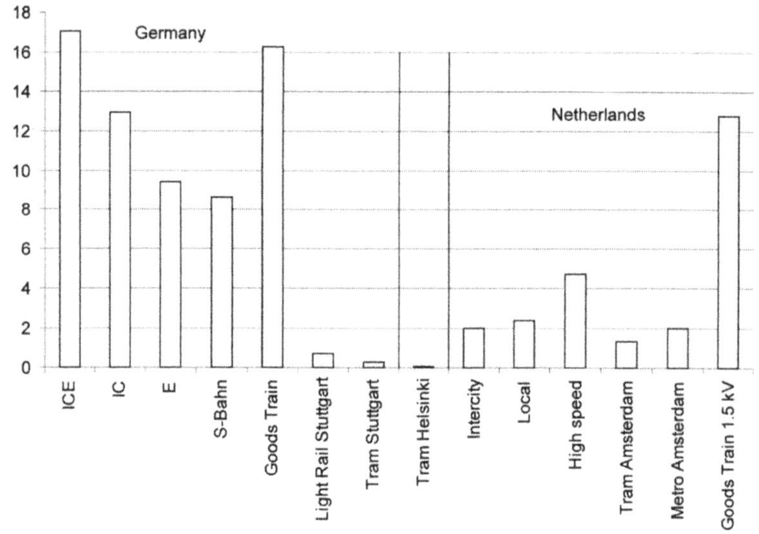


**Fig. 13.22.** Damage costs in Euro per 100 vkm due to electricity production for rail transport

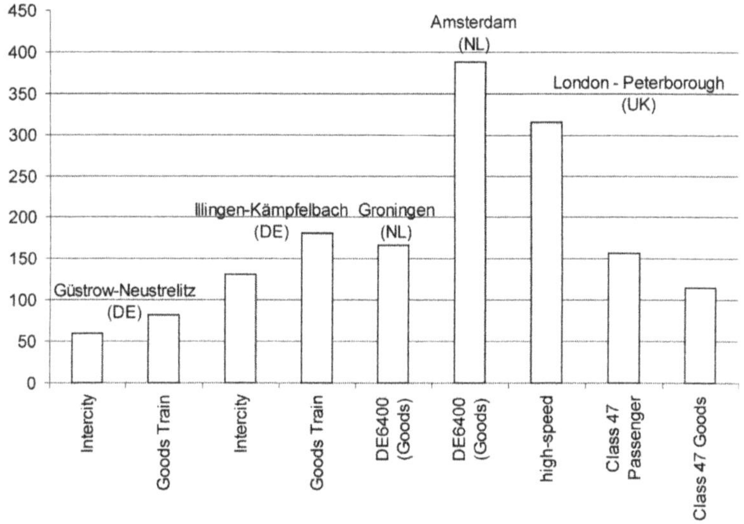


**Fig. 13.23.** Damage costs in Euro per 100 vkm for diesel trains

## 13.8.4 Other Modes

Figure 13.24 presents costs for container ships on different routes. The first three bars show the composition of damages in different locations in the Aegean sea. In

Piraeus, the harbour of Athens, the costs are the highest due to high local impacts. In Iraklio on the island of Crete, the local impacts are much smaller as the local population affected is only small compared to Athens. In the Aegean sea the local impacts are negligible. The costs of a trip from Felixstowe in the UK across the Channel to Rotterdam are in the same order of magnitude as for the inland waterway route from Rotterdam to Nijmegen.

Table 13.29 shows the quantifiable damages due to a landing and take-off (LTO) cycle at Berlin Tegel airport and at London Heathrow. The costs reflect the considerable variations in emission factors of the different engine types used. The costs for the 737-400 are lower at Berlin, mainly due to lower emissions in most components as a consequence of shorter times spent in most engine modes.

**Table 13.29.** Costs due to aircraft operation (LTO cycle) at Berlin-Tegel and London-Heathrow in Euro/LTO cycle.

| Location | Aircraft | Scale | $PM_{2.5}$ | $SO_2$ | Sulphates | Nitrates | Ozone | CO | Benzene | 1,3-Butadiene | Greenhouse gases | Total |
|---|---|---|---|---|---|---|---|---|---|---|---|---|
| Berlin | Boeing | Loc | 3.3 | 1.0 | | | | 0.003 | 0.003 | 0.09 | | 4.4 |
| Tegel | 737-300/400 | Reg | 1.9 | 1.1 | 1.8 | 29.3 | -7.3 | 0.002 | 0.002 | 0.07 | | 26.8 |
| Airport | CFM56 3C1 | Tot | 4.6 | 1.9 | 1.8 | 29.3 | -7.3 | 0.005 | 0.004 | 0.15 | 4.2 | 34.7 |
| London | Boeing | Loc | 24.6 | 2.0 | 0.0 | 0.0 | 0.0 | 0.006 | 0.014 | 0.49 | 0.0 | 27.2 |
| Heathrow | 737-200 | Reg | 8.3 | 0.4 | 2.5 | 27.4 | -9.2 | 0.002 | 0.004 | 0.15 | 0.0 | 29.7 |
| Airport | PW JT8D-15A | Tot | 32.9 | 2.4 | 2.5 | 27.4 | -9.2 | 0.008 | 0.018 | 0.64 | 6.8 | 63.7 |
| London | Boeing | Loc | 4.5 | 1.9 | 0.0 | 0.0 | 0.0 | 0.011 | 0.01 | 0.33 | 0.0 | 6.8 |
| Heathrow | 737-400 | Reg | 1.4 | 0.4 | 2.4 | 36.1 | -12.8 | 0.003 | 0.002 | 0.09 | 0.0 | 27.7 |
| Airport | CFM56-3C1 | Tot | 6.0 | 2.3 | 2.4 | 36.1 | -12.8 | 0.014 | 0.012 | 0.42 | 6.8 | 41.2 |
| London | Boeing | Loc | 128.9 | 8.1 | 0.0 | 0.0 | 0.0 | 0.021 | 0.024 | 0.82 | 0.0 | 137.8 |
| Heathrow | 747-400 | Reg | 46.6 | 1.8 | 10.6 | 367.8 | -134.6 | 0.005 | 0.008 | 0.29 | 0.0 | 293.7 |
| Airport | RB211-524H2 | Tot | 175.5 | 9.9 | 10.6 | 367.8 | -134.6 | 0.027 | 0.031 | 1.11 | 29.5 | 461.0 |
| London | Boeing | Loc | 6.4 | 2.8 | 0.0 | 0.0 | 0.0 | 0.008 | 0.004 | 0.13 | 0.0 | 9.4 |
| Heathrow | 757-200 | Reg | 2.4 | 0.6 | 3.7 | 87.9 | -32.4 | 0.002 | 8E-04 | 0.03 | 0.0 | 62.3 |
| Airport | RB211-535E4 | Tot | 8.8 | 3.4 | 3.7 | 87.9 | -32.4 | 0.010 | 0.004 | 0.16 | 10.3 | 81.9 |
| London | Boeing | Loc | 64.4 | 3.6 | 0.0 | 0.0 | 0.0 | 0.011 | 0.012 | 0.41 | 0.0 | 68.4 |
| Heathrow | 767-300 | Reg | 23.3 | 0.8 | 5.3 | 183.9 | -67.3 | 0.003 | 0.004 | 0.15 | 0.0 | 146.8 |
| Airport | RB211-524H | Tot | 87.7 | 4.4 | 5.3 | 183.9 | -67.3 | 0.013 | 0.016 | 0.56 | 14.7 | 229.9 |
| London | Boeing | Loc | 9.3 | 4.9 | 0.0 | 0.0 | 0.0 | 0.013 | 0.006 | 0.22 | 0.0 | 14.4 |
| Heathrow | 777-200 | Reg | 3.5 | 1.1 | 6.6 | 226.4 | -83.7 | 0.003 | 0.002 | 0.06 | 0.0 | 154.2 |
| Airport | GE GE90-92B | Tot | 12.8 | 6.0 | 6.6 | 226.4 | -83.7 | 0.016 | 0.008 | 0.28 | 18.3 | 186.9 |
| London | Boeing | Loc | 8.3 | 4.3 | 0.0 | 0.0 | 0.0 | 0.015 | 0.009 | 0.31 | 0.0 | 13.0 |
| Heathrow | 777-200 | Reg | 3.0 | 1.0 | 5.7 | 147.7 | -54.3 | 0.004 | 0.002 | 0.08 | 0.0 | 103.3 |
| Airport | GE90-76B | Tot | 11.3 | 5.3 | 5.7 | 147.7 | -54.3 | 0.019 | 0.011 | 0.39 | 15.8 | 132.0 |

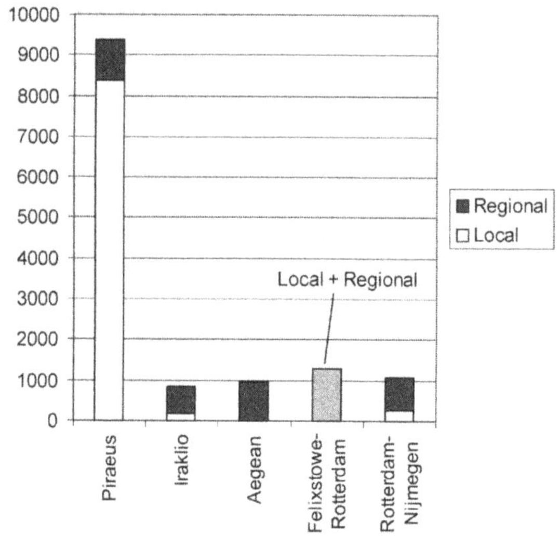


**Fig. 13.24.** Cost in Euro per 100 vkm due to airborne emissions of container ships

## 13.8.5 Costs Including Up- and Downstream Processes

The following section presents costs including vehicle use, vehicle production, fuel production and infrastructure use. The costs are related to the load of a vehicle to facilitate comparison between modes. Of course the load factors are very important for the results. The country-specific load factors used are given in Table 13.30 and Table 13.31. For illustration, selected transport technologies are compared for Germany, values for all countries are given in Table 13.30 and Table 13.31.

Figure 13.25 presents the damage costs due to passenger transport in the city centre of Stuttgart. In Stüttgart, rail vehicles for passenger transport are operated electrically only, that means there are no direct emissions from vehicle use. Of the remaining processes, fuel (electricity) production causes the highest share of costs. For vehicles with internal combustion engine, vehicle operation plays a very important role in the urban environment. The only exception is the petrol car complying with EURO II, where infrastructure use causes the highest cost per passenger kilometre (pkm). Costs per pkm due to vehicle production, fuel production and infrastructure use are much lower for busses and coaches than for cars. When looking at the total costs per pkm, the electric rail vehicles by far cause the smallest damage costs due to air pollution. They are followed by the coaches, but this is mainly due to the high load factor, which usually cannot be reached for inner-urban trips. So the ranking in terms of increasing damage costs for modern road passenger vehicles is modern urban bus, EURO II petrol car and EURO II diesel car.

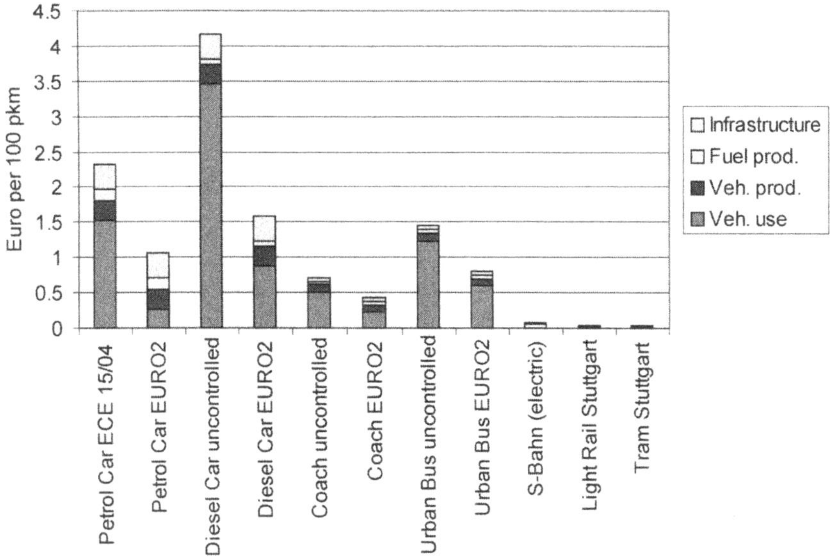


**Fig. 13.25.** Comparison of air pollution costs due to urban passenger transport in Stuttgart (pkm = passenger kilometre)

Figure 13.26 shows costs due to transport in extra-urban areas. It can be seen, that with lower population density (Güstrow-Neustrelitz), the importance of vehicle is reduced. For extra-urban areas the costs caused by direct emissions of a EURO II petrol car are almost negligible compared to the other processes. The diesel trains cause high costs, with the dominating contribution from direct emissions. The difference between the Intercity train and the Local train is mainly caused by the smaller capacity use of the Local train. By far the smallest costs are again caused by the trains with electric traction, the second best vehicle is the coach due to its high capacity use. The old vehicle technologies (incl. Local train) cause the highest damages due to their high emission factors.

The goods transport vehicles presented in Fig. 13.27 are run by diesel fuel or electricity. The highest costs per tonne kilometre (tkm) are caused by light duty vehicles (LDV) due to their comparably small load. The heavy duty vehicles (HDV) considered have a total maximum weight of 32 to 40 tonnes and a maximum load of about 25 tonnes (compared to 1 to 2 tonnes for LDVs). As they are used for long-distance transport, the average capacity use is higher than for LDVs. For this reason the cost per tkm is comparably small. The same is true for the goods trains considered. The costs of the electric train are hardly visible on the given scale.

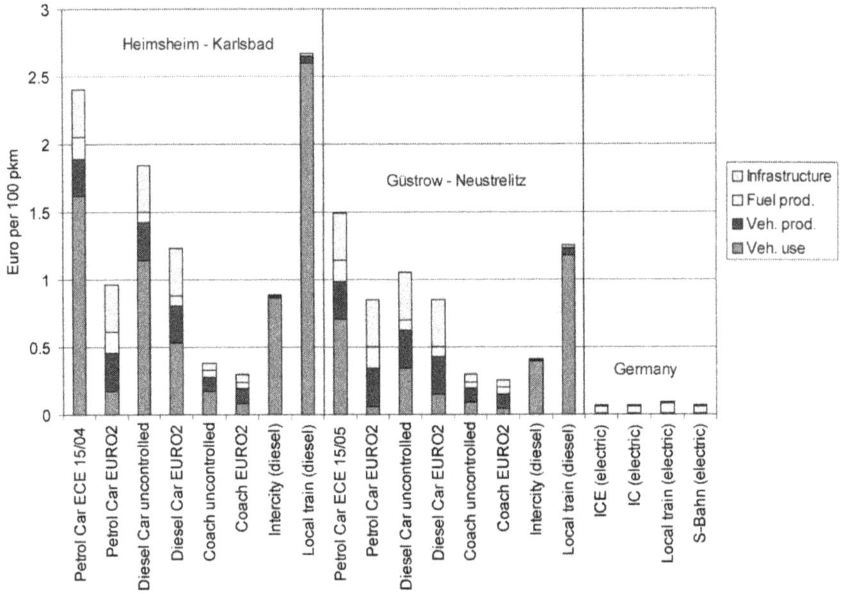


**Fig. 13.26.** Comparison of air pollution costs due to extra-urban passenger transport in Germany (pkm = passenger kilometre)

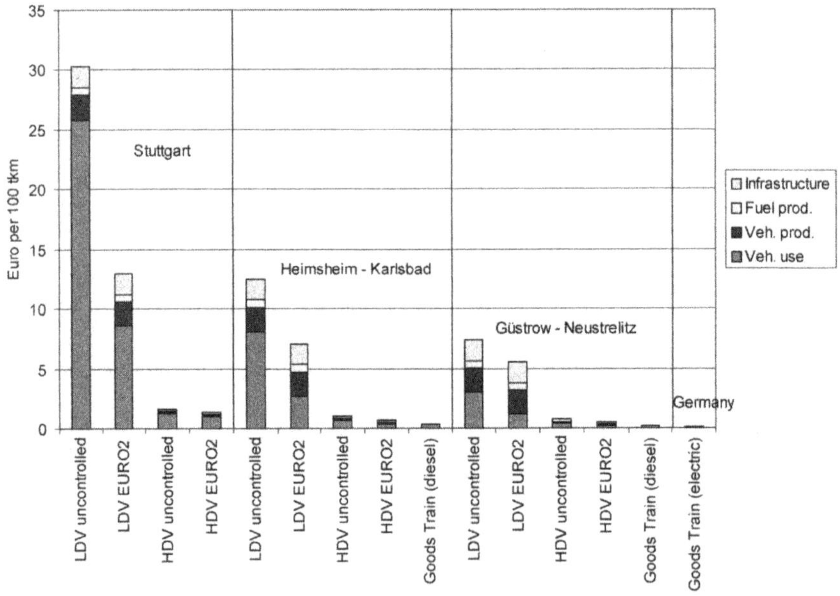


**Fig. 13.27.** Comparison of air pollution costs due to goods transport in Germany (tkm = tonne kilometre)

**Table 13.30.** Damage costs in Euro per 100 pkm due to air pollution including up- and downstream processes for selected passenger transport technologies

| Country | Vehicle type | Fuel | Technology | Average load | Location | Vehicle use | Vehicle production | Fuel production | Infra-structure | Total |
|---|---|---|---|---|---|---|---|---|---|---|
| BE | car | diesel | uncontrolled | 1.3 | Highway | 3.57 | 0.37 | 0.09 | 0.41 | 4.45 |
| BE | car | diesel | EURO II | 1.3 | Highway | 1.28 | 0.37 | 0.09 | 0.41 | 2.15 |
| BE | car | petrol | ECE 15/04 | 1.3 | Highway | 1.26 | 0.37 | 0.20 | 0.41 | 2.24 |
| BE | car | petrol | EURO II | 1.3 | Highway | 0.30 | 0.37 | 0.20 | 0.41 | 1.28 |
| BE | car | LPG | uncontrolled | 1.3 | Highway | 0.56 | 0.37 | 0.09 | 0.41 | 1.44 |
| BE | car | LPG | EURO I | 1.3 | Highway | 0.10 | 0.37 | 0.09 | 0.41 | 0.98 |
| BE | coach | diesel | uncontrolled | 50.0 | Highway | 0.21 | 0.04 | 0.02 | 0.02 | 0.28 |
| BE | coach | diesel | EURO II | 50.0 | Highway | 0.08 | 0.04 | 0.02 | 0.02 | 0.15 |
| BE | car | diesel | uncontrolled | 1.3 | Rural | 1.43 | 0.37 | 0.09 | 0.41 | 2.31 |
| BE | car | diesel | EURO II | 1.3 | Rural | 0.34 | 0.37 | 0.09 | 0.41 | 1.22 |
| BE | car | petrol | ECE 15/04 | 1.3 | Rural | 0.78 | 0.37 | 0.20 | 0.41 | 1.76 |
| BE | car | petrol | EURO II | 1.3 | Rural | 0.17 | 0.37 | 0.20 | 0.41 | 1.15 |
| BE | car | LPG | uncontrolled | 1.3 | Rural | 0.46 | 0.37 | 0.09 | 0.41 | 1.33 |
| BE | car | LPG | EURO I | 1.3 | Rural | 0.09 | 0.37 | 0.09 | 0.41 | 0.96 |
| BE | coach | diesel | uncontrolled | 50.0 | Rural | 0.16 | 0.04 | 0.02 | 0.02 | 0.23 |
| BE | coach | diesel | EURO II | 50.0 | Rural | 0.07 | 0.04 | 0.02 | 0.02 | 0.14 |
| BE | urban bus | diesel | uncontrolled | 15.0 | Rural | 0.39 | 0.13 | 0.06 | 0.06 | 0.64 |
| BE | urban bus | diesel | EURO II | 15.0 | Rural | 0.18 | 0.13 | 0.06 | 0.06 | 0.42 |
| BE | car | diesel | uncontrolled | 1.3 | Brussels | 9.62 | 0.37 | 0.09 | 0.41 | 10.49 |
| BE | car | diesel | EURO II | 1.3 | Brussels | 2.48 | 0.37 | 0.09 | 0.41 | 3.36 |
| BE | car | petrol | ECE 15/04 | 1.3 | Brussels | 2.18 | 0.37 | 0.20 | 0.41 | 3.15 |
| BE | car | petrol | EURO II | 1.3 | Brussels | 0.65 | 0.37 | 0.20 | 0.41 | 1.62 |
| BE | car | LPG | uncontrolled | 1.3 | Brussels | 0.63 | 0.37 | 0.09 | 0.41 | 1.50 |
| BE | car | LPG | EURO I | 1.3 | Brussels | 0.17 | 0.37 | 0.09 | 0.41 | 1.04 |
| BE | coach | diesel | uncontrolled | 50.0 | Brussels | 0.93 | 0.04 | 0.02 | 0.02 | 1.00 |
| BE | coach | diesel | EURO II | 50.0 | Brussels | 0.32 | 0.04 | 0.02 | 0.02 | 0.39 |
| BE | urban bus | diesel | uncontrolled | 15.0 | Brussels | 3.38 | 0.13 | 0.06 | 0.06 | 3.63 |
| BE | urban bus | diesel | EURO II | 15.0 | Brussels | 1.19 | 0.13 | 0.06 | 0.06 | 1.43 |
| FI | car | petrol | ECE 15/04 | 1.3 | Helsinki | 0.68 | 0.26 | 0.14 | 0.23 | 1.32 |
| FI | car | petrol | EURO II | 1.3 | Helsinki | 0.24 | 0.26 | 0.14 | 0.23 | 0.87 |
| FI | car | diesel | uncontrolled | 1.3 | Helsinki | 3.11 | 0.26 | 0.07 | 0.23 | 3.67 |
| FI | car | diesel | EURO II | 1.3 | Helsinki | 0.77 | 0.26 | 0.07 | 0.23 | 1.33 |
| FI | urban bus | diesel | uncontrolled | 14.0 | Helsinki | 0.36 | 0.09 | 0.05 | 0.04 | 0.54 |
| FI | urban bus | diesel | EURO II | 14.0 | Helsinki | 0.16 | 0.09 | 0.05 | 0.04 | 0.34 |
| FI | Tram | electric | Tram | 22.0 | Helsinki | - | 0.009 | 0.002 | 0.045 | 0.06 |
| FI | car | diesel | uncontrolled | 1.4 | T – K | 0.17 | 0.24 | 0.06 | 0.21 | 0.69 |
| FI | urban bus | diesel | uncontrolled | 14.0 | T – K | 0.11 | 0.09 | 0.05 | 0.04 | 0.29 |
| FI | urban bus | diesel | EURO II | 14.0 | T – K | 0.07 | 0.09 | 0.05 | 0.04 | 0.24 |
| DE | car | petrol | ECE 15/04 | 1.5 | Stuttgart | 1.56 | 0.28 | 0.16 | 0.35 | 2.34 |
| DE | car | petrol | EURO II | 1.5 | Stuttgart | 0.30 | 0.28 | 0.16 | 0.35 | 1.09 |

continued ...

**Table 13.30.** continued

| Country | Vehicle type | Fuel | Technology | Average load | Location | Vehicle use | Vehicle production | Fuel production | Infra-structure | Total |
|---|---|---|---|---|---|---|---|---|---|---|
| DE | car | diesel | uncontrolled | 1.5 | Stuttgart | 3.49 | 0.28 | 0.07 | 0.35 | 4.19 |
| DE | car | diesel | EURO II | 1.5 | Stuttgart | 0.91 | 0.28 | 0.07 | 0.35 | 1.61 |
| DE | coach | diesel | uncontrolled | 40.0 | Stuttgart | 0.51 | 0.10 | 0.05 | 0.05 | 0.72 |
| DE | coach | diesel | EURO II | 40.0 | Stuttgart | 0.22 | 0.10 | 0.05 | 0.05 | 0.44 |
| DE | urban bus | diesel | uncontrolled | 15.0 | Stuttgart | 1.25 | 0.10 | 0.05 | 0.05 | 1.46 |
| DE | urban bus | diesel | EURO II | 15.0 | Stuttgart | 0.60 | 0.10 | 0.05 | 0.05 | 0.81 |
| DE | train | electric | Light Rail | 34.2 | Stuttgart | - | 0.017 | 0.021 | 0.005 | 0.04 |
| DE | tram | electric | Tram | 18.9 | Stuttgart | - | 0.011 | 0.015 | 0.003 | 0.03 |
| DE | car | petrol | ECE 15/04 | 1.5 | H – K | 1.65 | 0.28 | 0.16 | 0.35 | 2.43 |
| DE | car | petrol | EURO II | 1.5 | H – K | 0.22 | 0.28 | 0.16 | 0.35 | 1.00 |
| DE | car | diesel | uncontrolled | 1.5 | H – K | 1.17 | 0.28 | 0.07 | 0.35 | 1.88 |
| DE | car | diesel | EURO II | 1.5 | H – K | 0.56 | 0.28 | 0.07 | 0.35 | 1.26 |
| DE | coach | diesel | EURO II | 40.0 | H – K | 0.09 | 0.10 | 0.05 | 0.05 | 0.30 |
| DE | coach | diesel | uncontrolled | 40.0 | H – K | 0.17 | 0.10 | 0.05 | 0.05 | 0.38 |
| DE | train | diesel | Intercity | 150.0 | I – K | 0.87 | 0.01 | 0.00 | 0.01 | 0.88 |
| DE | train | diesel | Local train | 45.9 | I – K | 2.60 | 0.05 | 0.00 | 0.02 | 2.67 |
| DE | car | petrol | ECE 15/04 | 1.5 | G – N | 0.73 | 0.28 | 0.16 | 0.35 | 1.52 |
| DE | car | petrol | EURO II | 1.5 | G – N | 0.09 | 0.28 | 0.16 | 0.35 | 0.88 |
| DE | car | diesel | uncontrolled | 1.5 | G – N | 0.37 | 0.28 | 0.07 | 0.35 | 1.07 |
| DE | car | diesel | EURO II | 1.5 | G – N | 0.17 | 0.28 | 0.07 | 0.35 | 0.87 |
| DE | coach | diesel | uncontrolled | 40.0 | G – N | 0.09 | 0.10 | 0.05 | 0.05 | 0.30 |
| DE | coach | Diesel | EURO II | 40.0 | G – N | 0.05 | 0.10 | 0.05 | 0.05 | 0.26 |
| DE | Train | diesel | Intercity | 150.0 | G – N | 0.39 | 0.01 | 0.00 | 0.01 | 0.41 |
| DE | Train | diesel | Local train | 45.9 | G – N | 1.18 | 0.05 | 0.00 | 0.02 | 1.25 |
| DE | train | electric | ICE | 313.2 | Germany | - | 0.008 | 0.054 | 0.003 | 0.066 |
| DE | train | electric | Intercity | 248.4 | Germany | - | 0.011 | 0.052 | 0.004 | 0.066 |
| DE | train | electric | Local train | 130.7 | Germany | - | 0.011 | 0.072 | 0.007 | 0.090 |
| DE | train | electric | S-Bahn | 157.6 | Germany | - | 0.006 | 0.055 | 0.004 | 0.065 |
| GR | car | petrol | ECE 15/04 | 1.4 | Athens centre | 3.29 | 0.14 | 0.11 | 0.21 | 3.75 |
| GR | car | diesel | EURO II | 1.4 | Athens centre | 4.55 | 0.14 | 0.06 | 0.21 | 4.96 |
| GR | car | petrol | EURO II | 1.4 | Athens centre | 1.10 | 0.14 | 0.11 | 0.21 | 1.55 |
| GR | car | diesel | uncontrolled | 1.4 | Athens centre | 18.35 | 0.14 | 0.06 | 0.21 | 18.76 |
| GR | car | LPG | uncontrolled | 1.4 | Athens centre | 0.71 | 0.14 | 0.07 | 0.21 | 1.12 |
| GR | car | LPG | EURO I | 1.4 | Athens centre | 0.17 | 0.14 | 0.07 | 0.21 | 0.58 |
| GR | car | petrol | ECE 15/04 | 1.4 | Athens suburbs | 1.76 | 0.14 | 0.11 | 0.21 | 2.22 |
| GR | car | diesel | EURO II | 1.4 | Athens suburbs | 1.99 | 0.14 | 0.06 | 0.21 | 2.41 |
| GR | car | petrol | EURO II | 1.4 | Athens suburbs | 0.50 | 0.14 | 0.11 | 0.21 | 0.96 |
| GR | car | diesel | uncontrolled | 1.4 | Athens suburbs | 7.62 | 0.14 | 0.06 | 0.21 | 8.03 |
| GR | car | LPG | uncontrolled | 1.4 | Athens suburbs | 0.71 | 0.14 | 0.07 | 0.21 | 1.12 |
| GR | car | LPG | EURO I | 1.4 | Athens suburbs | 0.17 | 0.14 | 0.07 | 0.21 | 0.58 |

continued ...

**Table 13.30.** continued

| Country | Vehicle type | Fuel | Technology | Average load | Location | Vehicle use | Vehicle production | Fuel production | Infra-structure | Total |
|---|---|---|---|---|---|---|---|---|---|---|
| GR | car | petrol | ECE 15/04 | 1.4 | Larissa centre | 1.25 | 0.14 | 0.11 | 0.21 | 1.70 |
| GR | car | diesel | EURO II | 1.4 | Larissa centre | 0.61 | 0.14 | 0.06 | 0.21 | 1.02 |
| GR | car | petrol | EURO II | 1.4 | Larissa centre | 0.19 | 0.14 | 0.11 | 0.21 | 0.65 |
| GR | car | diesel | uncontrolled | 1.4 | Larissa centre | 1.35 | 0.14 | 0.06 | 0.21 | 1.76 |
| GR | car | LPG | uncontrolled | 1.4 | Larissa centre | 1.11 | 0.14 | 0.07 | 0.21 | 1.53 |
| GR | car | LPG | EURO I | 1.4 | Larissa centre | 0.25 | 0.14 | 0.07 | 0.21 | 0.67 |
| GR | car | petrol | ECE 15/04 | 1.4 | Rural2 Motorw. | 2.00 | 0.14 | 0.11 | 0.21 | 2.46 |
| GR | car | diesel | EURO II | 1.4 | Rural2 Motorw. | 0.33 | 0.14 | 0.06 | 0.21 | 0.74 |
| GR | car | petrol | EURO II | 1.4 | Rural2 Motorw. | 0.18 | 0.14 | 0.11 | 0.21 | 0.64 |
| GR | car | diesel | uncontrolled | 1.4 | Rural2 Motorw. | 0.54 | 0.14 | 0.06 | 0.21 | 0.95 |
| GR | car | LPG | uncontrolled | 1.4 | Rural2 Motorw. | 1.59 | 0.14 | 0.07 | 0.21 | 2.01 |
| GR | car | LPG | EURO I | 1.4 | Rural2 Motorw. | 0.19 | 0.14 | 0.07 | 0.21 | 0.61 |
| NL | car | petrol | ECE 15/04 | 1.4 | Amsterdam | 0.69 | 0.26 | 0.14 | 0.29 | 1.38 |
| NL | car | petrol | EURO II | 1.4 | Amsterdam | 0.08 | 0.26 | 0.14 | 0.29 | 0.77 |
| NL | car | diesel | uncontrolled | 1.4 | Amsterdam | 0.79 | 0.26 | 0.07 | 0.29 | 1.41 |
| NL | car | diesel | EURO II | 1.4 | Amsterdam | 0.20 | 0.26 | 0.07 | 0.29 | 0.82 |
| NL | car | LPG | uncontrolled | 1.4 | Amsterdam | 0.50 | 0.26 | 0.07 | 0.29 | 1.12 |
| NL | car | LPG | EURO I | 1.4 | Amsterdam | 0.06 | 0.26 | 0.07 | 0.29 | 0.67 |
| NL | car | Ethanol | EURO I | 1.4 | Amsterdam | 0.07 | 0.26 | n.a. | 0.29 | 0.62 |
| NL | car | Methanol | EURO I | 1.4 | Amsterdam | 0.07 | 0.26 | n.a. | 0.29 | 0.62 |
| NL | car | CNG | EURO I | 1.4 | Amsterdam | 0.02 | 0.26 | 0.03 | 0.29 | 0.61 |
| NL | car | Hybrid | hybrid | 1.4 | Amsterdam | 0.01 | 0.26 | n.a. | 0.29 | 0.56 |
| NL | coach | diesel | uncontrolled | 13.7 | Amsterdam | 0.39 | 0.12 | 0.06 | 0.06 | 0.63 |
| NL | coach | diesel | EURO II | 13.7 | Amsterdam | 0.17 | 0.12 | 0.06 | 0.06 | 0.41 |
| UK | car | petrol | ECE 15/04 | 1.6 | London | 1.91 | 0.17 | 0.09 | 0.19 | 2.37 |
| UK | car | petrol | EURO II | 1.6 | London | 0.69 | 0.17 | 0.09 | 0.19 | 1.15 |
| UK | car | diesel | uncontrolled | 1.6 | London | 11.58 | 0.17 | 0.04 | 0.19 | 11.99 |
| UK | car | diesel | EURO II | 1.6 | London | 2.75 | 0.17 | 0.04 | 0.19 | 3.16 |
| UK | urban bus | diesel | uncontrolled | 12.0 | London | 4.19 | 0.09 | 0.04 | 0.04 | 4.36 |
| UK | urban bus | diesel | EURO II | 12.0 | London | 1.38 | 0.09 | 0.04 | 0.04 | 1.54 |
| UK | coach | diesel | uncontrolled | 10.0 | London | 5.64 | 0.11 | 0.05 | 0.05 | 5.85 |
| UK | coach | diesel | EURO II | 10.0 | London | 1.84 | 0.11 | 0.05 | 0.05 | 2.04 |
| UK | car | CNG | bi-fuel TWC | 1.6 | London | 0.06 | 0.18 | 0.02 | 0.19 | 0.45 |
| UK | car | LPG | bi-fuel TWC | 1.6 | London | 0.16 | 0.17 | 0.04 | 0.19 | 0.56 |
| UK | car | petrol | Reformer FC | 1.6 | London | 0.02 | 0.17 | 0.06 | 0.19 | 0.45 |
| UK | urban bus | CNG | lean burn | 12.0 | London | 1.48 | 0.09 | 0.02 | 0.04 | 1.63 |
| UK | urban bus | CNG | TWC | 12.0 | London | 1.41 | 0.09 | 0.02 | 0.04 | 1.56 |
| UK | urban bus | LPG | TWC | 12.0 | London | 2.19 | 0.09 | 0.03 | 0.04 | 2.35 |

continued ...

**Table 13.30.** continued

| Country | Vehicle type | Fuel | Technology | Average load | Location | Vehicle use | Vehicle production | Fuel production | Infra-structure | Total |
|---|---|---|---|---|---|---|---|---|---|---|
| UK | train | diesel | Pacer | 30.0 | L – P | 0.17 | 0.01 | 0.02 | 0.02 | 0.21 |
| UK | train | diesel | Sprinter | 40.0 | L – P | 0.16 | 0.004 | 0.02 | 0.01 | 0.20 |
| UK | train | diesel | Turbo | 47.0 | L – P | 0.15 | 0.004 | 0.01 | 0.01 | 0.17 |
| UK | train | diesel | Inter-City 125 | 336.0 | L – P | 0.93 | 0.005 | 0.03 | 0.002 | 0.97 |
| UK | train | diesel | Class 47 | 319.2 | L – P | 0.47 | 0.004 | 0.01 | 0.002 | 0.49 |
| UK | car | petrol | ECE 15/04 | 1.6 | Swindon | 0.87 | 0.17 | 0.09 | 0.19 | 1.33 |
| UK | car | petrol | EURO II | 1.6 | Swindon | 0.23 | 0.17 | 0.09 | 0.19 | 0.69 |
| UK | car | diesel | uncontrolled | 1.6 | Swindon | 2.43 | 0.17 | 0.04 | 0.19 | 2.84 |
| UK | car | diesel | EURO II | 1.6 | Swindon | 0.52 | 0.17 | 0.04 | 0.19 | 0.93 |
| UK | urban bus | diesel | uncontrolled | 12.0 | Swindon | 1.09 | 0.09 | 0.04 | 0.04 | 1.26 |
| UK | urban bus | diesel | EURO II | 12.0 | Swindon | 0.43 | 0.09 | 0.04 | 0.04 | 0.60 |
| UK | coach | diesel | uncontrolled | 10.0 | Swindon | 1.38 | 0.11 | 0.05 | 0.05 | 1.58 |
| UK | coach | diesel | EURO II | 10.0 | Swindon | 0.51 | 0.11 | 0.05 | 0.05 | 0.71 |
| UK | car | petrol | ECE 15/04 | 1.6 | Rural | 0.58 | 0.17 | 0.09 | 0.19 | 1.04 |
| UK | car | petrol | EURO II | 1.6 | Rural | 0.09 | 0.17 | 0.09 | 0.19 | 0.55 |
| UK | car | diesel | uncontrolled | 1.6 | Rural | 0.63 | 0.17 | 0.04 | 0.19 | 1.04 |
| UK | car | diesel | EURO II | 1.6 | Rural | 0.18 | 0.17 | 0.04 | 0.19 | 0.59 |
| UK | urban bus | diesel | uncontrolled | 12.0 | Rural | 0.49 | 0.09 | 0.04 | 0.04 | 0.66 |
| UK | urban bus | diesel | EURO II | 12.0 | Rural | 0.23 | 0.09 | 0.04 | 0.04 | 0.40 |
| UK | coach | diesel | uncontrolled | 10.0 | Rural | 0.47 | 0.11 | 0.05 | 0.05 | 0.67 |
| UK | coach | diesel | EURO II | 10.0 | Rural | 0.21 | 0.11 | 0.05 | 0.05 | 0.42 |

T – K = Tammisaari – Karjaa; H – K = Heimsheim – Karlsbad; G – N = Güstrow – Neustrelitz
I – K = Illingen – Kämpfelbach; L – P = London – Peterborough; n.a. = not available

**Table 13.31.** Damage costs in Euro per 100 tkm due to air pollution including up- and downstream processes for selected goods transport technologies

| Country | Vehicle type | Fuel | Technology | Average load | Location | Vehicle use | Vehicle production | Fuel production | Infra-structure | Total |
|---|---|---|---|---|---|---|---|---|---|---|
| BE | LDV | Diesel | uncontrolled | 2.0 | Highway | 4.35 | 0.36 | 0.11 | 0.27 | 5.09 |
| BE | LDV | Diesel | EURO II | 2.0 | Highway | 1.30 | 0.36 | 0.11 | 0.27 | 2.04 |
| BE | HDV | Diesel | uncontrolled | 16.0 | Highway | 0.79 | 0.09 | 0.03 | 0.12 | 1.03 |
| BE | HDV | Diesel | EURO II | 16.0 | Highway | 0.30 | 0.09 | 0.03 | 0.12 | 0.55 |
| BE | LDV | Diesel | uncontrolled | 2.0 | Rural | 1.69 | 0.36 | 0.11 | 0.27 | 2.43 |
| BE | LDV | Diesel | EURO II | 2.0 | Rural | 0.50 | 0.36 | 0.11 | 0.27 | 1.24 |
| BE | HDV | diesel | uncontrolled | 16.0 | Rural | 0.61 | 0.09 | 0.03 | 0.12 | 0.86 |
| BE | HDV | diesel | EURO II | 16.0 | Rural | 0.25 | 0.09 | 0.03 | 0.12 | 0.50 |
| BE | LDV | diesel | uncontrolled | 2.0 | Brussels | 6.51 | 0.36 | 0.11 | 0.27 | 7.24 |
| BE | LDV | diesel | EURO II | 2.0 | Brussels | 3.13 | 0.36 | 0.11 | 0.27 | 3.87 |
| BE | HDV | diesel | uncontrolled | 16.0 | Brussels | 2.56 | 0.09 | 0.03 | 0.12 | 2.81 |
| BE | HDV | diesel | EURO II | 16.0 | Brussels | 0.89 | 0.09 | 0.03 | 0.12 | 1.14 |
| FI | HDV | diesel | uncontrolled | 11.5 | Helsinki | 1.37 | 0.10 | 0.03 | 0.10 | 1.61 |
| FI | HDV | diesel | EURO II | 11.5 | Helsinki | 0.50 | 0.10 | 0.03 | 0.10 | 0.74 |
| DE | LDV | diesel | uncontrolled | 0.3 | Stuttgart | 26.05 | 2.04 | 0.65 | 1.70 | 30.44 |
| DE | LDV | diesel | EURO II | 0.3 | Stuttgart | 8.78 | 2.04 | 0.65 | 1.70 | 13.17 |
| DE | HDV | diesel | uncontrolled | 10.8 | Stuttgart | 1.31 | 0.11 | 0.05 | 0.18 | 1.64 |
| DE | HDV | diesel | EURO II | 10.8 | Stuttgart | 1.04 | 0.11 | 0.05 | 0.18 | 1.37 |
| DE | LDV | diesel | uncontrolled | 0.3 | H – K | 8.30 | 2.04 | 0.65 | 1.70 | 12.70 |
| DE | LDV | diesel | EURO II | 0.3 | H – K | 2.86 | 2.04 | 0.65 | 1.70 | 7.25 |
| DE | HDV | diesel | uncontrolled | 10.8 | H – K | 0.73 | 0.11 | 0.05 | 0.18 | 1.06 |
| DE | HDV | diesel | EURO II | 10.8 | H – K | 0.39 | 0.11 | 0.05 | 0.18 | 0.72 |
| DE | train | diesel | goods train | 557.6 | I – K | 0.32 | 0.01 | 0.0002 | 0.009 | 0.34 |
| DE | LDV | diesel | uncontrolled | 0.3 | G – N | 3.18 | 2.04 | 0.65 | 1.70 | 7.57 |
| DE | LDV | diesel | EURO II | 0.3 | G – N | 1.34 | 2.04 | 0.65 | 1.70 | 5.73 |
| DE | HDV | diesel | uncontrolled | 10.8 | G – N | 0.40 | 0.11 | 0.05 | 0.18 | 0.73 |
| DE | HDV | diesel | EURO II | 10.8 | G – N | 0.22 | 0.11 | 0.05 | 0.18 | 0.56 |
| DE | train | diesel | goods train | 557.6 | G – N | 0.15 | 0.01 | 0.0002 | 0.009 | 0.16 |
| DE | train | electric | goods train | 542.8 | Germany | - | 0.01 | 0.03 | 0.009 | 0.05 |
| GR | LDV | diesel | uncontrolled | 2.0 | Athens centre | 13.81 | 0.15 | 0.08 | 0.14 | 14.18 |
| GR | LDV | petrol | uncontrolled | 2.0 | Athens centre | 2.54 | 0.15 | 0.14 | 0.14 | 2.97 |
| GR | LDV | diesel | EURO II | 2.0 | Athens centre | 5.44 | 0.15 | 0.08 | 0.14 | 5.81 |
| GR | LDV | petrol | EURO I | 2.0 | Athens centre | 0.86 | 0.15 | 0.14 | 0.14 | 1.28 |
| GR | HDV | diesel | uncontrolled | 36.0 | Athens centre | 2.79 | 0.02 | 0.01 | 0.03 | 2.84 |
| GR | HDV | diesel | EURO II | 36.0 | Athens centre | 0.94 | 0.02 | 0.01 | 0.03 | 0.99 |

continued ...

**Table 13.31.** continued

| Country | Vehicle type | Fuel | Technology | Average load | Location | Vehicle use | Vehicle production | Fuel production | Infra-structure | Total |
|---|---|---|---|---|---|---|---|---|---|---|
| GR | LDV | diesel | uncontrolled | 2.0 | Athens suburbs | 6.00 | 0.15 | 0.08 | 0.14 | 6.38 |
| GR | LDV | petrol | uncontrolled | 2.0 | Athens suburbs | 1.43 | 0.15 | 0.14 | 0.14 | 1.86 |
| GR | LDV | diesel | EURO II | 2.0 | Athens suburbs | 2.37 | 0.15 | 0.08 | 0.14 | 2.75 |
| GR | HDV | diesel | uncontrolled | 36.0 | Athens suburbs | 1.28 | 0.02 | 0.01 | 0.03 | 1.33 |
| GR | LDV | petrol | EURO I | 2.0 | Athens suburbs | 0.44 | 0.15 | 0.14 | 0.14 | 0.87 |
| GR | HDV | diesel | EURO II | 36.0 | Athens suburbs | 0.47 | 0.02 | 0.01 | 0.03 | 0.53 |
| GR | LDV | diesel | uncontrolled | 2.0 | Rural2 Motorway | 0.91 | 0.15 | 0.08 | 0.14 | 1.28 |
| GR | LDV | petrol | uncontrolled | 2.0 | Rural2 Motorway | 1.49 | 0.15 | 0.14 | 0.14 | 1.91 |
| GR | LDV | diesel | EURO II | 2.0 | Rural2 Motorway | 0.45 | 0.15 | 0.08 | 0.14 | 0.83 |
| GR | LDV | petrol | EURO I | 2.0 | Rural2 Motorway | 0.24 | 0.15 | 0.14 | 0.14 | 0.67 |
| GR | HDV | diesel | uncontrolled | 36.0 | Rural2 Motorway | 0.23 | 0.02 | 0.01 | 0.03 | 0.29 |
| GR | HDV | diesel | EURO II | 36.0 | Rural2 Motorway | 0.14 | 0.02 | 0.01 | 0.03 | 0.19 |
| NL | LDV | diesel | uncontrolled | 0.3 | Amsterdam | 9.18 | 2.02 | 0.65 | 1.50 | 13.36 |
| NL | LDV | diesel | EURO I | 0.3 | Amsterdam | 3.21 | 2.02 | 0.65 | 1.50 | 7.38 |
| NL | LDV | diesel | EURO II | 0.3 | Amsterdam | 2.28 | 2.02 | 0.65 | 1.50 | 6.46 |
| NL | LDV | Hybrid | | 0.3 | Amsterdam | 0.03 | 2.02 | n.a. | 1.50 | 3.55 |
| UK | LDV | diesel | uncontrolled | 0.9 | London | 23.35 | 0.47 | 0.14 | 0.56 | 24.52 |
| UK | LDV | diesel | EURO II | 0.9 | London | 8.67 | 0.47 | 0.14 | 0.56 | 9.83 |
| UK | HDV | diesel | uncontrolled | 11.0 | London | 6.13 | 0.07 | 0.03 | 0.10 | 6.33 |
| UK | HDV | diesel | EURO II | 11.0 | London | 2.00 | 0.07 | 0.03 | 0.10 | 2.20 |
| UK | LDV | CNG | bi-fuel TWC | 0.9 | London | 0.21 | 0.48 | 0.07 | 0.56 | 1.31 |
| UK | LDV | LPG | bi-fuel TWC | 0.9 | London | 0.44 | 0.47 | 0.13 | 0.56 | 1.59 |
| UK | LDV | petrol | Reformer FC | 0.9 | London | 0.04 | 0.48 | 0.33 | 0.56 | 1.41 |
| UK | LDV | CNG | Reformer FC | 0.9 | London | 0.00 | 0.48 | 0.10 | 0.56 | 1.14 |
| UK | train | diesel | Class 47 | 355.0 | L – P | 0.31 | 0.004 | 0.02 | 0.001 | 0.34 |
| UK | train | diesel | Class 56 | 355.0 | L – P | 0.35 | 0.004 | 0.03 | 0.001 | 0.39 |
| UK | train | diesel | Class 58 | 355.0 | L – P | 0.34 | 0.004 | 0.03 | 0.001 | 0.37 |
| UK | train | diesel | Class 60 | 355.0 | L – P | 0.32 | 0.004 | 0.03 | 0.001 | 0.36 |
| UK | LDV | diesel | uncontrolled | 0.9 | Swindon | 7.76 | 0.47 | 0.14 | 0.56 | 8.92 |
| UK | LDV | diesel | EURO II | 0.9 | Swindon | 1.98 | 0.47 | 0.14 | 0.56 | 3.14 |
| UK | HDV | diesel | uncontrolled | 11.0 | Swindon | 1.53 | 0.07 | 0.03 | 0.10 | 1.73 |
| UK | HDV | diesel | EURO II | 11.0 | Swindon | 0.57 | 0.07 | 0.03 | 0.10 | 0.77 |
| UK | LDV | diesel | uncontrolled | 0.9 | Rural | 2.60 | 0.47 | 0.14 | 0.56 | 3.76 |
| UK | LDV | diesel | EURO II | 0.9 | Rural | 0.69 | 0.47 | 0.14 | 0.56 | 1.86 |
| UK | HDV | diesel | uncontrolled | 11.0 | Rural | 0.52 | 0.07 | 0.03 | 0.10 | 0.72 |
| UK | HDV | diesel | EURO II | 11.0 | Rural | 0.24 | 0.07 | 0.03 | 0.10 | 0.44 |

HDV = heavy duty vehicles with maximum total weight of 32 to 40 tonnes; LDV = light duty vehicle; H – K = Heimsheim – Karlsbad; G – N = Güstrow – Neustrelitz; I – K = Illingen – Kämpfelbach; L – P = London – Peterborough; n.a. = not available

# 14 Aggregated Costs

This chapter presents the results of the aggregation studies within the project. The aggregation task was only undertaken by a small number of teams, with the aim of developing and demonstrating the approach.

The results for the national level aggregation are presented first, undertaken for the UK, Belgium and the Netherlands. The results of an aggregation at a higher level of resolution, for a specific region of Germany, is also presented, showing the greater degree of sophistication as the corresponding data is available.

## 14.1 The External Costs of Road Transport in Belgium

This section outlines a summary of the method and results for the external costs of road transport in Belgium. It describes the data available on the composition of the Belgian car fleet, the data on general driving behaviour characteristics (vehicle usage, mileage distribution) and the additional assumptions used. The results identify the most important components of the total external cost in order to highlight their potential for policy options.

### 14.1.1 Vehicle Statistics

#### *14.1.1.1 Passenger Cars*

The Belgian passenger car fleet was split according to emission standard legislation, cylinder capacity and fuel. This data was taken from the Belgian National Institute of Statistics (NIS) which compiles the data from registration certificates. Vehicles were categorised according to their compliance with the different European emission standard from year of first registration.

In 1998 there were a total number of 4.47 million passenger cars registered in Belgium - a 5 % increase from 1996. Of these, a very high proportion are diesel cars and the share is increasing (34% in 1996, 37% in 1997-8). LPG cars make up less than 1%.

No extensive mileage statistics are available for Belgium and the Flemish Region. General information is available for goods transport at the Belgian Institute for National Statistics (NIS), but for the other vehicle categories estimations have been made, mostly based on methodologies and data from the years 1959 to 1987 (De Vlieger, 1992; De Vlieger & Debruyn, 1992). Where no new information was available, Vito's 1994 assessments were used (De Vlieger, 1996), based on the calibration of average annual mileage and fuel consumption from the energy balance for road transport (De Vlieger, 1995).

Average annual mileage was taken from the mobility enquiry of the Flemish Government (1994), which have a detailed vehicle category split (also used within the Flemish MIRA report, 1996). The study also gives a yearly growth in mileage per passenger car of 0.5% (Hajnal & Miermans, 1996). This percentage was used to extrapolate the mileage figures of 1994 to the 1993-1998 situation; the results for 1995 are shown in Table 14.1.

**Table 14.1.** Average mileage (km/year) for passenger cars according to cylinder capacity and fuel type in 1995 (Hajnal & Miermans, 1996).

| Fuel type | Cylinder capacity [cc] | | | Total |
|---|---|---|---|---|
| | < 1.400 | 1.400 – 2000 | > 2.000 | |
| Unleaded gasoline | 11 862 | 14 438 | 19 477 | 14 339 |
| Diesel | 22 200 | 22 119 | 22 837 | 22 416 |
| LPG | 20 000 | 19 944 | 31 642 | 24 262 |

### 14.1.1.2 Light & Heavy Duty Freight Vehicles

The NIS statistics show the average mileage driven in Belgium by heavy-duty freight vehicles was 38078 kilometres in 1996. Based on statistical figures over the period 1992-1997, an average yearly change in activity of 2.3% was found. The proposed mileage per weight class is shown in Table 14.2, based on the NIS statistics. As there is no uniformity in the statistics of heavy-duty vehicles (split in vehicle categories, in weight categories, coupling with mileage), only approximate estimations can be made.

**Table 14.2.** Average mileage (km/y) driven in Belgium by heavy duty freight vehicles in 1990, 1995 and 1998.

| Vehicle type | Mileage [km/y] | | |
|---|---|---|---|
| | 1990 | 1995 | 1998 |
| <3.5 ton | | 30 000 | |
| 3.5 – 7.5 ton | 27 790 | 31 140 | 33 340 |
| 7.5 – 16 ton | 31 500 | 35 290 | 37 780 |
| 16 – 32 ton | 36 310 | 40 670 | 43 540 |
| > 32 ton | 34 770 | 38 960 | 41 710 |

### 14.1.1.3
### Buses

The mileage of buses for public transport is based on information of the Brussels bus operating company. Values of the Flemish company 'De Lijn' show an average mileage of about 43000 km/y for there own buses. The mileage of buses of operators driving for De Lijn is not known (De Lijn, Annual Report 1996). The mileage driven in Belgium by coaches is taken from the Institute for coaches and buses (ICB). An overview of the average mileage per bus category for 1995 is given below in Table 14.3.

**Table 14.3.** Average mileage (km/y) for buses in Belgium 1995

| Vehicle type | Mileage [km/y] |
|---|---|
| Public transport | 46 240 |
| Coaches | 23 750 |

For coaches, mileage driven in Belgium for the years 1990 and 1998 are respectively 24710 km and 22940 km. Based on statistical figures over the period 1988-1998, an average yearly change in activity of -0.7% was found. Figures for the Flemish Region shows that the total amount of vehicle kilometres for public buses stays more or less constant during the period 1990-1996 (Source: Public Bus Company De Lijn, pers. com.).

### 14.1.1.4
### Mileage Distribution by Road Type

Several sources report the distribution of traffic over urban, rural and highway roads. The analysis was based on the work of two research studies, MIRA 96 (Flemish Environmental Report) and the Flemish working group 98 'Rational Energy Use–Traffic-Mobility'. Because of big differences between the mileage distributions of both teams, Vito used a distribution of 30 % urban, 35 % rural, and 35 % highway for passenger cars. Mileage distribution for other vehicles is given below in Table 14.4 based on approximate values from Vito.

**Table 14.4.** Overview of preliminary mileage distribution over different road types

| Vehicle type | Mileage distribution % | | |
|---|---|---|---|
| | urban | rural | highway |
| Passenger car | 30 | 35 | 35 |
| Public bus | 65 | 35 | 0 |
| Coach | 25 | 37.5 | 37.5 |
| Light duty | 35 | 35 | 30 |
| Heavy duty > 32 ton | 10 | 40 | 50 |

### 14.1.1.5
### *Factors Unaccounted For*

A number of important factors that influence the total external costs were not taken into account in this study. Congestion effects could not be included, as the average mileage driven in congested conditions on different road types is unknown. Cold start emissions could also not be included - although data on cold start emissions are available from some models, it very difficult to incorporate them for aggregation as data on the total mileage and spatial distribution were unavailable. There was also no statistical data that allowed the incorporation of driver behaviour. Finally, motorcycles and mini-buses were omitted because no reliable data on the composition of these fleet segments could be obtained.

For this study, no account was made of the impacts of Belgian cars driving on foreign roads and vice versa (though as mileage is calibrated on national fuel sales, and assuming a similar fleet composition and equal mileage, this should be a minor problem). The same is not true of heavy goods vehicles - it is clear that a significant proportion of trucks on Belgian roads are foreign. However, as there is no information on the fleet composition of foreign trucks (in terms of age and weight distribution) and no information on the mileage they drive in Belgium (let alone a road type distribution of the mileage) it was impossible to include them in the analysis.

### 14.1.1.6
### *Mileage as a Function of Vehicle Age*

Several European studies show that the mileage of vehicles depends strongly on the vehicle age for each category (André et al., 1998). In some European countries 'usage factors' are defined to quantify mileage as a function of the vehicle age (usually normalised to average annual mileage). For passenger cars the usage factors according to vehicle age of France were used as basic input (Hivert, 1996), but applied only by EURO category (rather than by year).

**Table 14.5.** Usage factors for passenger cars (1998)

| Vehicle type | Usage factor |
|---|---|
| EURO 0 | 0.83 |
| EURO I | 1.12 |
| EURO II | 1.19 |

## 14.1.2
## Data on External Costs

The aggregated value was calculated by adding all marginal external costs/km for all combinations of vehicle type, drive type and location. These values were weighted according to their annual mileage and the composition of the fleet. Data on costs were taken from the typical trajectories (urban, rural and highway) and represent the mid-estimate from a range of typical trajectories.

In all cases, the marginal external costs have been applied. Non-linearity in dispersion and chemical transformations is not accounted for, except for ozone (see marginal analysis for discussion) where a European average cost has been used.

### 14.1.3 Results and Conclusions

The total mileage driven in Belgium is 84 billion kilometres (1998). The resulting tailpipe emissions cause a total damage of 2.4 billion Euro or 1 % of GDP. This sum includes passenger cars, light duty vehicles, heavy-duty vehicles, buses and coaches. It excludes mini-buses, motorcycles and foreign trucks. Environmental impacts of other phases of the life cycle chain of transport (evaporative emissions, the construction of vehicles, fuel production and infrastructure) as well as non-environmental externalities of congestion, cold starts effects and accidents are also not included in this figure.

For passenger cars, this equates to an average of 2.1 Eurocent/km and for coach and buses an average of 28 Eurocent/vehicle km. Values have not been given by passenger km due to the variation on the seat occupancy rate, though it is known public buses perform better than passenger cars in large cities when seat occupancy rates are at least 25% (De Nocker et al., 1999).

The results show that two thirds of the external costs of road transport can be attributed to passenger cars, though this is no surprise given these vehicles account for more than 90% of all kilometres travelled. Of these, diesel cars are responsible for more than 80 % of the total environmental externalities, though for the base year, they only comprised half the vehicle km (and only 36% of the car fleet). The value for diesel was expected to be higher as they have higher external costs per km, though this is offset by the fact diesel cars are much newer and newer cars drive more km. Looking at the distribution by EURO standard, in 1998, 50% of the fleet still consisted of uncontrolled cars (pre- EURO). Although these cars are now relatively old and have a declining share in the total mileage, they still account for over 60% of total externalities.

**Table 14.6.** Total external costs and mileage for different segments of the fleet (in billion Euro and billion kilometres) for Belgium, 1998.

| | Billion km | % | Billion Euro | % |
|---|---|---|---|---|
| Passenger cars | 77.10 | 91.7% | 1.59 | 66.3% |
| Heavy Duty | 6.36 | 7.6% | 0.64 | 26.6% |
| Bus & Coach | 0.61 | 0.7% | 0.17 | 7.2% |
| Total | 84.1 | 100% | 2.4 | 100% |

Note value for HDV does not include Belgian trucks in other countries or damages from foreign trucks driving on Belgian roads.

**Table 14.7.** Total external costs for the use and non-use phase for different segments of the fleet (in billion Euro) for Belgium, 1998.

| | | Billion Euro | | | |
|---|---|---|---|---|---|
| External costs | Passenger cars | Heavy Duty Trucks | Bus & Coach | Total (all categories) | % |
| Use phase billion Euro | 1.59 | 0.64 | 0.17 | 2.4 | 66% |
| Non-use phase | | | | | |
| fuel production | 0.15 | 0.04 | 0.00 | 0.19 | 5% |
| vehicle production | 0.37 | 0.13 | 0.01 | 0.52 | 14% |
| infrastructure | 0.41 | 0.11 | 0.01 | 0.53 | 15% |
| Total (use + non-use) | 2.52 | 0.92 | 0.19 | 3.6 | 100% |
| % | 69% | 25% | 5% | 100% | |

The bulk of remaining externalities is attributable to heavy-duty vehicles. Indeed, although they only account for 6 % of total mileage, they cause 27% of the annual impacts of road transport. The remaining 7% of external costs result from buses and coaches (accountable for less then 1% of total mileage).

With respect to location, around two thirds of the total external costs of passenger cars can be attributed to urban traffic. For highways, the study shows that city ring roads are important – 20% of the environmental cost can be attributed to the city rings of Antwerp and Brussels though they only comprise 4% of the total length of Belgian highways.

From a policy point of view, a reduction of external costs from road transport should primarily be aimed at passenger cars (particularly certain types) and heavy-duty vehicles for freight. Targeting driving conditions where impacts are highest (urban areas) would yield the most efficient policy measures.

Finally, the study calculated the additional life-cycle costs of road transport. These have been calculated by means of multiplying impacts per vehicle km with the total mileage for the different categories. These add another 1.2 billion Euro to the aggregated external costs, taking the total road sector and upstream activities to 3.6 billion Euro or 1.6 % of GDP.

The total external costs for the road sector are presented in Table 14.7.

## 14.2
## The External Costs of Transport in the Netherlands

This section describes the estimation of the aggregate externalities of the whole transport sector in the Netherlands. The transport performance in the Netherlands in 1995 and the related direct emissions to air are presented in Tables 14.8 and 14.9 below.

**Table 14.8.** Transport Performance in the Netherlands in 1995.

| Type of transport | Transport performance |
|---|---|
| *Passenger transport* | *billion person km* |
| Private car | 146.8 |
| Motor | 1.6 |
| Moped | 1.2 |
| Bicycle | 13.3 |
| Train | 14.0 |
| Public bus | 6.0 |
| Private bus | 5.8 |
| Taxi | 2.4 |
| *Goods transport* | *million tonne km* |
| Road | 36,885 |
| Train | 721 |
| Ship | 6,899 |

Source: Central Bureau for Statistics, Statline database, www.cbs.nl., November 6 1999.

**Table 14.9.** Emissions from mobile sources in the Netherlands, incl. foreign mobile sources in the Netherlands in 1995 (ktonne/y)

| Transport type | | Pollutant | | | | | | | | |
|---|---|---|---|---|---|---|---|---|---|---|
| | | CO | $CO_2$ | $NO_x$ | $N_2O$ | $SO_2$ | VOC | NMVOC | $CH_4$ | $PM_{10}$ |
| Road Transport | Passenger cars | 408.40 | 17,191 | 113.08 | 4.40 | 4.79 | 99.48 | 95.64 | 3.84 | 4.34 |
| | Goods vehicles | 51.10 | 9,099 | 99.43 | 1.71 | 8.47 | 24.8 | 23.81 | 0.98 | 6.93 |
| | Other vehicles | 48.20 | 1,075 | 12.33 | 0.20 | 0.81 | 21.77 | 20.94 | 0.85 | 1.21 |
| | Total | 507.70 | 27,365 | 224.84 | 6.31 | 14.07 | 146.05 | 140.39 | 5.67 | 12.48 |
| Inland shipping | Passenger | 0.40 | 113 | 1.80 | 0.02 | 0.12 | 0.22 | 0.21 | 0.01 | 0.22 |
| | Goods | 1.30 | 1,349 | 25.87 | 0.30 | 2.7 | 1.29 | 1.24 | 0.05 | 1.72 |
| | Recreation | 6.90 | 92 | 1.12 | 0.03 | 0.11 | 1.57 | 1.49 | 0.08 | 0.11 |
| | Fishery | 0.00 | 33 | 0.62 | 0.01 | 0.07 | 0.03 | 0.03 | 0 | 0.04 |
| | Total | 8.60 | 1,587 | 29.41 | 0.36 | 3.0 | 3.11 | 2.97 | 0.14 | 2.09 |
| Rail transport (diesel) | Passenger | 0.20 | 47 | 0.52 | 0.01 | 0.09 | 0.07 | 0.07 | 0 | 0.04 |
| | Goods | 0.20 | 86 | 1.87 | 0.02 | 0.17 | 0.04 | 0.04 | 0 | 0.03 |
| | Total | 0.40 | 133 | 2.39 | 0.03 | 0.27 | 0.11 | 0.11 | 0 | 0.07 |
| Air transport | | 5.20 | 405 | 2.48 | 0.04 | 0.21 | 1.11 | 0.90 | 0.09 | 0.15 |
| Others | Agriculture | 10.50 | 1,100 | 17.56 | 0.24 | 1.2 | 3.51 | 3.37 | 0.14 | 1.76 |
| | Construction | 3.50 | 367 | 5.85 | 0.08 | 0.40 | 1.17 | 1.12 | 0.05 | 0.59 |
| | Total | 18.30 | 1,906 | 30.44 | 0.42 | 2.08 | 6.09 | 5.85 | 0.24 | 3.04 |
| Total | Traffic & Transport | 521.90 | 29,490 | 259.12 | 6.74 | 17.55 | 150.38 | 144.37 | 5.90 | 14.79 |
| | Mobile sources | 540.20 | 31396 | 289.56 | 7.16 | 19.63 | 156.47 | 150.22 | 6.14 | 17.83 |

Source: Central Bureau for Statistics, Statline database, www.cbs.nl, 1999.

The externalities of direct emissions to air from road, rail, ship and air transport have been estimated using different methods. The aggregated externalities of direct emissions to air from mobile sources in the agricultural and construction sectors were estimated by using the road transport approach. The approaches and results are presented in the following sub-sections. The externalities of non-operation stages of road, rail and ship transport are included and have been derived by multiplying the average externalities of these phases by the total number of vehicle kilometres driven (see Table 14.9 and Table 14.10). For more information on the external costs of the non-operational life cycle stages the reader should refer to the methodology and marginal cost sections.

## 14.2.1 Road Transport

A three-step approach was followed. The first step in the road transport aggregation was to estimate the average externalities per tonne pollutant emitted in the Netherlands. This was done by first applying the EcoSense Transport model to straight vertical and horizontal trajectories spanning the full width and height of each province and crossing through the centre of the province. Then, the average

**Table 14.10.** Average and Total External Costs of the Non-Operational Life Cycle Stages of Transport in the Netherlands.

| Type of transport | Average external costs non-operational life cycle stages | | | Total external cost of the non-operational life cycle stages |
|---|---|---|---|---|
| | Manufacture | Fuel production | Infrastructure | |
| Passenger transport | Eurocent/pkm | Eurocent/pkm | Eurocent/pkm | (million Euro) |
| Private car[a] | 0.259 | 0.119 | 0.289 | 980.2 |
| Motor | n.a. | n.a. | n.a. | n.a. |
| Moped | n.a. | n.a. | n.a. | n.a. |
| Bicycle | n.a. | n.a. | n.a. | n.a. |
| Train[b] | 0.007 | 0.024 | 0.033 | 9.0 |
| Public bus | 0.119 | 0.062 | 0.055 | 14.1 |
| Private bus | 0.119 | 0.062 | 0.055 | 13.7 |
| Taxi[c] | 0.259 | 0.069 | 0.289 | 14.8 |
| Goods transpor | Eurocent/tkm | Eurocent/tkm | Eurocent/tkm | Million Euro |
| Road[d] | 0.193 | 0.078 | 0.267 | 198.6 |
| Train[b,e] | 0.004 | 0.006 | 0.013 | 0.2 |
| Ship[f] | 0.031 | 0.018 | 0.067 | 8.0 |
| | | | Total | 1,238.6 |

[a] 13% LPG, 19% diesel and 68% gasoline.

[b] External costs of non-operational life cycle stages of electricity production (0.15 Eurocent/kWh) not included.

[c] Assumed 100% diesel

[d] Assumed 100 % diesel fuelled HDV

[e] Assumed average of diesel and electric

[f] Assumed average of bulk, container and tanker

**Table 14.11.** Average externalities per air pollutant for the different transport types. The standard deviations for road transport are given in brackets[a].

| Pollutant | Road vehicles/Diesel trains | | Ships | Electricity production[b] | |
|---|---|---|---|---|---|
| | Euro/t | s.d. | Euro/t | Euro/t | Eurocent/kWh |
| $NO_x$ (Nitrate) | 4464 | (256) | 4,218 | 4,755 | 0.19 |
| $NO_x$ (Ozone) | -2020 | | -2,020 | -2,020 | -0.080 |
| NMVOC (Ozone) | 1,376 | | 1,376 | 1,376 | 0.0010 |
| PM | 87,671 | (27,976) | 107,801 | 23,686 | 0.014 |
| $SO_2$ (direct) | 2,895 | (676) | 3,383 | 1,582 | 0.016 |
| $SO_2$ (Sulphate) | 3,552 | (577) | 3,878 | 5,641 | 0.052 |
| $CO_2$ | 2.4 | | 2.4 | 2.4 | 0.095 |
| $N_2O$ | 748 | | 748 | n.a. | n.a. |
| Methane | 45 | | 45 | 45 | 2.8E-08 |
| Benzene | 557 | (170) | 678 | n.a. | n.a. |
| Butadiene | 21,867 | (6,402) | 26,442 | n.a. | n.a. |
| CO | 0.7 | (0.2) | 0.9 | n.a. | n.a. |
| DME | 731 | (128) | 801 | n.a. | n.a. |

[a] If no standard deviation is presented this implies that the damage estimate used is not depending on the location of the emission.

[b] The average externalities of the non-operational stages of power production in the Netherlands in 2000 are estimated at 0.15 Eurocent/kWh.

**Table 14.12.** Externalities of Road Transport in the Netherlands in 1995 (million Euro)

| Pollutant | Passenger cars | Goods vehicles | Other vehicles |
|---|---|---|---|
| $NO_x$ (Nitrate) | 504.84 | 443.90 | 55.05 |
| $NO_x$ (Ozone) | -228.42 | -200.85 | -24.91 |
| NMVOC (Ozone) | 131.60 | 32.76 | 28.81 |
| $PM_{2.5}$ | 380.49 | 607.56 | 106.08 |
| $SO_2$ (direct) | 13.87 | 24.52 | 2.34 |
| $SO_2$ (Sulphate) | 17.01 | 30.08 | 2.88 |
| $CO_2$ | 41.26 | 21.84 | 2.58 |
| $N_2O$ | 3.29 | 1.28 | 0.15 |
| Methane | 0.17 | 0.04 | 0.04 |
| Benzene | 2.58 | 0.64 | 0.57 |
| Butadiene | 8.88 | 2.21 | 1.94 |
| CO | 0.29 | 0.04 | 0.03 |
| DME | 3.17 | 5.07 | 0.88 |
| Nitrification | -2.63 | -2.31 | -0.29 |
| Non-operational life cycle stages | 980.22 | 198.60 | 42.63 |
| Total | 1856.63 | 1165.38 | 218.80 |
| % GDP | 0.52 | 0.33 | 0.06 |

externalities of a 1-km drive were estimated by weighting by the length of the two trajectories for each province. The provincial averages were arithmetically averaged to obtain the national average. The results are presented in Table 14.12.

In the second step these national average estimates were multiplied by the total emission due to passenger and goods road transport in the Netherlands. The final step was to include the life cycle externalities of the non-operational life cycle stages.

## 14.2.2 Rail Transport

In rail transport, electric trains and diesel trains for passenger and goods transport were treated separately. For diesel rail transport the total emissions from rail transport (see Table 14.9) were multiplied by the average externalities of road transport on a cost per tonne pollutant basis. For electric driven trains the average externalities for electricity production in the Netherlands were estimated by taking the average externalities per tonne pollutant emitted and per kWh produced from power plants in the Netherlands in the year 2000. A discussion of the methodology used is given in the marginal cost section on the Netherlands. In the second step the average externalities of electricity production (see Table 14.11) were multiplied by the total electricity consumption for passenger and goods rail transport (1.15 billion and 101.5 million kWh respectively). In the final step the life cycle externalities of the non-operational life cycle stages were included.

**Table 14.13.** Externalities of Rail Transport in the Netherlands in 1995 (million Euro)

| Pollutant | Passenger (diesel) | Goods (diesel) | Passenger (electric) | Goods (electric) |
|---|---|---|---|---|
| $NO_x$ (Nitrate) | 2.32 | 8.35 | 2.16 | 0.19 |
| $NO_x$ (Ozone) | -1.05 | -3.78 | -0.92 | -0.08 |
| NMVOC (Ozone) | 0.10 | 0.06 | 0.01 | 0.00 |
| $PM_{2.5}$ | 3.51 | 2.63 | 0.16 | 0.01 |
| $SO_2$ (direct) | 0.26 | 0.49 | 0.17 | 0.01 |
| $SO_2$ (Sulphate) | 0.32 | 0.60 | 0.60 | 0.05 |
| $CO_2$ | 0.11 | 0.21 | 1.10 | 0.10 |
| $N_2O$ | 0.01 | 0.01 | n.a. | n.a. |
| Methane | 0.00 | 0.00 | 0.00 | 0.00 |
| Benzene | 0.00 | 0.00 | n.a. | n.a. |
| Butadiene | 0.01 | 0.00 | n.a. | n.a. |
| CO | 0.00 | 0.00 | n.a. | n.a. |
| DME | 0.03 | 0.02 | n.a. | n.a. |
| Nitrification | -0.01 | -0.04 | -0.01 | 0.00 |
| Non-operational life cycle stages | n.a. | n.a. | 1.83 | 0.16 |
| Total | 5.60 | 8.56 | 5.10 | 0.45 |
| % GDP | 0.00 | 0.00 | 0.00 | 0.00 |

### 14.2.3 Inland Shipping

Inland shipping mainly consists of goods transport by barges, tankers, container ships and push tugs. To estimate the externalities of this transport type the same approach as with road transport was used. In the first step the average externalities per tonne pollutant emitted from a ship sailing from Rotterdam to Nijmegen were estimated (while for road and diesel train transport an emission height of 2.5 meters is assumed, for shipping an emission height of 15 meters is used). The average external cost per tonne of pollutant emitted is presented in Table 14.11. In the second step these average externalities were multiplied by the total emission due to shipping (see Table 14.9). Thus, the assumption is made that this task is representative for shipping in the whole of the Netherlands. In the final step the life cycle externalities of the non-operational life cycle stages were included.

### 14.2.4 Aircraft Transport

In the marginal cost analysis for the Netherlands only the externalities of 1 specific take-off on one of the airstrips at Schiphol airport were estimated. These estimates can hardly be seen as representative for all aeroplane transport in the Netherlands. As the externalities of greenhouse gas emissions are not site specific only these externalities were estimated. This was undertaken by multiplying the external costs of greenhouse gases by the total emissions from this sector (see Table 14.9). The results are given in Table 14.15.

**Table 14.14.** Externalities of Inland Shipping in the Netherlands in 1995 (million Euro)

| Pollutant | Passenger | Goods | Recreation | Fishery |
|---|---|---|---|---|
| $NO_x$ (Nitrate) | 7.59 | 109.12 | 4.72 | 2.62 |
| $NO_x$ (Ozone) | -3.64 | -52.26 | -2.26 | -1.25 |
| NMVOC (Ozone) | 0.29 | 1.71 | 2.05 | 0.04 |
| $PM_{2.5}$ | 23.72 | 185.42 | 11.86 | 4.31 |
| $SO_2$ (direct) | 0.41 | 9.13 | 0.37 | 0.24 |
| $SO_2$ (Sulphate) | 0.47 | 10.47 | 0.43 | 0.27 |
| $CO_2$ | 0.27 | 3.24 | 0.22 | 0.08 |
| $N_2O$ | 0.01 | 0.22 | 0.02 | 0.01 |
| Methane | 0.00 | 0.00 | 0.00 | 0.00 |
| Benzene | 0.01 | 0.04 | 0.05 | 0.00 |
| Butadiene | 0.02 | 0.14 | 0.17 | 0.00 |
| CO | 0.00 | 0.00 | 0.01 | 0.00 |
| DME | 0.18 | 1.38 | 0.09 | 0.03 |
| Nitrification | -0.04 | -0.62 | -0.03 | -0.01 |
| Non-operational life cycle stages | n.a. | n.a. | n.a. | n.a. |
| Total | 29.28 | 267.99 | 17.70 | 6.33 |
| % GDP | 0.01 | 0.08 | 0.00 | 0.00 |

### 14.2.5 Other Mobile Sources

The externalities of direct emissions to air from mobile sources in the agricultural and construction sector were also estimated, using average road transport external costs estimates (see Table 14.11) and the emissions from these sources (see Table 14.12). This assumes that all externalities from these sources are comparable to the average externalities of road transport. This assumption is justified for construction emissions, as these are likely to be centred in and around urban areas. It probably leads to an overestimation of the externalities from agricultural emissions, as road transport externalities include urban and non-urban areas, while agricultural emissions mainly arise in non-urban areas. The results are presented in Table 14.15.

### 14.2.6 Conclusions

A summary table of the total life cycle externalities due to mobile sources in Netherlands in 1995 is shown below.

The total externalities from mobile sources are estimated at 3.9 billion Euro, which is almost 1.2% of the Gross Domestic Product (GDP) of the Netherlands in 1995. The main transport externalities are related with the operation and non-operation life cycle stages of road transport by passenger cars and freight vehicles. Table 14.12 shows that in 1995, the externalities from tail-pipe emissions from

**Table 14.15.** Total Externalities of Mobile Sources in the Netherlands in 1995 (million Euro)

| Pollutant | Road | Shipping | Rail | Air | Total transport | Agricul-ture | Construc-tion | Total mobile sources |
|---|---|---|---|---|---|---|---|---|
| $NO_x$ (Nitrate) | 1003.8 | 124.05 | 13.02 | n.a. | 1,141 | 78.40 | 26.12 | 1,277 |
| $NO_x$ (Ozone) | -454.18 | -59.41 | -5.83 | n.a. | -519 | -35.47 | -11.82 | -581 |
| NMVOC ($O_3$) | 193.17 | 4.09 | 0.17 | n.a. | 197 | 4.64 | 1.54 | 205 |
| $PM_{2.5}$ | 1094.1 | 225.31 | 6.31 | n.a. | 1,326 | 154.30 | 51.73 | 1,592 |
| $SO_2$ (direct) | 40.73 | 10.15 | 0.93 | n.a. | 52 | 3.47 | 1.16 | 58 |
| $SO_2$ (Sulphate) | 49.97 | 11.64 | 1.57 | n.a. | 63 | 4.26 | 1.42 | 71 |
| $CO_2$ | 65.68 | 3.81 | 1.52 | 0.97 | 72 | 2.64 | 0.88 | 77 |
| $N_2O$ | 4.72 | 0.26 | 0.02 | 0.03 | 5.0 | 0.18 | 0.06 | 5.4 |
| Methane | 0.25 | 0 | 0 | 0.00 | 0.26 | 0.01 | 0.00 | 0.28 |
| Benzene | 3.79 | 0.1 | 0 | n.a. | 3.9 | 0.09 | 0.03 | 4.1 |
| Butadiene | 13.03 | 0.33 | 0.01 | n.a. | 13 | 0.31 | 0.10 | 14 |
| CO | 0.36 | 0.01 | 0 | n.a. | 0.37 | 0.01 | 0.00 | 0.38 |
| DME | 9.12 | 1.68 | 0.05 | n.a. | 11 | 1.29 | 0.43 | 13 |
| Nitrification | -5.23 | -0.7 | -0.06 | n.a. | -6.0 | -0.41 | -0.14 | -6.7 |
| Non-operational | 1221.5 | 0 | 1.99 | n.a. | 1,223 | n.a. | n.a. | 1,223 |
| Total | 3240.8 | 321.3 | 19.71 | 1.0 | 3583 | 213.71 | 71.52 | 3952 |
| % GDP | 0.91 | 0.09 | 0 | 0 | 1.02 | 0.06 | 0.02 | 1.13 |

n.a. = not available.

passenger and goods transport by road in the Netherlands are comparable. However, the externalities from non-operational life cycle stages of passenger road transport and roughly five times those of goods transport by road.

## 14.3 The External Costs of Road Transport in the UK

The UK aggregation assessment assessed the road transport sector. The approach used was outlined in Chap. 12 (and shown in Fig. 12.2), estimating national emissions from the sector (dis-aggregated by location of emissions) and using relevant unit costs per tonne of pollutant (derived from marginal runs) to produce total aggregated values. The study also looked at current and historical damages from the sector to show how road transport externalities have changed over time with changes in the vehicle fleet.

**Table 14.16.** Proportion of UK Vehicle Stock and Vehicle km (1997 fleet and km)

| Category | Technology | % fleet by category | % km by category |
|---|---|---|---|
| Gasoline car | Pre-EURO | 60.9 | 49.6 |
| | EURO I | 30.5 | 37.9 |
| | EURO II | 8.6 | 12.4 |
| Diesel car | Pre-EURO | 29.6 | 22.2 |
| | EURO I | 57.2 | 61.5 |
| | EURO II | 13.2 | 16.3 |
| Gasoline LDV | Pre-EURO | 92.6 | 86.3 |
| | EURO I | 6.3 | 11.4 |
| | EURO II | 1.2 | 2.3 |
| Diesel LDV | Pre-EURO | 54.9 | 44.6 |
| | EURO I | 37.9 | 46.0 |
| | EURO II | 7.2 | 9.4 |
| Rigid HDV | Old | 34.9 | 25.4 |
| | Pre-EURO | 28.3 | 29.6 |
| | EURO I | 26.0 | 31.8 |
| | EURO II | 10.8 | 13.3 |
| Articulated HDV | Old | 19.4 | 13.5 |
| | Pre-EURO | 25.8 | 24.1 |
| | EURO I | 38.5 | 43.8 |
| | EURO II | 16.3 | 18.6 |
| Bus | Old | 50.5 | 33.5 |
| | Pre-EURO | 19.5 | 23.6 |
| | EURO I | 19.6 | 28.0 |
| | EURO II | 10.4 | 14.8 |

### 14.3.1 Road Transport Statistics

The emissions and data for the study were taken from the base 1997 National Atmospheric Emissions Inventory (NAEI). The inventory has detailed data on fleet composition, based on vehicle licensing statistics and emission standards in place when the vehicles were first registered.

10.7% of the 1997 passenger car fleet is diesel (although they comprise 20% of new car sales). 66% of the LDV fleet is diesel. The database also takes account of decreasing annual mileage with increasing age of vehicle using National data on annual mileage by age of vehicle. The inventory also includes road statistics, showing the breakdown of vehicle km by location (urban, rural and motorways), shown in Table 14.17.

### 14.3.2 Road Emission Factors and Damage Costs

Vehicle emission factors for the UK road sector were also taken from the National Atmospheric Emissions Inventory. Emission factors for each EURO standard, speed and location were used. Thus, for each individual vehicle category above, there were emission factors for pre- EURO, EURO I, and EURO II vehicles, each at representative speeds for urban, rural and motorway driving.

These emission factors were multiplied by the vehicle km (adjusted for fleet composition, age distribution, and mileage distribution) to produce estimates of the total tonnes of pollutants from each vehicle category and from the entire transport sector. Data on car catalyst failure rates (based on a failure rate of 5% per annum, with 95% of failed catalysts being repaired after the vehicle is 3 years old) and cold start emissions (based on the COPERT II methodology) were also included.

**Table 14.17.** Estimated Vehicle Mileage Breakdown in the UK in 1997.

| | Billion km | | | | |
|---|---|---|---|---|---|
| Vehicle | urban | rural | motorway | Total | % |
| Gasoline car | 149.78 | 134.86 | 54.46 | 339.1 | 73.88 |
| Diesel car | 17.95 | 16.16 | 6.53 | 40.64 | 8.8 |
| Gasoline LDV | 5.82 | 6.05 | 2.26 | 14.13 | 3.1 |
| Diesel LDV | 11.4 | 11.85 | 4.43 | 27.68 | 6.0 |
| Rigid HDV | 6.52 | 9.01 | 4.9 | 20.43 | 4.5 |
| Artic. HDV | 1.26 | 5.15 | 6.12 | 12.53 | 2.7 |
| Bus | 3.09 | 1.43 | 0.55 | 5.07 | 1.1 |
| TOTAL | 195.82 | 184.51 | 79.25 | 459.58 | |
| % | 42.6 | 40.2 | 17.2 | | |

Total externalities for the sector were then estimated by multiplying emissions for urban, rural, and motorway locations by the relevant unit pollution costs (costs per tonne for each pollutant) derived from the urban, rural and motorway assessments of the UK marginal costs assessment. For this initial assessment, a single urban value was used. In reality, a significant range exists in the costs per tonne from large urban areas (particularly London) down to small urban areas (almost an order of magnitude difference). It is stressed that a greater uncertainty exists for the urban value used and more accurate aggregation values could be produced by the use of a greater dis-aggregation for UK urban locations.

### 14.3.3 Results

The results show that in 1997, transport sector externalities (tailpipe emissions) were estimated at 8.7 billion, which is around 1% of UK GDP. The results do not include upstream emissions. The breakdown of these total externalities, and the comparison with vehicle km, are shown below.

The results clearly indicate that diesel fuel use generates much higher external costs than gasoline. They also show that urban traffic generates much higher external costs. In terms of the split by vehicle category, the results show that although cars have a very high share of the vehicle km driven (>80%) they only account for around 40% of the damage. Goods vehicles, especially HDVs, have a disproportionately large contribution to damages (33% of damages from 7% of vehicle kms).

**Table 14.18.** % of UK Road Sector Damages by Location, Vehicle and Fuel (1997).

| | % vehicle km | % damage |
|---|---|---|
| By location | | |
| Urban | 43.1 | 64.8 |
| Rural | 39.8 | 21.2 |
| Motorway | 17.1 | 14.0 |
| By vehicle | | |
| Gasoline cars | 73.1 | 34.3 |
| Diesel cars | 8.8 | 7.4 |
| Gasoline LDVs | 3.0 | 2.1 |
| Diesel LDVs | 6.0 | 12.5 |
| Rigid HDV | 4.4 | 23.3 |
| Articulated HDV | 2.7 | 10.3 |
| Bus | 1.1 | 8.9 |
| Motorcycle | 0.9 | 1.2 |
| By fuel | | |
| Diesel | 22.9 | 62.4 |
| Gasoline | 77.1 | 37.6 |

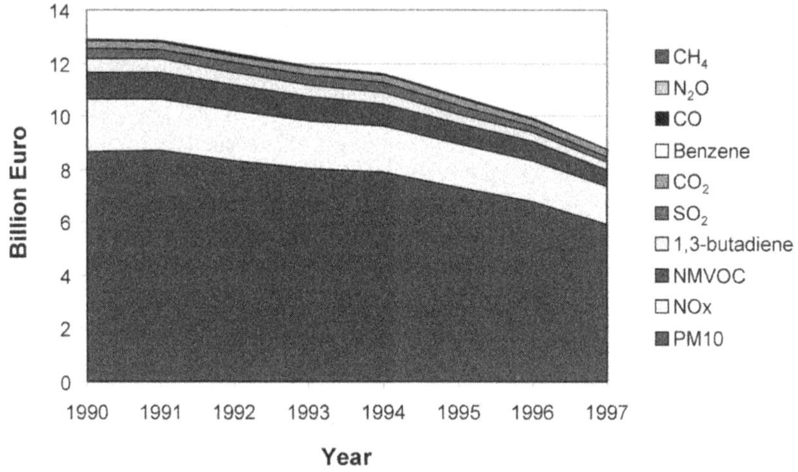


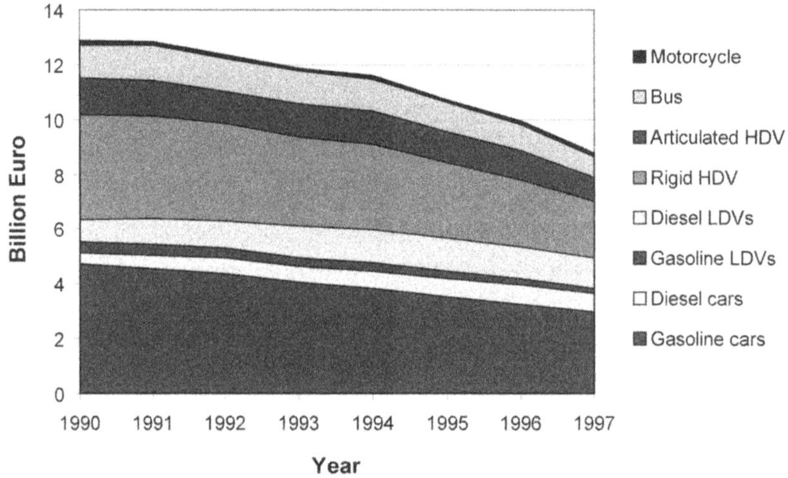


**Fig. 14.2.** Total UK Road Sector Externalities 1990-1997, by Vehicle

As well as 1997 data, the analysis was repeated using the same approach for the years 1990-1996. This provides some interesting information on the benefits of the introduction of tightening vehicle emissions standards in reducing road sector externalities over time. Figures 14.1 to 14.3 show national road sector externalities by pollutant, by fuel, by vehicle type and by location.

The results show the clear downwards externalities from the transport sector, arising from the introduction of EURO I and EURO II vehicles into the fleet. Indeed, the overall road transport externalities have declined by almost 4 billion Euro from 1990 to 1997, even though total vehicle km have risen over this same time period from 419 billion km to 464 billion km.

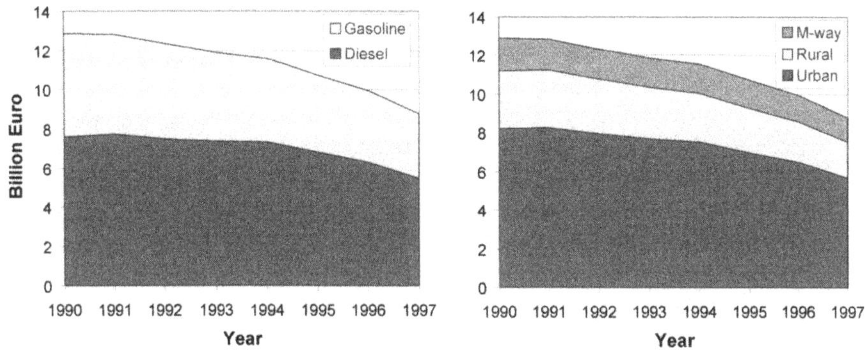


**Fig. 14.3.** Total UK Road Sector Externalities 1990-1997 by Fuel and Location

# 14.4 The External Costs of Road Transport in the Federal State of Baden-Württemberg, Germany

This section outlines the methodology and results for a regional level aggregation, looking at the road transport system of the federal state of Baden-Württemberg, a region of Germany. This study is particularly interesting as it provides an example of how a greater level of resolution can be brought to the local scale impact analysis.

## 14.4.1 Emission Scenario

The emission scenario was based on vehicle fleet and emission factor data given by the MEET project (Samaras et al., 1998, MEET, 1999). The spatial transport activity data was based on the 1995 road traffic census, including federal highways, rural and district roads. It is given in terms of mean daily traffic volumes (Lensing, 1995, adopted by Wickert, 1999).

**Table 14.19.** Representative vehicle speeds for different vehicle categories on different types of roads (MEET, 1999).

| Vehicle category | Urban | Extra-urban | Highway |
|---|---|---|---|
| Car, gasoline | 36.9 | 75 | 105.9 |
| Car, diesel | 36.9 | 75 | 125.1[a] |
| LGV | 36.9 | 75 | 115.5 |
| HDV | 20.0 | 60 | 90.0 |
| Motorcycles | 36.9 | 75 | 105.9 |

[a] The speed for diesel passenger cars is different from the gasoline cars because the larger cylinder capacity class was taken as representative for diesel vehicles.

Traffic volume data were summarised under 5 vehicle categories: passenger cars (PC), light duty vehicles (LDV), heavy duty vehicles (HDV), coaches, urban buses and two-wheelers (2W). Three road classes were distinguished: highways, major extra-urban roads (federal roads) and urban roads.

Emission factors were calculated for each vehicle technology, e.g. passenger car, gasoline, cylinder capacity <1.4 l, EURO I, using representative speeds (see Table) and the corresponding emission formulas. These emission factors were aggregated to give average emission factors per vehicle category by weighting with the annual total mileage of each vehicle technology. Due to the availability of data, the fleet composition data was taken from 1995, annual mileage per vehicle was taken from 1990.

Emission factors were calculated for CO, $CO_2$, $NO_x$, $PM_{2.5}$, $SO_2$, Benzene and 1,3 Butadiene. The latter 3 were derived following the procedure described in Chap. 3.

The Tables 14.20 and 14.21 show the total annual mileage for 1995 in Baden-Württemberg (about 12% of the total annual mileage driven in Germany) split by type of road and vehicle category and the resulting total emissions.

**Table 14.20.** Annual mileage for 1995 in Baden-Württemberg on different road classes by vehicle categories (Mio vkm/a).

| Mio vkm/a | Highways | Federal roads | Other roads | Urban roads | Total |
|---|---|---|---|---|---|
| Passebger Car | 15526 | 16404 | 24483 | 20767 | 77180 |
| Motorcycles | 173 | 432 | 788 | 523 | 1916 |
| LDV | 513 | 466 | 717 | 563 | 2260 |
| HDV | 2406 | 1241 | 1199 | 943 | 5789 |
| Bus | 143 | 112 | 222 | 174 | 651 |

**Table 14.21.** Annual emissions for 1995 in Baden-Württemberg on different roads, given per pollutant (kt).

| Pollutant | Highway | Federal roads | Other roads | Urban roads | Total |
|---|---|---|---|---|---|
| PM | 1.1 | 0.6 | 0.7 | 1.1 | 4 |
| $SO_2$ | 0.8 | 0.5 | 0.6 | 0.7 | 2 |
| NOX | 21.1 | 12.5 | 17.0 | 17.2 | 68 |
| VOC | 4.7 | 3.7 | 4.8 | 8.8 | 22 |
| CO | 59.2 | 26.3 | 40.3 | 70.8 | 197 |
| Benzene | 0.2 | 0.1 | 0.2 | 0.3 | 1 |
| Butadiene | 0.0 | 0.0 | 0.0 | 0.0 | 0 |
| Ethene | 0.7 | 0.5 | 0.5 | 1.0 | 3 |
| Formaldehyde | 0.6 | 0.4 | 0.5 | 0.9 | 2 |
| $CO_2$ | 4866.8 | 3192.7 | 4533.1 | 5154.7 | 17747 |

## 14.4.2 Local Damage Estimation

The next step in the analysis was to calculate the damage costs for the above emissions. Consistent with discussion in Chap. 12, a high level of resolution was adopted for the local level analysis. The approach used was to use simple geo-spatial and social-spatial parameters and to examine a large number of case studies by regression analysis, to find the relationship between exposure and parameters such as meteorology and population distribution.

Given the exposure-response functions (ERF) are linear (see Chap. 12), it is possible to simplify the local damage analysis, by splitting the accumulated exposure and the slope and evaluating separately. The general ERF structure is shown in Equation 14.1. The damage (D) is the sum of damage in each grid cell, each of which are derived from multiplication of the impact and monetary value. Given linearity, the monetary value and slope are independent of the grid cell (e.g. background concentrations), and so Equation 14.1 can be transformed to Equation 14.2, where the product of slope and monetary value is called the costfactor (CF) and DeltaConcentration times population is called the accumulated exposure (ACCEXP).

$$D = \sum_{i=0}^{n} MonetaryValue \cdot \underbrace{slope \cdot \Delta_{Conc,i} \cdot Population_i}_{impact_i} \tag{14.1}$$

$$D = \underbrace{MonetaryValue \cdot slope}_{\text{costfactor}} \cdot \underbrace{\sum_{i=0}^{n} \Delta_{Conc,i} \cdot Population_i}_{\text{accumulated exposure}} \tag{14.2}$$

The local dispersion modelling in the marginal analysis uses a Gaussian approach. This also assumes a linear relationship between emissions and concentration, and therefore the "normalised accumulated exposure" (NACCEXP) can be introduced. The NACCEXP equates to the accumulated exposure due to one gramme of pollutant emitted. Multiplication of the NACCEXP with the actual emission results in the actual accumulated exposure. The damage is then calculated with Equation 14.3.

$$D = \text{costfactor}_p \cdot NACCEXP \cdot Emission_p \tag{14.3}$$

There is a costfactor for each primary pollutant p for which one or more ERFs exist. The total damage of one vehicle trip is the sum of the damages per pollutant. The costfactors for the current set of primary pollutants and monetary values are summarised in Table 14.22.

**Table 14.22.** Cost factors (Euro/ACCEXP, prices 2000)

| Pollutant | Costfactor (Euro/ACCEXP) |
|---|---|
| $PM_{2.5}$ | 34.99 |
| 1,3 Butadiene | 6.76 |
| $SO_2$ | 0.89 |
| Diesel particles (via cancer effects) | 0.60 |
| Benzene | 0.18 |
| Formaldehyde | 0.13 |
| Ethene | 0.08 |
| CO | 0.001 |

### 14.4.3
### Local Exposure Estimates

As damage estimates are dominated by health impacts, the key parameter to assess is the exposure of the population, i.e. how many people are exposed and to what concentration. The local dispersion modelling, together with high-resolution population data, allows us to assess the population weighted exposure level.

In order to substitute the dispersion modelling step, the driving parameters for exposure estimation were studied, separating modeling parameters and input data. The first category comprises the parameterisation schemes of the gaussian dispersion model, which have to be specified according to the environment where the model is applied, and the resolution of the adaptive grid. The quality of meteorological data and the quality of population data belong to the second category.

The ROADPOL-model includes the parameterisation schemes: Pasquill-Gifford (PG), Benson (B), Briggs-rural (BR) and Briggs-urban (BU). The PG-scheme was derived from near-surface release in flat terrain (roughness length $z_0$ = 0.03). The Benson scheme agrees with PG in the downwind range from 100 m to 10 km, except $\sigma_z$ values for A and B stability classes approximate the B2 and B1 Brookhaven curves. The roughness length of the Brookhaven experiment was $z_0$ = 1m. The BU-scheme, also called McElroy-Pooler sigmas, were derived from several urban dispersion experiments with low level tracers. The U.S. EPA recommends these sigma values as the one most appropriate for dispersion simulations in urban areas. For most atmospheric conditions, BU gives higher concentrations than those for rural conditions, reflecting the influences of increased mechanical turbulence during night-time hours as the results of the released urban heat stored in the structures and pavement during the daytime (Turner, 1994).

The given dispersion schemes were compared to German measurement data for low level releases (Panitz et al., 1989), referred to as *KFK* sigmas. These sigmas where derived from measurement over flat terrain (Rhine valley), characterised by open spaces, built-up and wooded areas. It was found, that KFK sigmas were more similar to BU than BR. Considering, that roughness lengths $z_0$ = 0.03 are rarely found in Germany, BU sigmas were therefore applied in all cases. Studying the effect of the minimal grid cell size, no effect for rural areas were found, however an increase of exposure estimates from 75% to 105% with decreasing grid width could be observed (125m refer to 100%). Therefore, 125 m minimal grid width was applied in all cases.

To account for meteorological characteristics, 9 data sets of statistical meteorological data were used. The data was calculated from 10 years continuous measurements in different, characteristic regions all over Germany (Deutscher Wetterdienst, 1999). The yearly average wind speed was identified as most important parameter.

To find a relationship between NACCEXP and simple parameters, such as population density, 70 model runs were conducted with the detailed local range model. The model parameters were set as described above. The road sections were 1 km long and selected randomly within the nine regions of the meteorological data sets. Different locations e.g. city centres, city suburbs, villages, outside populated areas and far from populated areas were covered. Thereupon, a regression analysis was undertaken to estimate a simple relationship between population parameters, wind speed and exposure estimates. The population density (PD) parameters tested were: PD1, PD3, PD10, PD20, PD1_3, PD3_10, PD10_20 (where the numbers refer to radius in km around the road segment, or as with 1_3, the PD in the circle between 1 and 3 km radius).

The best results were achieved when emissions within and outside built-up areas were distinguished. The results from the regression analysis are given in Equations 14.4 for urban areas and 14.5 for extra-urban areas. As implicitly the characteristic population distribution and meteorological conditions are contained in these equations, they are strictly only valid for application in Germany.

$$NACCEXP_u = 1.68 * 10^{-4} + \frac{(1.28 * PD1 + 1.79 * PD20)}{WINDSPEED} * 10^{-6} \qquad (14.4))$$

$$NACCEXP_{e-u} = -2.6 * 10^{-6} + \frac{8.6 * 10^{-7} * (PD10 + PD10\_20)}{WINDSPEED} + 3.9 * 10^{-7} * PD1 \qquad (14.5)$$

### 14.4.4 Classification of Road Network Links

As stated above, the damage due to emissions on a certain road segment can readily be calculated by population parameters and wind speed. Figure 14.4 illustrates the calculation of the population parameters PD1 and PD10. A buffer is drawn around the road segment and linked to the population map. The population within a certain radius is identified and the average population density within this buffer area is calculated. In this way, a whole road network can be classified.

For the local scale damages, each road segment was characterised according to the "local damage potential": 184 road segments for highways, 848 road segments

**Fig. 14.4.** Classification of road sections by population parameters: PD1 (buffer with radius 1km) and PD10 (buffer with radius 10km around road section).

for extra-urban (federal) roads, and 3988 polygons, representing the urban roads, were analysed throughout Baden-Württemberg. Wind speed was interpolated based on measurement data of meteorological stations in Baden-Württemberg, provided by the German Meteorological Service (Traub and Kruse, 1996). The interpolation was performed on the EMEP 50 x 50 km grid, the value was assigned to each road segment in the respective grid cell. 'Kriging' was used as interpolation algorithm.

## 14.4.5 Regional Range Impacts

Regional range impacts can be best quantified with the use of the 'multi-source' EcoSense model, which uses the CORINAIR 1994 emissions scenario. Two model runs are performed: one with the baseline emission scenario, and a second

run with the road traffic emissions (i.e. total traffic emissions in Baden-Württemberg) subtracted from the baseline scenario.. The difference in impacts is attributed to the non-marginal impact of road transport emissions. Note, population data of Baden-Württemberg was set to zero for the evaluation of regional impacts in order to avoid double counting with the local analysis. Global range impacts were assessed using the marginal cost estimates in Chap. 9.

### 14.4.6 Results

The aggregated costs of road transport in Baden-Württemberg in 1995 were calculated as 756.5 Million Euro. This amounts to 0.29 % of GDP of Baden-Württemberg (Federal Statistical Office, 1999). Detailed results are given in Table 14.23.

**Table 14.23.** Damage due to direct emissions from road transport in Baden-Württemberg 1995

| Pollutant | Highway | Federal roads | Million Euro<br>Other roads | Urban roads | Total |
|---|---|---|---|---|---|
| Local scale | | | | | |
| $PM_{2.5}$ | 29.89 | 18.62 | 23.36 | 76.63 | 148.51 |
| $SO_2$ | 0.52 | 0.37 | 0.51 | 1.25 | 2.65 |
| CO | 0.05 | 0.03 | 0.04 | 0.16 | 0.28 |
| Benzene | 0.03 | 0.02 | 0.03 | 0.11 | 0.19 |
| Butadiene | 0.17 | 0.14 | 0.16 | 0.65 | 1.12 |
| Ethene | 0.04 | 0.03 | 0.04 | 0.16 | 0.27 |
| Formaldehyde | 0.06 | 0.05 | 0.06 | 0.24 | 0.40 |
| Total (local) | 30.76 | 19.26 | 24.20 | 79.20 | 153.41 |
| Regional scale | | | | | |
| $PM_{2.5}$ | 28.51 | 15.14 | 18.99 | 27.72 | 90.36 |
| $NO_x$[a] | 131.02 | 77.59 | 105.59 | 106.53 | 420.74 |
| NMVOC[b] | 9.48 | 7.30 | 9.60 | 17.56 | 43.94 |
| $SO_2$[c] | 1.62 | 0.97 | 1.31 | 1.48 | 5.40 |
| Sum (regional) | 170.63 | 101.00 | 135.31 | 153.29 | 560.44 |
| $CO_2$ | 11.68 | 7.66 | 10.88 | 12.37 | 42.59 |
| TOTAL | 213.07 | 127.92 | 170.39 | 244.86 | 756.44 |

[a] via nitrates and ozone; [b] via ozone; [c] direct and via sulphates

## 14.5 Conclusions

This chapter presents the results of the aggregation studies. It demonstrates the application of the approaches proposed in Chap. 12 for aggregation at a national/regional level and shows the potential usefulness of the numbers in policy and strategy studies.

The national level results are particularly interesting. Although the national aggregation assessments use different methods (and so some care must be taken in directly comparing results), they have all used a higher dis-aggregation for local impacts as recommended in Chap. 12. The results show that the aggregated costs for the road transport sector are significant. The externalities of exhaust emissions from the road sector are estimated at 4.2 billion Euro (1.8% of GDP) for Belgium and 8.7 billion Euro (1% of GDP) for the UK. These values are interesting as they are higher than many estimates in the literature (which are of the order of 0.5% of GDP (European Commission, 1995b)). This is to be expected, as these are the first national aggregation studies to take local impacts into account properly when estimating the external costs of the road transport sector (i.e. specifically taking into account the higher external costs in urban locations). All studies show that passenger cars are the dominant source of road transport sector damages, though heavy goods vehicles give rise to a dis-proportionately large share of damages relative to their share of vehicle kilometres.

The analysis of the entire transport sector in the Netherlands shows that road transport is also the dominant source of damages from all transport and mobile sources, responsible for around 82% of the total transport externalities of 3.9 billion (1.1% of GDP). Finally, the results show that life-cycle emissions are important: the aggregation studies in Belgium and the Netherlands show emissions from fuel production, vehicle manufacture and infrastructure are 23% (Belgium) and 40% (Netherlands) of total road sector externalities.

The regional analysis for Germany presents an elaborated approach to increase the resolution of local scale impact analysis. The central feature is the use of cost factors which were derived per exposed receptor. With these factors, aggregated costs can be calculated using a GIS (Geographic Information System) up to national or the European level. Thus, together with the multi-source EcoSense model this approach allows to calculate damage costs on the national level for all European countries.

# 15 Policy Case Studies

This chapter presents the results of the policy case studies. These studies demonstrate the potential application of the ExternE accounting framework in policy and decision making. The studies have included a wide range of examples at different levels (local to national), covering different issues. Together they provide a rich source of information on the practical use of the ExternE Transport values in providing decision support for transport policy issues.

## 15.1 Belgium: Do Policies Cope with Transportation Trends?

New cars (EURO III) have significantly lower costs per vehicle km compared to older cars (an improvement of 75 % to 90 %), thanks to technological improvements but mainly because of more stringent environmental standards at the EU level. These policies are mainly tail pipe oriented and do not affect the trends in passenger and goods transportation, which are mainly determined by technological and socio-economic developments. In this case study it will be examined to what extent the standards set address transport growth and the trend towards more diesel cars for private transport in Belgium.

### 15.1.1 The Evolution of the Total Environmental Costs in Belgium

The previous chapter explained how the total externalities for the Belgian road transport sector were obtained. This methodology was repeated for all years between 1993 and 1998. In this way the net result of all transport related policy measures (European and national) in this time frame can be evaluated. During this period the changing composition of the fleet and the mileage were taken into account, but no changes to the composition of fuels or mobility distribution data were included. The result for all transport is presented in Table 15.1 and shown in Fig. 15.1.

The results indicate that the cost of air pollution from Belgian vehicles has only slightly decreased between 1993 and 1998. This is entirely due to the reduction of environmental impacts from passenger cars. Total externalities from HDV and buses have risen during this period.

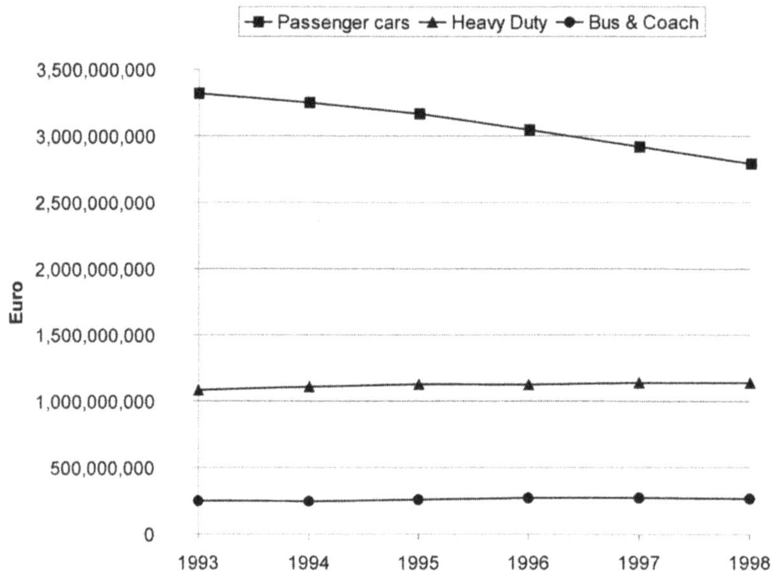


**Fig. 15.1.** Total air pollution costs caused by Belgian passenger cars, heavy-duty vehicles, buses and coaches between 1993 and 1998.

#### *15.1.1.1*
#### *Passenger Cars*

Although the important pollutants such as $NO_x$, PM and VOC from new passenger cars have been cut to very low levels, the rate at which old cars are replaced by new models imposes a time lag of several years before the implementation of new regulations can be observed. For Belgium, even over a time frame of 6 years, the reduction is modest and falls short of expectations based on the impressive technological progress from uncontrolled engines to EURO II.

**Table 15.1.** Total external costs for the use phase for different segments of the fleet (in billion Euro) for Belgium, 1998.

| | External costs | | | Mileage | | |
|---|---|---|---|---|---|---|
| Year | Passenger cars | Heavy Duty | Bus & Coach | Passenger cars | Heavy Duty | Bus & Coach |
| 1993 | 3.31 | 1.08 | 0.25 | 67.03 | 5.29 | 0.51 |
| 1994 | 3.25 | 1.11 | 0.25 | 69.14 | 5.48 | 0.51 |
| 1995 | 3.16 | 1.13 | 0.26 | 70.87 | 5.67 | 0.55 |
| 1996 | 3.04 | 1.13 | 0.27 | 72.53 | 5.82 | 0.58 |
| 1997 | 2.91 | 1.14 | 0.27 | 74.76 | 6.10 | 0.60 |
| 1998 | 2.79 | 1.14 | 0.27 | 77.10 | 6.36 | 0.61 |

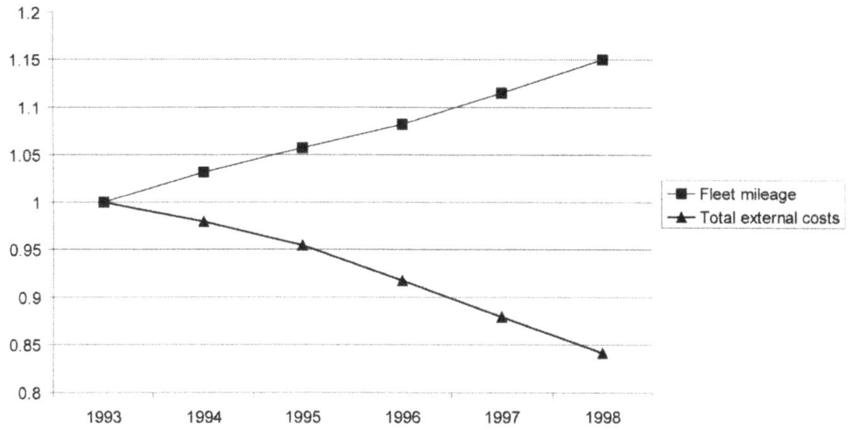


**Fig. 15.2.** Evolution of total air pollution costs compared to fleet mileage (Belgian passenger cars, 1993=1)

This is due to the combined effect of three factors:

- The number of registered passenger cars in Belgium keeps increasing (+12% since 1993).
- The annual average mileage per car keeps increasing (+15%, see Fig. 15.2)
- There is a clear trend towards more diesel cars. Old petrol cars are often replaced by diesel cars that have higher external costs.

The lower annual mileage of older (EURO 0) cars can explain part of the decrease. This effect (see Chap. 14, Table 14.5) even seems sufficient to counteract the increased fleet growth in 1997 and 1998. It has to be noted however that the distribution of mileage for older and newer cars is subject to relatively high uncertainty as no exact figures exist for Belgium.

Because of the reduction of external costs from EURO 0 cars (as opposed to the increase for other vehicles, see below), the importance of passenger cars in the total costs has also declined. In 1993, 71% of the externalities were due to passenger cars and this number has fallen to 66% in 1998.

#### *15.1.1.2*
#### *Heavy Duty Vehicles, Buses and Coaches*

The trends that were calculated for heavy-duty vehicles and buses are different to those discussed in the previous paragraph for passenger cars. Instead of a decline, an increase in the externalities of trucks and buses since 1993 is calculated. Whereas technological progress for passenger cars was sufficient to compensate for the increased mileage growth of the fleet and the trend towards diesel, this is not the case for trucks nor buses (Fig. 15.3 and Fig. 15.4).

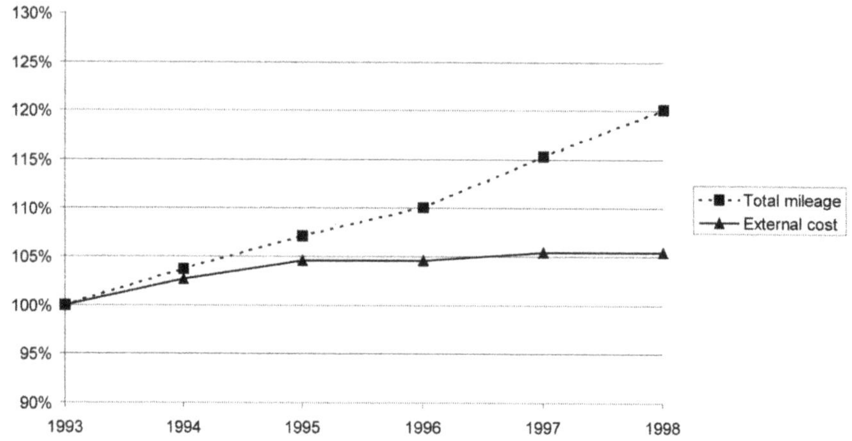


**Fig. 15.3.** Evolution of the mileage and externalities of Heavy Duty vehicles (1993=100%)

For trucks there has been an considerable growth, both in terms of the number of vehicles and in average mileage per vehicle. This combined increase leads to a rise of the total fleet mileage that is much higher than for passenger cars. Two phases can be distinguished in Fig. 15.3. Between 1993 and 1996 the growth of the fleet was rather slow, but external costs were rising. In 1997 and 1998 the growth of the fleet was much more important, explaining the accelerating increase of the fleet mileage. On the other hand, the fast growing fleet has now acquired a significant proportion of newer, environmentally better, EURO II trucks. Since 1995 this technological progress seems to compensate for the growth in the road transport sector.

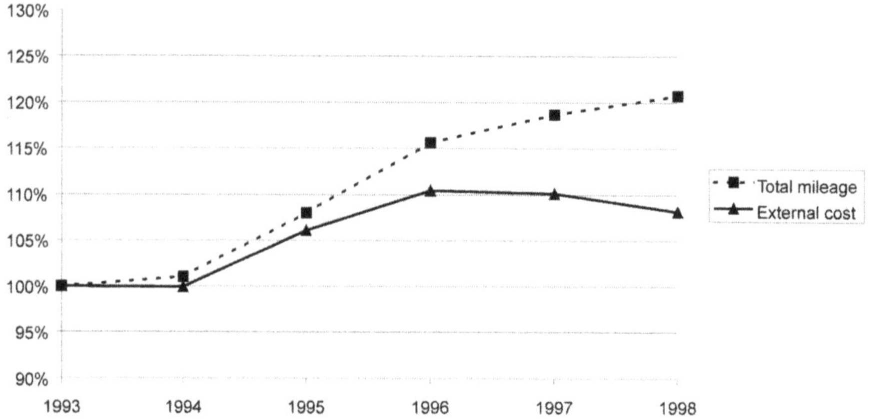


**Fig. 15.4.** Evolution of the mileage and externalities of Belgian buses and coaches (1993=100%)

A similar trend in external cost was found for buses and coaches (Fig. 15.4). After an initial increase between 1993 and 1996, externalities are now decreasing. The turning point coincides with a reduced growth of the fleet mileage in 1997 and 1998. Therefore the pattern of changing external costs completely reflects the important changes in new bus sales.

### 15.1.2 Concluding Remarks

Our analysis shows that total environmental externalities from Belgian vehicle transport have declined by 9% over the period 1993-1998. Transport growth and the shift to diesel vehicles have partly compensated the benefits from the introduction of much cleaner EURO I and II vehicles. For private transport, the figures indicate a declining trend, but for trucks and busses, total externalities did not decline over the period. Future work will check these trends up to 2005.

## 15.2 Belgium: Taxation of Transportation Fuels and External Costs per Litre of Fuel

This section looks at what lessons the ExternE data could provide for internalising energy related environmental externalities in transport fuel prices. The case study expressed the external costs (in Eurocent per litre of transportation fuel) for all the different reference technologies and trajectories for the Belgium implementation. The results for gasoline are shown in Fig. 15.5 and those for diesel cars in Fig. 15.6 and Fig. 15.7.

### 15.2.1 External Costs per Litre of Petrol and Diesel

The first obvious lesson is that the external costs vary a lot between the different case studies. This is due to a combination of emission legislation (or vintage year), traffic conditions (speed) and location. For all categories, impacts are the highest if the fuel is burnt in EURO 0 cars in urban traffic, and the lowest for new EURO III cars driving in rural areas. The difference between the maximum and minimum value is a factor 10 for gasoline and a factor of up to 30 for diesel.

The ranking in between these extremes is based on a mixture of different factors and cannot be easily summarised. For gasoline, the vintage year is the dominant factor. For EURO I vehicles onwards, the differences per litre of gasoline are more limited (a factor 2.5). For the pre-catalyst cars (EURO 0) traffic conditions are also important because high speeds (highway) lead to high $NO_x$ emissions and related impacts. For diesel cars, location is more important than for gasoline cars, because particle emissions have a higher share in total impacts. It has to be noted that this ranking is different from the results per vehicle-km, because the latter ranking also reflects the fuel efficiency of different technologies and traffic situations.

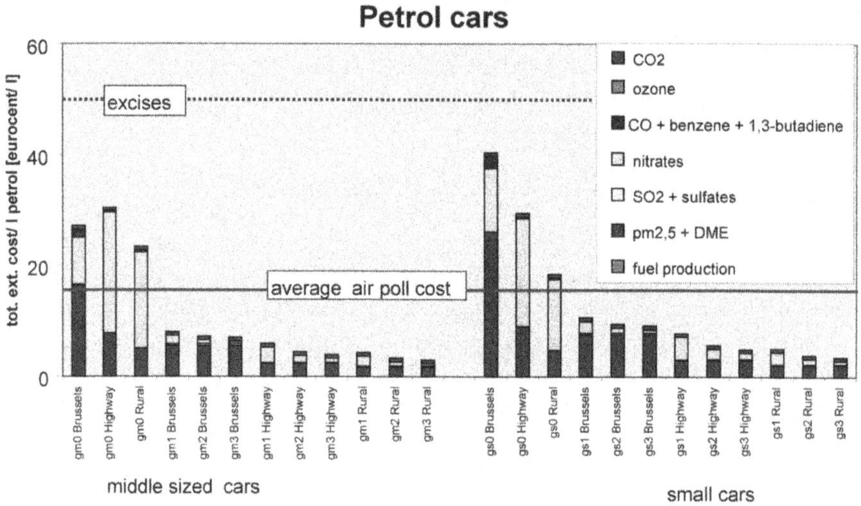


**Fig. 15.5.** External costs per litre of petrol used in small and middle sized petrol cars in Belgium, (EURO 0 to EURO III), for urban, highway and rural traffic reference trajectories; and compared with consumer prices and excises for petrol (all in Eurocent /litre).

For diesel vehicles, the external costs per litre of fuel were also compared for passenger cars, buses, heavy and light-duty vehicles (Fig. 15.7). All vehicle categories show the same pattern of external costs - varying widely for vehicle vintage, location and traffic situation, although the ranges are a little bit smaller compared to passenger cars.

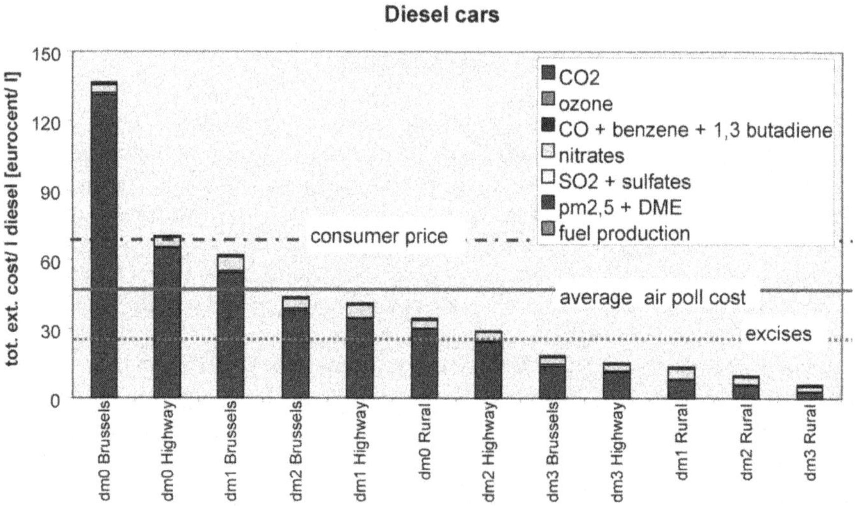


**Fig. 15.6.** External costs per litre of diesel used in diesel cars in Belgium, (EURO 0 to EURO III), for urban, highway and rural traffic reference trajectories; and compared with consumer prices and excises for diesel (all in Eurocent /litre).

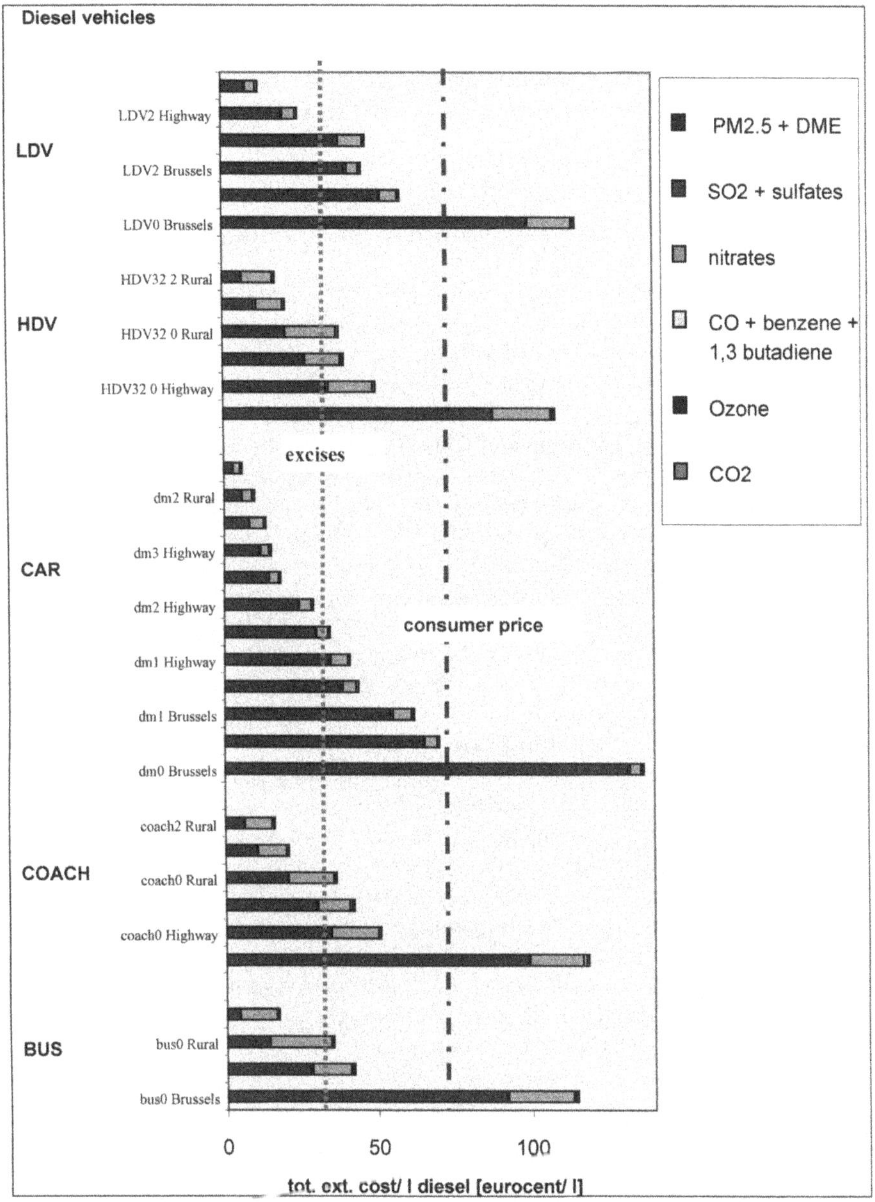


**Fig. 15.7.** External costs per litre of diesel used in different diesel vehicles in Belgium, (EURO 0 to EURO III), for urban, highway and rural traffic reference trajectories and compared with consumer prices and excises for diesel (all in Eurocent /litre)

Because EURO vehicles still have an important share in total mileage, and with mileage more or less equally distributed over urban, highway and rural traffic, one cannot recommend a single 'environmental tax' per litre of fuel to reflect the environmental damage costs. Furthermore, in reality the external costs will differ even more, because the analysis was based on the mid-estimate from ‘reference trajectories’ covering trajectories with external costs that are a factor higher or lower. Finally, cold start, differences in maintenance and driving behaviour were not considered.

### 15.2.2 Comparing External Air Pollution Costs with Prices and Excises

The average externality per litre of fuel – for 1998 – amounts to 0.5 Euro/litre of diesel. On average, air pollution externalities associated with diesel are about two thirds of the consumer price for diesel and amount to almost double the excise for diesel. Once again, this lesson is not valid for all vehicle and transport categories. For diesel, all pre EURO III vehicles have higher external air pollution externalities for urban and highway drive then the excise tax on diesel used for transportation. Only for EURO I and II rural and for EURO III vehicles, these externalities (at least as quantified here) are lower then the excise.

The picture is different for petrol (Fig. 15.5). The average externality for petrol is around 16 Eurocent /litre (for 1998), which is about 15 percent of the consumer price for petrol and about one-third of the excise on petrol. For pre-catalyst cars (EURO 0) the externalities are far above the average, but the external costs do not amount to the level of excise for any of the categories.

Because the external costs vary so much for different vehicles, traffic situations and locations, it is hard to draw policy conclusions from the comparison with the current consumer prices and excise taxes. Moreover, our estimates are a minimum estimate in the sense that some impacts cannot be quantified (ecological impacts from acidification and eutrophication, some global warming impacts, etc). Nonetheless, even with these caveats, it is clear from an environmental point of view that there is little to justify the lower excise on diesel relative to petrol. This issue is especially important as the shift from petrol to diesel cars is one of the reasons why the total impacts from passenger cars have not dropped significantly over time, even with the introduction of stricter environmental standards for new cars. Finally, the total externalities from diesel fuelled buses, coaches and light and heavy-duty vehicles have all increased over the period 1993-1998 indicating consumer prices may be giving the wrong signals.

## 15.3 Finland: Alternative Fuels in Helsinki City Buses

The use of the ExternE-methodology in decision making for environmental and transport policy was assessed with a case study on city buses in Helsinki. The study assessed the air pollution related externalities on the central city bus lines and suburban bus with the use of natural gas instead of diesel fuelled vehicles.

## 15.3.1 Background Data on the Helsinki City Bus Traffic

The existing Helsinki internal bus fleet travels 34 million kilometers per year and uses low-sulphur city-diesel as a fuel. There are currently a number of buses that use gaseous fuel, though for central city traffic the decision has been made to gradually change completely from diesel to gaseous fuels. The use of natural gas would improve air quality in the centre by reducing the emissions of carbon monoxide, nitrogen oxides and particles. The data on Helsinki city-buses is presented in Table 15.2.

The map of bus routes, covering all the central city bus lines and a few examples of the suburban lines was provided by Helsinki City Transport (HCT) in MapInfo format, shown in Fig. 15.8.

Measurement data on the speeds per bus line section (for each operating hour) and data on the departures per line section (again for each operating hour) were also available. The length of the bus lines varied from 2.5 to 17 km. Each line, divided into 4 - 13 sections, was translated in EcoSense as seven trajectories.

The average speeds (in each operating hour) for the central city lines varied from 15 - 35 km/hour and for the suburban lines from below 30 to 60 km/hour. The speeds in the separate sections (in each operating hour) varied from below 10 km/h to over 50 km/h for central city lines, and from below 10 to over 100 km/h on suburban lines (see Table 15.3). The speeds for each line section were chosen as follows: average speed for the bus line in each operating hour was calculated by weighting the speeds per line section with the length of the line section. The operating hour for the bus line with the lowest speed was chosen. The operating hour with the lowest speed usually corresponds to the operating hour with the most departures - the rush hour. Average speeds in the rush hour varied from 15 to 25 km per hour on central city lines and from 30 to 50 km per hour on suburban lines. The variation within line sections varied from 8 to 60 km per hour (central city lines) and from 7 to 68 km per hour (suburban lines).

**Table 15.2.** Data on the Helsinki City-buses

| | |
|---|---|
| Bus network (Helsinki internal transport 1997) | |
| Number of bus lines | 84 |
| Total network length (km) | 890 |
| Distance traveled | 34 million line-km/year |
| Bus fleet (1997) | |
| Uncontrolled | 55 |
| EURO I | 138 |
| EURO II | 358 |
| CNG | 18 |
| LPG | 7 |

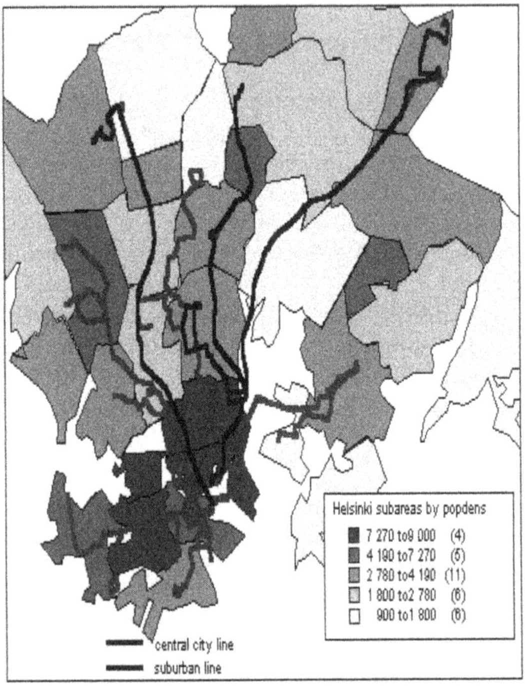


**Fig. 15.8.** Helsinki Internal Bus Traffic Line Data

The bus lines are located in Helsinki, in an area of 185 km². The average population density in Helsinki is 2 961/km². The average population density in 33 Helsinki sub-areas varies from just under 1000 to 13 500 inhabitants per km². The local meteorological data used for the EcoSense calculations was produced in the Helsinki-Vantaa meteorological station.

**Table 15.3.** Speed Data on the Bus Lines and the Line Sections

| | Central city lines | Suburban lines |
|---|---|---|
| Variation in each operating hour | | |
| Average line speed (km/h) | 15 – 35 | 30 – 60 |
| Section speeds (km/h) | 7 – 50 | 7 – 100 |
| Variation in rush hour | | |
| Average line speed (km/h) | 15 - 25 | 30 - 50 |
| Section speeds (km/h) | 8 - 60 | 7 - 68 |

## 15.3.2 Emission Factors

The speed dependent natural gas emission factors applied in the calculations were based on data from several sources. The Helsinki Metropolitan Area Council (YTV) provided the original emission factors. These factors were formed for the traffic planning studies for the Helsinki Metropolitan Area originally for the year 1995. The factors were compared to the latest data available for gas emission factors from Finnish sources (Laurikko and Nylund, 1998 and to data from the UK (ETSU, 1996). The only speed dependent data available for the gas buses was that of YTV for 1995: other data were average estimates of emissions per km. The emission factors for the diesel buses applied were the MEET functions taken from the EcoSense model. Table 15.4 compares the speed dependent emission factors of natural gas buses used in this calculation with the emission factors for diesel buses.

**Table 15.4.** Emission Factors for Natural Gas Buses

| Emission component | (% of gas bus emissions from diesel bus emissions) | Emission Factor Source |
|---|---|---|
| CO | 50 – 80 % | YTV speed dependent function |
| $CO_2$ | 100 % | MEET function for diesel buses |
| $NO_x$ | 70 – 90 % | YTV speed dependent function |
| PM | 15 – 17 % | YTV speed dependent function, multiplied with a factor of 0.1 to correspond to other available data (ETSU, 1996; Laurikko and Nylund, 1998 |
| VOC | 300 - 500 % | YTV speed dependent function multiplied with a factor of 0.1 to correspond to other available data (ETSU, 1996; Laurikko and Nylund, 1998 |

## 15.3.3 Results

### *15.3.3.1 Concentrations*

The results of the EcoSense calculations, for single central city bus lines, showed that local $PM_{2.5}$ concentrations in the Helsinki area varied from 1E-011 to 4E-07 $\mu g/m^3$/line-km with diesel fuel. The average local concentration (the concentration per grid cell weighted by the population density of the grid cell) was of the order of 3-4 E-09 $\mu g/m^3$/line-km. For the total 34 line-km of the Helsinki internal bus traffic in 1997, the average local $PM_{2.5}$ concentration caused by bus traffic with diesel fuel in the Helsinki area would be just over 0.1 $\mu g/m^3$ (517 000 people exposed). The highest local concentrations of $PM_{2.5}$ (at least 25 % of the maximum concentration) were formed within a distance of about 100 metres of the road. At distance of about 1 to 2 km from the road, the concentrations were less than 1% of the maximum concentrations. The biggest regional nitrate concentration increments are caused in the area of Finland, Sweden and Norway, and in Russia, east of Finland.

The concentrations calculated by the EcoSense for one bus trajectory were compared with the concentrations estimated by the Finnish Meteorological Institute for bus traffic in the Helsinki Metropolitan Area in an ongoing study by Ekono Energy (Ekono, 1999). In the Finnish model the concentrations were estimated with the help of a line source model, based on an analytical solution and with chemical transformation of emissions based on the "discrete parcel" method, corresponding to the Caline model. The maximum concentrations of $PM_{2.5}$ within the Helsinki area estimated by the Finnish model and with the EcoSense model seemed to be in about the same range.

### 15.3.3.2 Environmental Impacts and Costs

In the case of diesel buses, most of the externalities were caused by the impact of local $PM_{2.5}$ concentrations on human health (comprising around 60-70% of total costs). Other significant costs were from health impacts due to regional concentrations of nitrates (10%), climate change (5%) and crop losses due to ozone (5%). In the case of natural gas buses, nearly 50% of externalities were due to nitrates. Climate change accounted for about 15% and the remainder was due to particles and ozone.

For central city lines, the environmental costs of diesel buses (EURO II vehicles) varied from 25 - 40 Finnish pennies/km. For suburban lines the results were of the order of just over 20 pennies/km. For the technologies with higher emission rates (EURO I and uncontrolled vehicles) the results in each case were 2 or 3 times higher. The results for natural gas buses on the different lines varied from 10 to 15 pennies/km.

The environmental life-cycle costs (for fleet manufacture, fuel production and road infrastructure) were calculated over a 720 000 vehicle km lifetime and increased the costs for diesel buses by 10 - 40% and natural gas buses by 60-80%. The total life-cycle costs for natural gas buses are of the order of 30 - 35% of those for diesel buses. No information was available on the life-cycle costs of the extra infrastructure needed for the gas system.

### 15.3.3.3 Comparison to the Results of a Previous Study

The results calculated in this case study were compared with the results of a study by Ekono Energy (Otterström, 1998) on Helsinki bus traffic, which used Finnish CVM-based WTP-values for health impacts. The two studies show very different results, with the earlier study showing the benefits of switching from diesel to natural gas to be FIM 31.5 million (based on the same line kilometres). The difference is explained by the differences in the emissions data, estimated concentrations and exposure-response functions. The nitrogen oxides and particle emissions in the earlier study seem to be somewhat higher. The $PM_{2.5}$ concentrations estimated by the Finnish Meterorological Institute for the 1998 study were about 10 times as high as those calculated by EcoSense. The estimation of the concentrations for the 1998 study was based on measured concentrations near the traffic routes. The nitrate concentrations estimated in 1998 were also relatively high. The slope of the health impact functions by particles has decreased by about 70% since 1998. Regarding the unit values, in the case of serious health impacts, the Finnish values are higher than those applied within ExternE. In the case of minor health impacts, the reverse is true.

### 15.3.4 Comparison of Total Costs for Diesel and Natural Gas Buses

The total operating costs of the buses in the Helsinki Metropolitan Area (Laurikko and Nylund, 1998) are FIM 11.4/km for diesel buses (EURO II) and FIM 12.4/km for gas buses, an 8% difference in favour of diesel buses (note 5.95 FIM = 1 Euro). As much as FIM 7.9/km of the total costs, consisting of salaries and other staff expenses, tyres, insurance and general expenses is estimated to remain constant in both fuel alternatives.

The costs of the fleet and maintenance (20 - 24% of total operating costs) are slightly higher for the gas buses and in addition, the refuelling infrastructure needed for the gas system introduces some extra costs compared with the diesel bus system. The total fuel costs of the diesel buses are higher. This is caused by the relatively heavy taxation of diesel fuel in Finland - fuel costs account for about 10% of the total operating costs. If the total costs of diesel and gas buses were compared without taxes, the difference in total operating costs would be 17% in favour of diesel buses.

The total operating costs of Helsinki bus traffic (34 million km in 1998) for the current diesel bus fleet is FIM 389 million per year. The equivalent value for a gas bus fleet is FIM 420 million per year. The associated environmental costs are FIM 12 million per year (FIM 16 million for the total life cycle) for diesel and FIM 3 million per year (FIM 5 million for the total life cycle) for natural gas. Thus, for all costs, the values are FIM 405 million for diesel and FIM 425 million for natural gas. However, the change from diesel to natural gas is more beneficial for central city lines, where the emissions are higher due to lower speeds and higher population weighted exposure. Therefore, the case study also looked at the replacement of the diesel buses with natural gas buses in the central city traffic only (leaving diesel buses on other lines). This policy would increase total operational costs from FIM 389 to 397 million per year. The associated environmental costs (including life cycle) for the centre city lines would fall from FIM 5.2 million (for diesel) to FIM 1.7 million (for natural gas) – an environmental benefit of FIM 3.5 million per year (including environmental life-cycle costs). Therefore total costs would go from FIM 405 million per year to FIM 410 million per year

The study shows that the environmental benefits gained with natural gas buses do not fully cover their higher operating costs. However, it is important to stress that not all environmental impacts were quantified and valued in this study. Additional benefits from gas buses include lower noise levels and, because of cleaner burning, less odour. The lower particle emissions from gas buses also reduce building soiling. Taking these effects into account would all increase the benefits of natural gas buses. Finally different hydrocarbons are emitted from diesel and natural gas though due to a lack of data it was not possible to take these effects into consideration.

The uncertainties in the emission data also cause uncertainty in the results. There was considerable variation in the data reviewed in this study. In addition, the uncertainty of the environmental cost estimates might be important. The confidence intervals estimated by Rabl (1999) for the environmental costs of different pollutants indicate the uncertainty with particle emissions is slightly higher than the uncertainty of nitrates (these two pollutants making the greatest contribution to the total results). While the proportion of particles in the total costs of diesel buses is slightly higher than that of natural gas, the uncertainty of the environmental costs in the case of diesel buses is slightly greater.

### 15.3.5 Conclusions

This case study has assessed the introduction of alternative fuel buses in Helsinki, relative to the existing diesel fleet. The results show significant benefits from the use of natural gas in terms of the reduction in environmental costs. The change from diesel to gas buses would lead to slightly higher operating costs, though it is stressed that the benefits gained with the gas buses might be greater than quantified in this study due to a number of additional benefits.

The environmental benefits estimated support the decision to replace diesel with natural gas buses in Helsinki, though the benefits are not as high as might be expected due to the existing use of lower polluting city-diesel fuel in the existing bus fleet.

## 15.4 France: Electric versus Conventional Vehicles: Social Costs and Benefits

Urban transport is one of the largest contributors to urban air pollution as well as to anthropogenic greenhouse gas emissions. Concern over these issues has propelled the electric vehicle (EV) to the forefront of the debate concerning technologies and strategies for reducing the environmental and health impacts of transport. With zero tailpipe emissions, compact size, and an essentially silent motor, electric vehicles could be the urban commuter's panacea were it not for its high purchase price. However, they have been criticised as "pollution elsewhere vehicles" because of the emissions by the power plants which supply the electricity.

There are few regions in Europe where the emissions trade-off is as favourable to the EV as in Ile-de-France (Greater Paris). 90 to 95% of the electricity of France comes from sources with negligible air pollution, nuclear providing almost 80% and hydro about 12% (for an explanation of the damage costs of nuclear, see the earlier section on nuclear electricity). Furthermore, because of the large population (about 11 million inhabitants) and population density, any air pollutants that are emitted here can have considerable health impacts.

The French policy case study assesses those external costs that differ between electric and conventional vehicles, namely those due to the emission of pollutants. The study is based on a paper published by Funk and Rabl (1999), but with damage costs adjusted with the new ExternE values. Accidents, congestion or noise were not addressed, although we note in passing that lower noise could be an appreciable external benefit of electric vehicles. Since the uncertainties are large, the sensitivity of the results is also considered to uncertainties in the damage cost estimates.

## 15.4.1 Vehicle Technologies and Costs

Table 15.5 presents key figures and hypotheses used to calculate the total cost for each vehicle. Since costs are realised throughout the lifetime of the vehicle, the choice of discount rate becomes an issue. Here 5% real are chosen, i.e. above inflation. All costs are in constant currency, without any escalation beyond general inflation.

The new vehicles studied here are representative of four-passenger subcompact cars popular in France. They are electric, gasoline and diesel versions of the same model, the Peugeot 106. The prices are for the model with electric windows, electric locks and power steering in 1998. The electric car has a Nickel-Cadmium battery because that is the preferred choice for vehicles of this size in France. Lead-Acid batteries contain nearly half the energy density, and are not favoured in a detailed life-cycle assessment comparison conducted by Electricité de France (Ollivier, 1996). It was assumed that the battery is leased rather than bought.

Table 15.5 Key data and assumptions for the economic analysis. These costs include taxes: approximately 80% of gasoline and 72% of diesel fuel are a fuel tax, and 17% of all other costs are a value added tax. Values for existing ("<97") cars are in ().

| | Electric | Gasoline | Diesel |
|---|---|---|---|
| Lifetime | 10 yr | 10 yr | 10 yr |
| Discount rate (above inflation) | 5% | 5% | 5% |
| km/yr | 9,125 | 9,130 | 16,425 |
| Purchase price, without battery, Euro[a] | 14,204 | 10,400 for >97 (0 for <97) | 11,577 for >97 (0 for <97) |
| Lease of battery, Euro/yr[b] | 1122 | | |
| Registration, Euro | 89 | 149 | 149 |
| Insurance, Euro/yr | 336 | 437 | 437 |
| Maintenance, repairs + technical control, Euro/yr | 89 | 200 | 200 |
| Electricity or fuel per km, urban driving[c] | 0.25 kWh/km | 0.070 l/km >97 (0.192 l/km <97) | 0.060 l/km >97 (0.113 l/km <97) |
| Energy price | 0.0696 Euro/kWh | 0.955 Euro/l | 0.649 Euro/l |
| Energy cost Euro/yr | 159 | 611 for >97 (1676 for <97) | 640 for >97 (1205 for <97) |

[a] Peugeot, 1998
[b] Club du Véhicule Electrique de Paris, 1998
[c] Electricité de France

For existing cars (labelled "<97" in our figures and tables), a purchase price of 0 is taken because from the point of view of society it is a sunk cost. For simplicity the same number of km/yr and the same costs for maintenance, insurance etc are assumed. While real repair costs are higher, this is compensated by the fact that the numbers for <97 cars represent an average over the current fleet and include models that are much larger than the subcompact Peugeot 106. That can be seen from the fuel consumption data in Table 15.5.

This point illustrates the difficulty of obtaining representative data about the current vehicle fleet. There is continual evolution in technologies, tastes and driving patterns, and reliable data about the current situation are hard to get. The evolution has been particularly rapid with regard to emissions. Furthermore, a single average number fails to capture the variability. For instance, a key parameter for the life cycle cost per km is the distance driven per year, which varies greatly between different owners. Rather than trying to analyse a large number of cases, the results are presented in a graphical format that we hope enables the reader to visualise how a change in one of the cost components would affect the ranking of choices.

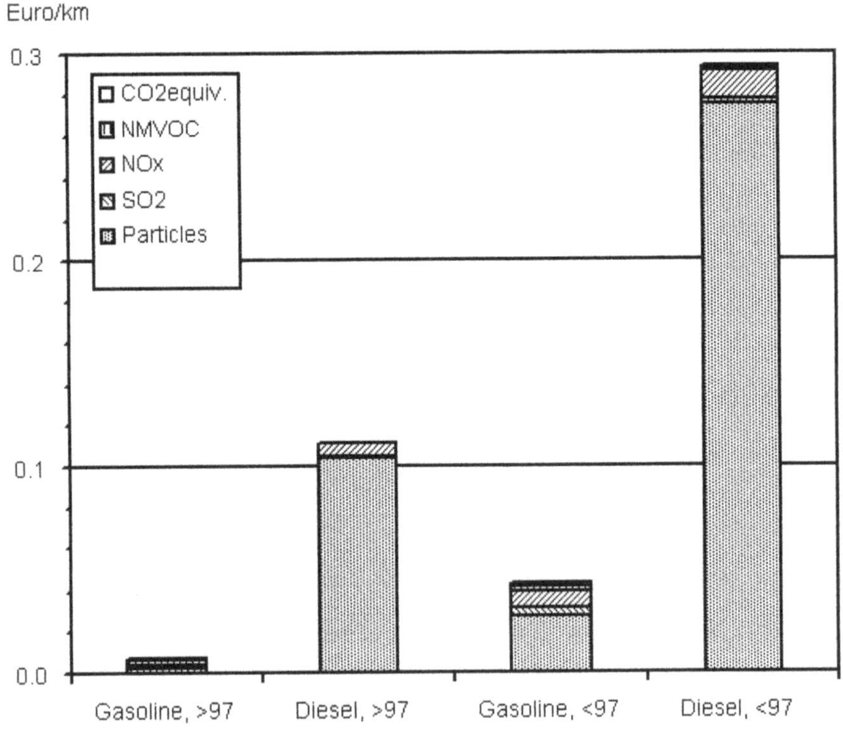


**Fig. 15.9.** Damage cost due to emissions from tail pipe and from refinery, for diesel and gasoline cars driven in the Paris metropolitan area. "<97" = pre 1997, ">97" = post 1997.

## 15.4.2
## Emissions and Damage Costs per km

The damage costs per km are shown in Fig. 15.9. The contribution of CO, according to the assumptions of ExternE, is so small that it is not even shown in the figures. We emphasise that they are for the Paris metropolitan area. Because of the high population density, the damage is dominated by particle emissions. For more rural driving the damage of the primary pollutants, especially the particles, would be one to two orders of magnitude smaller, while damage from $NO_x$ would change very little.

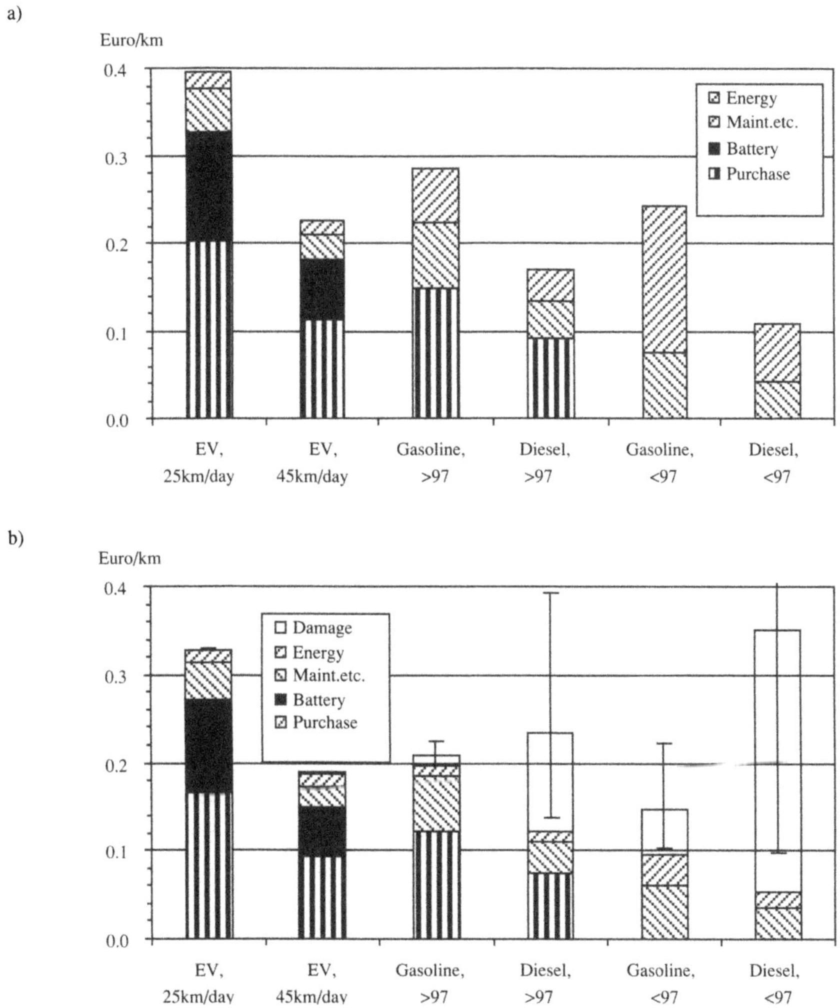


**Fig. 15.10.** Life cycle costs per km* **a)** private costs (incl. general taxes), **b)** social costs (excl. general taxes). *Vehicle lifetime for all vehicles is 10 years, but annual mileage is 9125 km/yr for electric and gasoline and 16425 km/yr for diesel. "<97" = pre 1997, "new" = post 1997.

In Fig. 15.9 the emissions from the following phases are taken into account: emissions from electricity production, emissions from fuel production, and tailpipe emissions. Emissions for other life-cycle phases, such as vehicle production and disposal, are more difficult to estimate given the multitude of options and actors involved throughout the lifecycle. Since the data provided by Lewis and Gover (1995) suggest that emissions from the production of cars is small compared to those from the tailpipe, they will not be considered in this analysis, implicitly assuming that their external costs are equal among EVs and ICEs.

### 15.4.3 Life Cycle Costs/Benefits

The results of the life cycle costs per kilometre are shown in Fig. 15.10. As diesel and gasoline cars are used differently, the EV are shown both for 25 km/day and 45 km/day utilisation. This allows comparisons to be made on the basis of equal driving distance, but neglects that higher utilisation is likely to increase repair costs or decrease vehicle life-time. Part a) shows the private costs, part b) the social costs. From the point of view of society, general taxes represent a transfer of funds rather than a cost. Therefore all general taxes were removed from part b), keeping only vehicle registration fees because they are paid for expenses related to road transport (the Vignette, a national auto tax based on vehicle size, age, and location of registration, and the carte grise, a car registration fee).

The removal of taxes between parts a) and b) is not perfect. Fuel taxes could be considered as payment for public expenditures for transport, but their level is set by politics without direct relation to specific expenditures. In any case, the resulting uncertainties are small compared to the other items in Fig. 15.10 b).

Likewise subsidies should be included in part a) but not in part b) of Fig. 15.10. Currently the EV benefits from a subsidy by Electricité de France to car manufacturers (10,000 F = 1,524 Euro, i.e. 11% of purchase price) and a subsidy by the State to consumers (5,000 F = 762 Euro, i.e. 5.5% of purchase price). It is not clear how much of the Electricité de France subsidy is passed on to the consumer. For simplicity these subsidies were not considered in the Figure; in any case they would only change the contribution of the purchase price by a small amount.

In terms of private costs the EV is about 30 to 40% more expensive than a new ICE car, comparing respectively the gasoline car with an EV driven 25 km/day and the diesel with an EV driven 45 km/day. For existing cars the comparison is even less favourable for the EV; this would not change appreciably if one were to include the resale value (highly variable) which was neglected because it is a sunk cost from the perspective of society.

In terms of total social cost the EV is at least 50% more expensive than the gasoline car, both <97 and new, but about 20 to 40% less expensive than diesel cars. The latter is due to the high damage cost of primary particle emissions in a metropolis like Paris.

These estimates are based on currently available EV technology and battery usage over 10 years. This lifetime requires the equivalent of two battery packs (leased or purchased), which represent almost as much cost as the purchase of the vehicle itself. Costs are certain to decrease as technology improves (with production economies of scale, higher autonomy, advanced batteries, etc.). It remains to be seen if this results in more rapid reductions in total cost than those from incremental improvements in advanced ICE vehicles themselves.

The error bars in Fig. 15.10 b) indicate the uncertainty of the damage costs only, as estimated by Rabl and Spadaro (1999). A cost-benefit analysis of a policy to encourage the use of EVs involves many other uncertainties, especially driving patterns and the evolution of cost and performance of the technologies. Can improvements in battery technology keep pace with emission reductions of the ICE? This kind of uncertainty is best addressed by considering different scenarios, but it is beyond the scope of the present analysis. However, the magnitude of the cost components in Fig. 15.10 gives an idea of the sensitivity to different assumptions.

### 15.4.4 Conclusions

Using the damage cost estimates of the ExternE Project, the external cost of air pollution were calculated for electric vehicles and for conventional cars fuelled by gasoline or by diesel. It was found that the cost of air pollution from cars in Paris is large, even for cars that meet the rather strict emission limits of the latest regulations. The damage cost amounts to approximately 4% of the private life cycle cost for new gasoline cars, and 70% for new diesel cars. If these costs were internalised, the EV would be competitive at current prices against diesel cars but not against gasoline (the precise comparison depends on the individual situation, especially the number of km travelled per year). In terms of total social cost the EV is at least 50% more expensive than the gasoline cars, both <97 and new, but about 20 to 40% less expensive than diesel cars.

It is ironic that the diesel looks so poor in this comparison; after all it has been encouraged on the grounds of higher energy efficiency and lower emissions. Indeed, except for particle emissions the diesel is relatively clean, especially compared to gasoline cars without catalyst (introduced in France since 1993). The perspective presented here results from the recent epidemiological studies that have highlighted the health effects of particles, coupled with the fact that primary pollutants can have large impacts if emitted in the streets of large cities.

However, the cost of air pollution is not enough to give the EV a clear advantage against all conventional cars, even in Paris. If the EV can barely be justified in a large metropolitan area where the external costs of electricity are small, the concept loses much of its appeal. This conclusion comes on top of questions about the appeal of vehicles whose range and performance is still significantly below their conventional counterpart.

For policy decisions it would be helpful if the uncertainties could be reduced, but that is unlikely in the near future, at least as far as the damage costs are concerned. A key consideration is the evolution of the respective technologies. One can expect significant cost reductions in a new technology such as the EV (including development of the infrastructure). On the other hand, the emission control of the internal combustion engine has been making impressive progress in recent years and technologies for the reduction of diesel particle emissions are under development. It may be more cost-effective to curb pollution by technologies other than the EV (e.g. hybrid or fuel cell vehicles) or by policies such as scrapping older highly polluting cars.

## 15.5 Germany: Benefits of Introducing CNG-Fuelled Vehicles in the Federal State of Baden-Württemberg

Due to the increasing knowledge of the adverse effects of air pollutants, it is becoming more and more interesting to compare alternative technological strategies for reducing emissions in an efficient and cost-effective way. The introduction of new vehicle technologies with alternative fuel systems is one such option for emissions reductions, but requires considerable investment. It is therefore very desirable to compare these additional costs against the environmental benefits they achieve – the reduction in external costs, as calculated within ExternE, is an excellent way of putting these costs and benefits in context.

The German policy case study has looked at such an assessment for the road transport sector in the federal state of Baden-Württemberg. The study looked at the introduction of alternative fuelled vehicles, building on a project ("strategies to reduce particulate and other emissions of road transport due to the improved use of CNG- and fuel cell vehicles" (Krüger et al., 1998)) supported by the federal government under the "Umweltplan Baden-Württemberg".

The federal funded project aimed to analyse technological options and potentials to reduce emissions due to a change in the vehicle fleet by the (supported) introduction of CNG fuelled or fuel cell powered vehicles. The time scale under assessment was from 1995 to 2020. As future scenarios depend heavily on the assumptions of introduction and fleet turn-over, the potentials with respect to technology availability and infrastructure needs (e.g. fuel stations etc.) of the use of CNG vehicles were discussed in detail. The limited radius of action of CNG fuelled vehicles together with the insufficient network of refuelling stations (at present) were discussed as particularly important issues. The study assessed direct emissions and also emissions of up- and downstream processes were taken into account. These are emissions used as the starting point for this policy case study, applying the ExternE methodology to assess the environmental benefits (in economic terms) of the CNG-scenario for 2010.

### 15.5.1 Emission Scenarios

Table 15.6 gives the emissions of the reference scenario in the year 2010 and the scenario with enhanced use of CNG vehicles (the CNG scenario). Both scenarios were developed taking into account fleet turnover (usage of modern technology), improved fuel quality (e.g. sulphur content corresponding to EU regulations) and increased mileage due to higher demand. In addition to the reference scenario, the CNG scenario considered the substitution of conventional fuelled vehicles with CNG-fuelled vehicles. The actual substitution is mainly determined by mobility restrictions due to the technically and economically limited CNG storage capacities, as well as assumptions for the time needed for the introduction in the vehicle fleet. The theoretical potential for the substitution of private conventional passenger cars with CNG fuelled cars is calculated as 24.7% (for the year 1995). The potential for commercial used passenger cars is found to be 20% at maximum. For the year 2010, it is assumed that 80% of the theoretical potential has been put into practice.

From a technical perspective, it is assumed that all urban buses can be converted to CNG (it is assumed that 80% are converted by 2010). In contrast, coaches have a low potential for conversion to CNG due to their larger travel distances. It is therefore assumed that only 10% are converted to CNG by 2010. Heavy and light duty vehicles are distinguished into short- and long-distance traffic (assuming of 80% of short-distance 10% of long-distance for total freight transport mileage). No substitution is assumed for truck trailers and articulated trains.

**Table 15.6.** Direct and evaporative air pollutant emissions and emissions from up-stream processes for the reference and the CNG scenario of the year 2010 (kt) (Based on handbook of emission factors, taken from (Krüger et al., 1998)).

| Pollutant | | Reference Scenario 2010 | CNG Scenario 2010 |
|---|---|---|---|
| VOC | direct | 12.428 | 12.192 |
| | evaporative | 2.738 | 2.343 |
| | up-stream | 4.579 | 10.807 |
| | *total* | *19.745* | *25.341* |
| NMVOC | direct | 11.649 | 10.321 |
| | evaporative | 2.738 | 2.343 |
| | up-stream | 2.814 | 2.372 |
| | *total* | *17.202* | *15.036* |
| $CH_4$ | direct | 0.779 | 1.870 |
| | up-stream | 1.765 | 8.435 |
| | *total* | *2.544* | *10.305* |
| CO | direct | 59.120 | 53.998 |
| | up-stream | 1.402 | 1.858 |
| | *total* | *60.523* | *55.855* |
| $NO_x$ | direct | 49.511 | 44.387 |
| | up-stream | 5.579 | 5.243 |
| | *total* | *55.090* | *49.630* |
| $SO_2$ | Direct | 0.662 | 0.561 |
| | up-stream | 10.325 | 8.781 |
| | *total* | *10.987* | *9.342* |
| $PM_{2.5}$ | Direct | 1.914 | 1.704 |
| | up-stream | n.a. | n.a. |
| | *total* | *1.914* | *1.704* |
| $CO_2$ | Direct | 20828 | 20520 |
| | up-stream | 2035 | 1936 |
| | *total* | *22863* | *22456* |

## 15.5.2 Implementation and Results

To determine the local effects of air pollution, it is important to consider the environment where the emission takes place. In addition to variations in population density in general, it is important to distinguish between emissions within and outside built-up areas. Furthermore, meteorological parameters play an important role for the dispersion of pollutants in the boundary layer of the earth.

The source study provided the emissions by pollutant and road category (urban, extra-urban, highway) as a total sum for the whole federal-state of Baden Württemberg, which were calculated by means of emission factors and total annual mileage. Therefore it was necessary to model the spatial distribution of these emissions. Based on the data used in the aggregation case study, described in Chap. 14, the emissions were spatially dis-aggregated assuming the same mileage and emission characteristics distribution. The % share of total emissions was calculated for each road section, vehicle category and pollutant. Evaporative emissions were added to the direct emissions, whereas emissions from up-stream processes were treated as described in Chap. 10.

The externalities of the reference scenario as well as the CNG scenario were then calculated, using the methodology described in Chap. 14. The external costs for the reference scenario were estimated as 551 Million Euro. Of this, the damages (55.7%) were dominated by nitrate and ozone impacts at the regional scale from direct emissions of $NO_x$. The remaining major impacts were from local particle damages to human health (15.5%), regional damages from fuel production emissions (13%) and global warming (10%).

The external costs of the CNG scenario totalled 494 Million Euro. Thus, the total benefits of the direct emissions reductions amounted to 56 Million Euro, 10% of the external costs of the Reference Scenario. These benefits were due to reductions in $NO_x$ emissions (responsible for 55.6% of total benefit), particle emissions (15.2% of benefit) and emissions from up-stream processes (16.9% of which more than ¾ are due to reduced $SO_2$ emissions). The results are given in Table 15.7 and Table 15.8.

**Table 15.7:** External costs due to direct emissions (Million Euro).

| Pollutant | Reference scenario 2010 | | | CNG scenario 2010 | | |
|---|---|---|---|---|---|---|
| | local | Regional | Global | local | Regional | global |
| PM | 85.64 | n.a | -- | 75.54 | n.a. | -- |
| $SO_2$[a] | 0.75 | 1.43 | -- | 0.63 | 1.21 | -- |
| CO | 0.09 | n.a. | -- | 0.08 | n.a | -- |
| $NO_x$[b] | -- | 307.02 | -- | -- | 275.25 | -- |
| NMVOC[c] | -- | 28.75 | -- | -- | 25.30 | -- |
| $CO_2$ | -- | -- | 49.99 | -- | -- | 49.25 |
| Methane | -- | -- | 0.03 | -- | -- | 0.08 |

[a] direct and via sulphates; [b] via nitrates and ozone; [c] via ozone

**Table 15.8:** External costs due to emissions from up-stream processes (Million Euro).

| Pollutant | Referencescenario2010 | CNGscenario2010 |
|---|---|---|
| PM[a] | n.a | n.a |
| $SO_2$[b] | 51.46 | 43.77 |
| CO | n.a | n.a |
| $NO_x$[c] | 16.30 | 15.32 |
| NMVOC[d] | 4.65 | 3.92 |
| $CO_2$ | 4.88 | 4.65 |
| Methane | 0.079 | 0.379 |

[a] no emissions available [b] direct and via sulphates; [c] via nitrates and ozone; [d] via ozone

### 15.5.3 Concluding Remarks

Quantifiable environmental damages due to road transport in Baden-Württemberg could be reduced by 10% by replacing conventional fuels by CNG. Damages are reduced for all species considered, except for methane. But results show that the increased costs due to methane emissions are negligible compared to the overall reductions.

## 15.6 Greece: Benefits from Introducing CNG Buses in Athens

Although the operation of the bus network is a minor source of total emissions in the city of Athens, people consider it a serious pollution problem because of the age of the vehicles. The objective of the Greek policy case study was to look at this issue and estimate the environmental impacts caused by the operation of the bus network, and to assess the benefits from replacing old diesel buses with CNG buses. The study also set out to assess the optimal number of CNG buses through the use of cost-benefit analysis.

### 15.6.1 The Public Transport Network of Athens

The public transport network of Athens consists of an extensive bus network and one metro line serving 410 and 95 million passengers per year respectively. Public transport accounts for 31.7% of the total travel demand, while private vehicles have 54.5% of the modal share.

The bus network covers the entire greater Athens area and has 280 bus lines with an overall length of about 6,000 kilometres. The operation of this network requires the use of 1,500 buses and 290 trolley buses, running about 95 million vehicle kilometres per year. The bus speeds range from 8-12 km/h in the urban core of Athens to 24 km/h on suburban lines. In 1998 about 40% of the buses complied with the EURO I standard, while the rest of the vehicles had an average age of 16 years. The use of these old vehicles causes severe problems to the air quality of Athens due to the high pollutant emissions. Recently, in an effort to improve the atmospheric environment and the image of public transport, the authorities decided to start the renewal of the bus fleet by purchasing a significant number of new buses, including 290 CNG buses. It was estimated that the use of these CNG buses alone would result in a reduction of CO emissions by 41.16 tonnes/year, $NO_x$ emissions by 327.1 tonnes/year, PM emissions by 17.05 tonnes/year, $SO_2$ by 7.4 tonnes/year and VOC by 37.1 tonnes/year.

### 15.6.2 CNG Buses

The use of CNG as a fuel for road vehicles is already a widespread technology - more than two million vehicles operate in about 40 countries around the world. The main differences between a CNG and a diesel bus are located in the vehicle engine and fuel tanks. CNG buses use spark ignition engines, which like all Otto-cycle engines have a performance about 20% below that of diesel engines. The fuel is stored in high-pressure tanks (about 200 bars) which are usually located on the vehicle roof, making coach-work reinforcements necessary. The resulting increase in bus weight due to CNG tanks is around 1,000 kg. Two types of compression stations can be used for refuelling: fast filling taking 10-15 minutes and slow filling taking 6-8 hours (usually overnight). The reliability of CNG buses has proved to be similar to diesel vehicles as are their maintenance costs. The use of CNG buses results in significantly reduced emissions of PM and $SO_2$. Hydrocarbon emissions mainly consist of $CH_4$, which has no harmful effects on the local environment. However, $CO_2$ emissions are higher than those of a diesel bus, due to the increased fuel consumption of CNG engines.

### 15.6.3 The Environmental Cost of the Bus Network of Athens

The transport externalities of the entire bus network of Athens were estimated by adding the costs of each one of the 280 lines calculated with the ExternE methodology, using bus lines maps and data concerning the daily number of trips and the average bus velocities. The environmental cost of each line was assessed for four different bus technologies:

- Early 90's diesel bus
- Diesel bus EURO I
- Diesel bus EURO II
- CNG bus

The results showed that there is a great variability of the resulting damages per vkm depending on the location of the bus line. Some indicative results for three specific bus lines are given in Fig. 15.11.

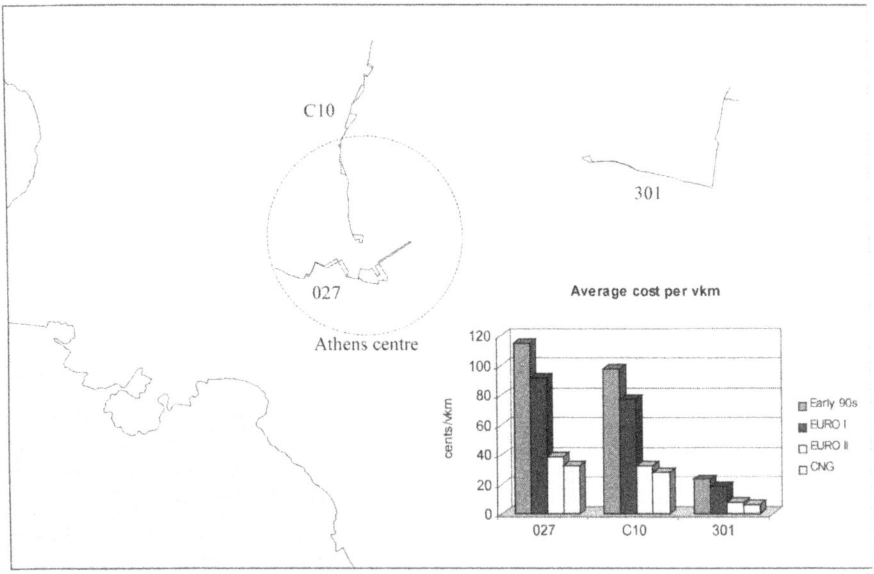


**Fig. 15.11.** Location and cost per vkm of bus lines "027", "301" and "C10"

It is obvious that line 027, located in the Athens centre, has the highest costs per vkm, followed by line C10, connecting Athens centre with a suburban area. Line 301, being a suburban line, has the lowest damages per vkm. The average and the maximum damages per vkm of all the bus lines and all bus technologies are listed in Table 15.9.

The externalities caused by the operation of the entire Athens bus network were calculated using the resulting damage values per vkm for each bus line and data concerning each line's length, the number of trips per day and the bus types used. The total cost under the current status and the costs that would result if all bus lines were using a specific bus technology are displayed in Fig. 15.12.

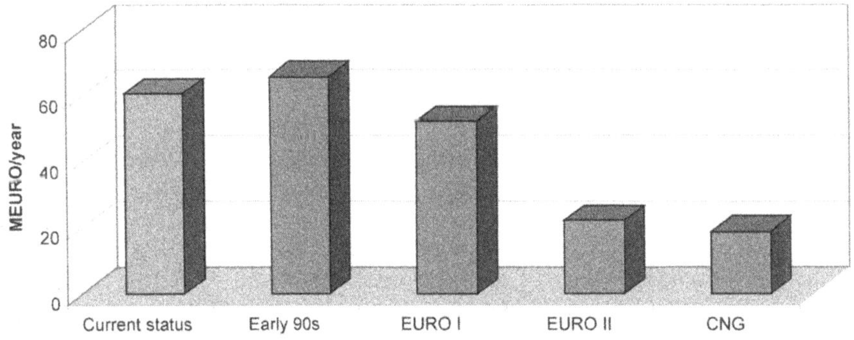


**Fig. 15.12.** Total damage cost from the operation of the bus network

**Table 15.9.** Average and maximum damages of all bus lines

| Damages [Eurocents/vkm] | Average | Maximum |
|---|---|---|
| Early 90s diesel bus | 67.99 | 115.49 |
| Diesel bus EURO I | 53.97 | 91.68 |
| Diesel bus EURO II | 22.95 | 38.98 |
| CNG bus | 19.49 | 33.11 |

### 15.6.4 Evaluation of the CNG Buses Use

In order to estimate the benefits that would result from the introduction of the 290 CNG buses in the Athens bus fleet, it was assumed that these buses would operate in the lines having the highest damages per vkm. It was found that the CNG buses could serve about 50 lines, most of which are located in the city centre, and that they would travel in total about 65,000 vkm/day. The total daily damages resulting from the operation of the CNG buses were estimated to be 16,390 Euro, while the average damage per vkm was 25.35 Eurocents/vkm. The benefits due to the substitution of the old technology buses were 39,360 Euro/day, and were calculated by comparing the total daily cost of the whole network before and after the operation of the CNG buses.

Besides the environmental benefits, the operation of the CNG buses requires significant capital costs for the construction of the refuelling stations and the purchase of the vehicles, as well as increased operational costs mainly due to the higher fuel consumption. From the perspective of the bus operator, such an investment would not be seen positively. However, from a social perspective, it is possible that the external benefits resulting from the use of CNG buses could exceed the capital and the increased operational costs. In that case, the state, being the conveyor of social interest should provide the means (subsidies, low interest loans) in order to convert a socially interesting plan into a profitable investment. In order to evaluate such an investment, the resulting costs and benefits, including the social ones, must be specified. The basic possible elements of cost and benefit of an investment plan concerning the substitution of diesel with CNG buses are:

- Capital costs: Refuelling station, Maintenance facilities, CNG buses
- Operational costs: Fuel cost, Labour cost, Maintenance cost
- Operator benefits: Trip time decrease, Increase of passenger volume
- Social benefits: Trip time decrease, Environmental benefits

In the following analysis the costs and benefits resulting from the operation of 50 to 550 CNG buses in the public transport network of Athens are estimated. The various costs of the investment plan are listed in Table 15.10. It is assumed that the use of CNG buses will not cause changes in trip times and passenger volumes and that the labour cost (driver and administrative) will remain the same. The

**Table 15.10.** Fixed and operational costs for CNG buses [Euro]

| | |
|---|---|
| Refuelling and maintenance facilities | 3,800,000 |
| CNG bus price | 265,000 |
| CNG bus fuel cost per vkm | 0.138 |
| Diesel bus fuel cost per vkm | 0.111 |
| CNG bus maintenance cost per vkm | 0.083 |
| Diesel bus maintenance cost per vkm | 0.051 |

Source: Chandler et al. (1996)

lifetime of the project is considered to be 10 years. The social benefits resulting from the operation of the CNG buses were calculated over the lifetime of the project by comparing the total daily cost of the whole network with and without their operation.

The resulting benefits as a function of the number of CNG buses used are shown in Fig. 15.13. As the number of CNG buses in the network increase the resulting environmental benefits per bus decrease, as they have to operate in lines outside the city centre where the benefits per vkm are lower.

The impact of this decrease to the total net present value of the investment is shown in Fig. 15.14, where there appears to be an optimal number of CNG buses. The net present value and the optimal number of buses are highly dependent on the rate used for discounting. Two different rates of time preference have been used for comparison, 0% and 3%. As shown in Fig. 15.14, increased time preference rates decrease the optimal number of buses and the efficiency of the investment, which in both cases produces positive results.

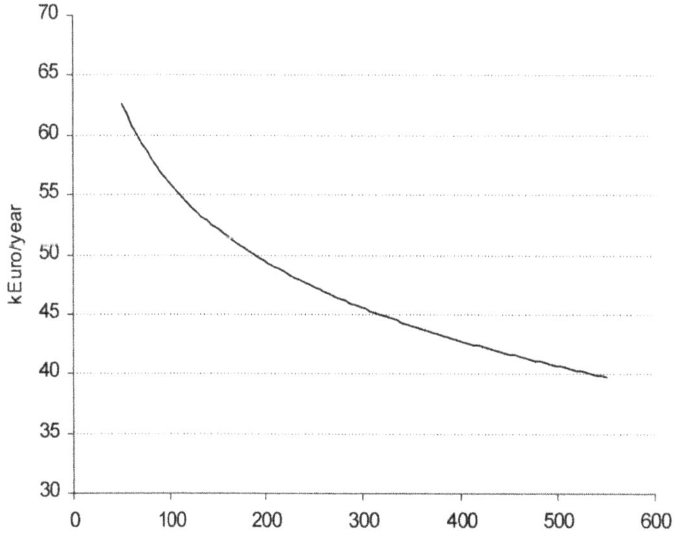


**Fig. 15.13.** External benefit per CNG bus

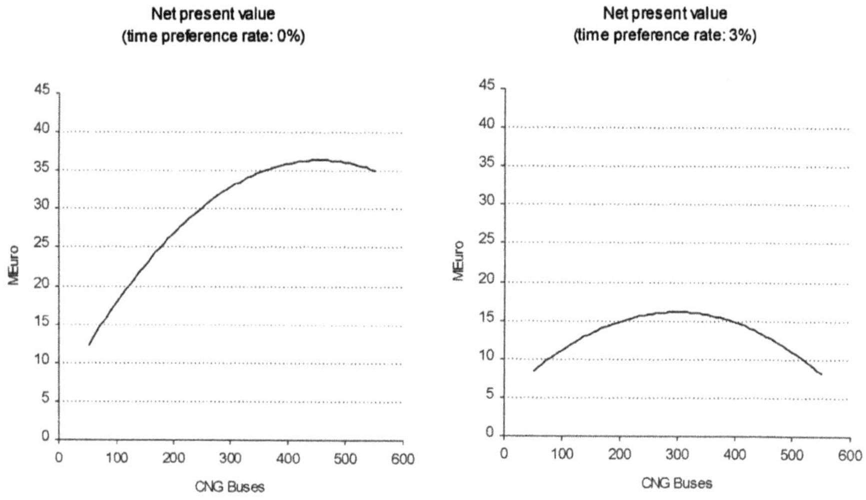


**Fig. 15.14.** Net present value, including social benefits, as a function of the number of CNG buses

### 15.6.5 Concluding Remarks

The bus network of Athens is the main public transport system of the city. It is characterised by the existence of old buses, which cause significant environmental burdens to the local environment and deteriorate the image of public transport. The assessment of the externalities resulting from the operation of the bus network, using different bus technologies, revealed a great potential for the substitution of the old buses with new technology or alternative fuelled vehicles. A cost-benefit analysis was conducted for various scenarios of substitution of old vehicles with CNG buses. The analysis showed that such an investment would have positive results, if the resulting social benefits are included, and that the optimal number of CNG vehicles varies between 300 and 450 vehicles depending on the time preference rate.

## 15.7 A Life Cycle Analysis of Tyres in Western Europe

A system dynamics model was developed to analyse the life cycle of truck tyres in Western Europe (van Beukering and Janssen, 2000). The main purpose of the model is to determine the most effective measures to improve the environmental performance of the truck tyre lifecycle. The model is calibrated for the period 1990-1999 and predicts developments for the period 2000-2020. The region selected includes Germany, Netherlands, United Kingdom, Belgium, France, and Luxembourg. As shown in Fig. 15.15 the model incorporates the complete life cycle of truck tyres. The main stages identified include production, consumption, transport and solid waste management (SWM). The model allows for technological and behavioural changes over the long term and takes into account both private and external effects as monetary costs for each stage in the lifecycle. The sum of these costs is defined as the social cost. To calculate the external costs, the standard values for human health, crops, forests, material damage and global warming from the aggregation analysis for the Netherlands were used (see Chap. 14). External values for amenity and ecosystems were excluded from the analysis.

Four scenarios reflecting issues in the current tyre lifecycle were tested:

- In the base case, technological improvements are extrapolated based on the current rate of change: the durability of tyre improves from 250 000 kilometres in 1990 to 300 000 kilometres in 2020 and fuel efficiency improves by 1.5 percent annually.
- In the Eco-Tyre Scenario, tyres are introduced that are more expensive – in the order of 15 to 25 Euro per tyre – but consume up to 5 percent less fuel. The market share of the eco-tyre increases from 0 percent in 2000 to 40 percent by 2020.
- The majority of the drivers under-inflate their tyres. This leads to higher fuel consumption and a more rapid wearing out of the tyre. In the 'pressure scenario' consumers reduce under-inflation from 10 percent in 2000 to 0 percent by 2020.
- Increased lifetime of tyres decreases the overall demand for new tyres. The 'lifetime scenario' simulates a gradual increase in the lifetime of truck tyre from 250 000 in 1990 to 400 000 km by 2020.

Several conclusions can be drawn from the model simulations (Fig. 15.16 and Fig. 15.17). Almost 95 percent of external costs in the lifecycle of truck tyres are from human health damage arising in the consumption stage. The main pollutants are $PM_{10}$ and $NO_x$, which account for 58 and 31 percent of human health costs, respectively. The contribution of global warming to the external costs is only 2 percent. Damage to crops has a significant negative value because the ozone depleting effect of $NO_x$ has a positive impact on crop yields and exceeds the negative impact of VOC and $SO_2$ on crop yields.

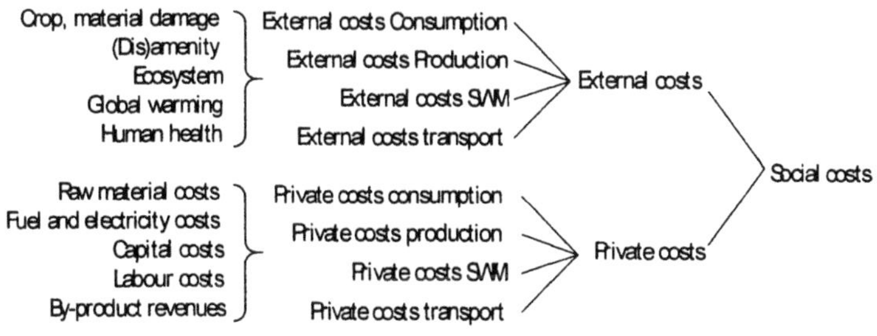


**Fig. 15.15.** Cost structure of the model

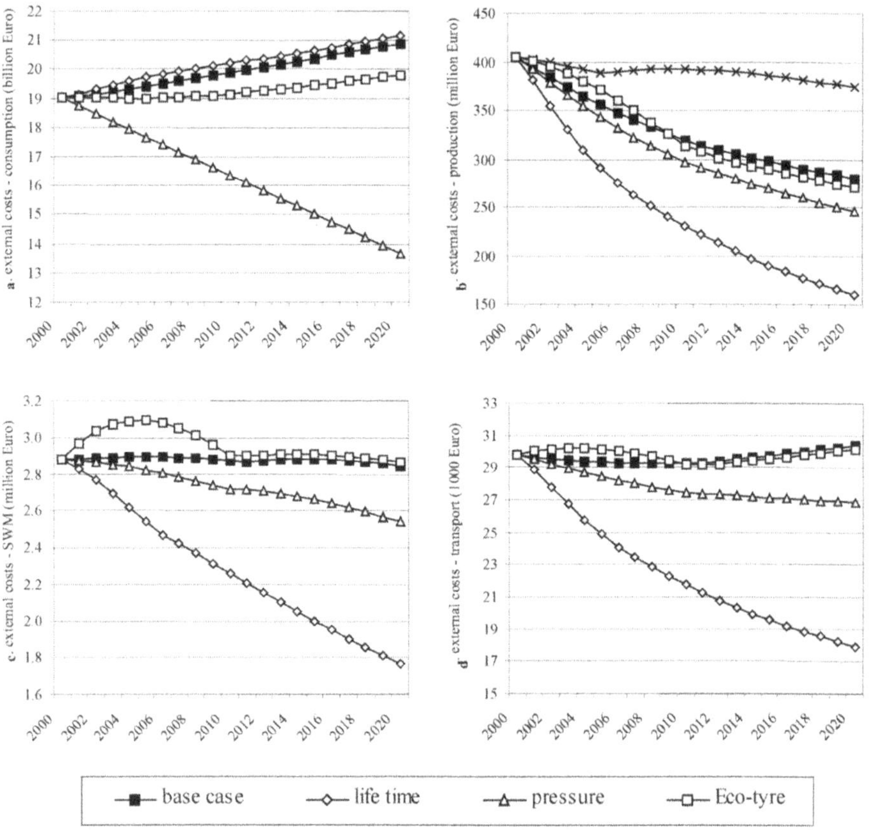


**Fig. 15.16.** Aggregated external costs in separate stages of the lifecycle of truck tyres

Although most policy interventions in the tyre lifecycle currently focus on waste managers and tyre-manufacturers, the consumer is the most powerful stakeholder when it comes to improving the environmental performance. By improving the monitoring and management of the tyre pressure, environmental gains can be achieved that are not achievable by technological or economic changes in the other stages. The second best option – utilising eco-tyres rather than the normal tyre – also lies within the domain of the consumer. Typically, both options can be achieved at negative private costs. In other words, the environmental improvements in the tyre cycle with the highest potential are both win-win options. Deficiency in driver-awareness and lack of user-friendly pressure equipment may be reasons why consumers are not already using these options.

The role of tyre-manufacturers is still of crucial importance in creating a more sustainable tyre lifecycle. By improving the durability and fuel efficiency of tyres, significant improvements can be achieved. From the perspective of the tyre manufacturer it is more rational to focus on fuel efficiency because this avoids

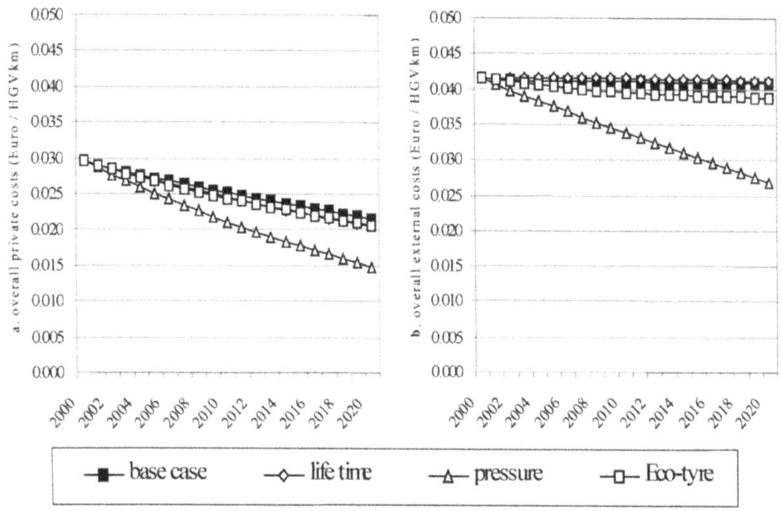


**Fig. 15.17.** **a)** Private and **b)** external costs aggregated over the lifecycle of truck tyres. Note: The unit Euro/HGVkm expresses the overall private and external cost per kilometre of a Heavy Good Vehicle (HGV) with a capacity of 10 tonnes and a loading efficiency of 75 percent.

shrinking the sales market. In introducing eco-tyres in the traditional market, however, it is important that the retreading industry simultaneously makes the shift to these new tyres. Retreading techniques require certain modifications that are only economically feasible if the retread market for eco-tyres has sufficient volume. Therefore, the government could play a role in developing and promoting new retreading techniques for eco-tyres.

The waste stage of tyres also deserves additional attention. Eventually every tyre will reach the end of its functional life and therefore has to be utilised as efficiently as possible. To create incentives for the individual member countries, the European Commission defined strict targets that need to be met in the next few decades. Most standards simulated in the model are in line with these targets.

It can be concluded that a model has been developed for the lifecycle of truck tyres in Western Europe. This model is dynamic in nature and incorporates environmental and economic criteria. Tentative results indicate that more than 95 percent of the overall environmental impact during the life of a tyre occurs before disposal. This result implies that the emphasis in environmental policies related to tyres should shift from the waste stage to the use stage (i.e. additional fuel consumption due to operation at too low tire pressure).

## 15.8 Spain: Assessment of Environmental Externalities of the Use of Biofuels in Transport in a Mediterranean Region

The Spanish Promotion Plan for Renewable Energies (IDAE, 1999) has a strategic objective to introduce 300 000- 500 000 toes of biofuels for transport. This figure represents 1.8% of the overall gasoline and diesel demand in the transport sector in Spain. The Spanish Policy Case Study assessed the environmental benefits from the introduction of biofuels for transport in a region of Spain.

Bio-diesel fuel is included within the objectives of the Plan but at a low level and therefore has not been included in this analysis. Production of ETBE from bio-ethanol is constrained by the isobuthilene production capacity of Spanish refineries. Therefore, although in a first phase of the Plan, ethanol would be used as raw material for the production of ETBE, the final objective in a second phase resides in the direct use of ethanol mixed with gasoline.

The case study area was the region of Castellón de la Plana, an area of Mediterranean coast that suffers from traffic jams in summer due to tourism. The area includes the city of Castellón, which is also an important population and industrial centre which itself generates significant transport activity. The area has specific topological and meteorological features, which combined with local emissions, result in persistent tropospheric ozone concentrations. This case study example is therefore particularly relevant in reducing atmospheric emissions in the region.

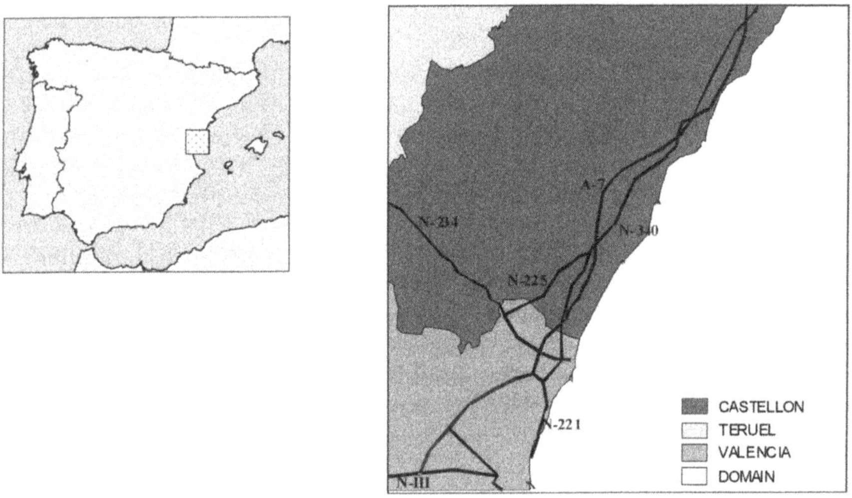


**Fig. 15.18.** Case study domain (left) and daily average traffic intensities (right)

## 15.8.1
## Routes and Vehicle Technologies

The domain under study is a 100km x100km square with the centre in Burriana (Castellón de la Plana). Six different motorways and roads within the domain have been analysed. This includes the N-221, a small but very congested ring road around Valencia. Figure 15.18 shows the domain area and the roads and motorway under study. The length of these roads is shown Table 15.11.

Transport activities in the region were characterised gathering data on the number of vehicles depending on their category, class, fuel and cylinder capacity, and daily average traffic intensity in each road. Eight vehicle technologies have been studied from the many types of vehicles driving along these roads, eight fed on fossil fuels and one on biofuels. The fossil fuel vehicles are:

*Passenger cars:*

1. Gasoline (95-97) 1.2-2.0 l. and ECE 15/04, representing 38% of passenger car activity and 26% of the total transport activities in the region.
2. Gasoline (95-97) <1.2 l. and ECE 15/04, representing 14% of passenger car activity and 9.5% of the regional total.
3. Diesel 96 <2.0 l., representing 15% of car activity and 10% of the regional total.

*HDV:*

4. Diesel 96 3.5-7 Tn, representing 43% of HDV activity and 2% of the regional total.
5. Diesel 96 7-16 Tn, representing 25% of HDV activity and 1.2% of the regional total.

*LDV:*

6. Diesel 96, representing 58% of LDV activity and 10% of the regional total

*Buses:*

7. Early 90's, diesel 96, representing 69% of bus activity and 0.09% of the regional total.
8. EURO I, diesel 96, representing 24% of buses and 0.03% of the total

Together these represent 75.3% of the transport activities in the region. The final technology (9) were buses running on an E95 mixture, comprising ethanol (95%) and gasoline (5%).

**Table 15.11.** Main roads and motorways length in kilometres

| A-7 | N-340 | N-III | N-234 | N-221 | N-225 |
|---|---|---|---|---|---|
| 129.84 | 88.31 | 23.88 | 50.77 | 16.73 | 22.49 |

### 15.8.2 Bioethanol Introduction Scenario

The objective of the Spanish Promotion Plan for Renewable Energies is set at 300,000-500,000 toes/year of biofuels by the year 2010.

Biodiesel is considered to be less competitive. Therefore, most of the effort is being directed to the introduction of bioethanol, initially used as raw material for the production of ETBE and later mixed in different proportions for captive fleets and eventually for all vehicles.

The scenario analysed looks at the introduction of ethanol mixed in a high proportion (95% of ethanol and 5% gasoline (E95)) in a fleet of buses. These would displace the existing vehicles in the area assessed. The reference year for the evaluation is 1996. The current average distance driven per vehicle per year is 36500 km and the average fuel consumption is 40.5 litres per 100 km. The scenario assumes 87% of the bus fleet would be converted to E95. This percentage translates into a fleet of 272 E95 buses, replacing the same number of the oldest buses. The amount of ethanol required for the scenario is 3815 $m^3$ per year - representing around 1940 toes.

The raw materials for the production of ethanol considered in the Promotion Plan are cereals and sugar beet. In this policy case, wheat has been selected as a representative cereal crop for the production of ethanol. With the current wheat yields in the area and considering a conversion factor of 2.6 t of wheat per $m^3$ of ethanol, around 6200 hectares of wheat would be needed to produce the required amount of ethanol.

### 15.8.3 Results

#### *15.8.3.1 Ethanol Production from Wheat*

In the scenario, ethanol is produced from wheat grown in set-aside cereal areas in the Castilla-La Mancha region. Wheat yield in the area is 1.6 tonnes per ha with a 12% moisture content. A Life Cycle assessment of the agricultural production of winter wheat in this area of Spain has been performed following the methodology developed by Cowell (1998).

Data on emissions from the processes involved in the conversion of wheat to ethanol were taken from existing lifecycle assessments of alternative fuels (ETSU, 1995).

The fuel consumption of E95 coaches is similar to that of diesel buses as the lower energy content of ethanol is counterbalanced by the greater fuel efficiency and power density of ethanol due to its higher octane rating (Pitstick, 1993). Total emissions per vehicle km are shown in Table 15.12

Emissions of the wheat production stage are very high, mainly due to the low yield of wheat in this area of Spain (1.6 t/ha). Transformation of wheat to ethanol is a very energy intensive process and therefore the associated atmospheric emissions from this stage are also very high. As a consequence, the emissions of

**Table 15.12.** Life cycle emissions of the production of ethanol from wheat (g/GJ) and corresponding emissions of the mixture fuel E95 (g/vkm)

| | g/GJ | | | | | |
|---|---|---|---|---|---|---|
| | Wheat production | Transport of wheat / straw | Wheat to ethanol process | Distribution and storage | Total ethanol | Total E95 (g/vkm) |
| $NO_x$ | 1089.1 | 20.3 | 159.9 | 8.6 | 1277.9 | 11.20 |
| $SO_2$ | 362.4 | 0.5 | 303.0 | 0.2 | 666.1 | 5.85 |
| VOC | 226.0 | 0.7 | 341.9 | 8.6 | 577.2 | 5.24 |
| $PM_{10}$ | 315.2 | 1.6 | 20.2 | 0.7 | 337.7 | 2.96 |
| CO | 346.5 | 6.1 | 85.3 | 2.6 | 440.5 | 3.86 |
| $CO_2$ | 88634.5 | 1322.0 | 54460.0 | 561.0 | 144977.5 | 1271.83 |
| $N_2O$ | 378.8 | | 3.1 | | 381.9 | 3.34 |
| $CH_4$ | 101.5 | | | | 101.5 | 0.90 |

the upstream stages of the production of bioethanol from wheat in Spain are very high compared to those of fossil fuels. This shows how important the correct selection of the raw material for the production of ethanol is and it is stressed that raw materials with higher yields per ha are needed in order to enhance the environmental performance of the agricultural stages of bioethanol production. A promising alternative for Spain is the crop Helianthus tuberosum, which is highly adapted to the agronomic conditions in Spain, with low requirements and high yields per ha. The energetic optimisation of the ethanol transformation process is also a crucial point in the environmental balance of the whole process. This could be done by burning the residual biomass in the boilers of the transformation plant at a higher rate than assumed in this study or from growing energetic crops, such as Cynara cardunculus, to meet the thermal requirements of the transformation process. In these conditions the energetic and $CO_2$ balances are optimised. However, in the conditions analysed in this study, the production of ethanol from wheat for bus fuel does not have many environmental benefits.

### 15.8.3.2 Marginal Costs

***Damages due to vehicle use.*** The impacts from the vehicle technologies previously defined were assessed on each road and motorway. The resulting damages (average figures for all the trajectories) in EUR/100 vkm are shown in Fig. 15.19.

Gasoline cars show higher total damages than diesel cars, although diesel cars have higher local higher damages (due to higher particulate emissions). The highest values per vkm are from buses and especially the early 90's vehicles. Buses running on bioethanol have lower damages per vehicle km than older buses and some benefits over EURO I buses.

Global warming damages were estimated for each type of vehicle in each of the trajectories analysed. Results are show in Table 15.13 $CO_2$ emissions from E95 buses are low (the main advantage of biofuel vehicles) – the emissions from the

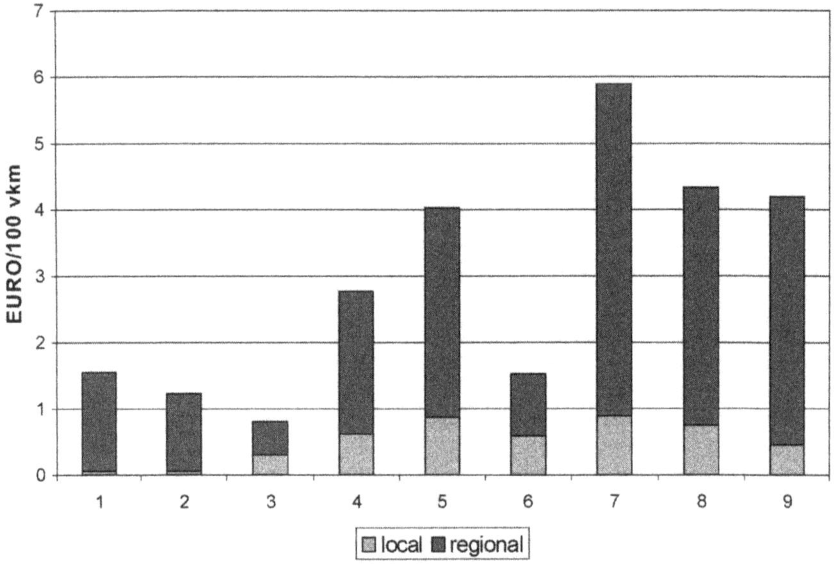


**Fig. 15.19.** Local and regional damages per vehicle km for each technology
1. Gasoline cars 1.2-2.0 l
2. Gasoline cars <1.2 l
3. Diesel cars < 2.0 l
4. Diesel HDV 3.5-7 t
5. Diesel HDV 7-16 t
6. Diesel LDV
7. Diesel buses Early 90's
8. Diesel buses EUROI
9. E95 buses

ethanol component are zero, as the emissions are from carbon fixed by the growing crop, and so only emissions from the fraction of gasoline have been considered.

***Damages per tonne of pollutant emitted.*** Damages per tonne of pollutant were estimated for three different trajectories. A71 and N340 are trajectories crossing a mainly rural territory while N221 is a ring road around the city of Valencia and as such it can be considered as suburban. Results are shown in Table 15.14.

**Table 15.13.** Total damages of the vehicle use stage (Euro /100 vkm)

| Vehicle | Damages | Global warming damages | Total |
|---|---|---|---|
| 1. Gasoline cars1.2-2.0 l | 1.546 | 0.040 | 1.586 |
| 2. Gasoline cars <1.2 l | 1.226 | 0.035 | 1.261 |
| 3. Diesel cars < 2.0 l | 0.798 | 0.035 | 0.833 |
| 4. Diesel HDV 3.5-7 t | 2.762 | 0.090 | 2.852 |
| 5. Diesel HDV 7-16 t | 3.997 | 0.119 | 4.116 |
| 6. Diesel LDV | 1.509 | 0.061 | 1.570 |
| 7. Diesel buses Early 90's | 5.857 | 0.164 | 6.021 |
| 8. Diesel buses EURO I | 4.305 | 0.161 | 4.466 |
| 9. E95 buses | 4.183 | 0.008 | 4.191 |

**Table 15.14.** Damages in € per tonne of pollutant emitted in different trajectories

| | Scale | $PM_{2.5}$ | $SO_2$ - direct | $SO_2$ - $2^{ndary}$ particles | $NO_x$ - $2^{ndary}$ particles | CO | Benzene | 1,3-Butadiene |
|---|---|---|---|---|---|---|---|---|
| A7 | Loc | 4274 | 114 | | | 0.033 | 33 | 787 |
| | Reg | 15634 | 181 | 3433 | 4257 | 0.194 | 151 | 4728 |
| | Total | 19627 | 294 | 3433 | 4257 | 0.225 | 181 | 5461 |
| N340 | Loc | 5834 | 149 | | | 0.043 | 35 | 1311 |
| | Reg | 16213 | 179 | 3422 | 4245 | 0.189 | 123 | 5963 |
| | Total | 21575 | 327 | 3422 | 4245 | 0.228 | 155 | 7166 |
| N221 | Loc | 50911 | 1300 | | | 0.374 | 305 | 11445 |
| | Reg | 16735 | 181 | 3413 | 4152 | 0.190 | 124 | 5999 |
| | Total | 66933 | 1480 | 3413 | 4152 | 0.559 | 425 | 17279 |

The results show the influence of the density of receptors in the damages. Thus, for the N221 located near a highly populated city, total results are around threefold those obtained for the A71 and N340 located in rural areas. These effects are especially important for local damages, which are 10-15 times higher in N221 than in the other trajectories.

***Total marginal damages.*** The total marginal damages take into account the damages produced in the upstream stages of vehicle production, fuel production, and provision of infrastructure. Emissions of different pollutants for these stages and damages factors per tonne of pollutant for Spain have been used to calculate damages for these upstream stages. Average occupancy rates for buses have been used to estimate damages per passenger km, and average load rates for both categories of HDV vehicles have been used to estimate damages per tonne and km.

**Table 15.15.** Total marginal damages for the vehicle technologies analysed in Castellón (Euro/100pkm and Euro/100 tkm)

| Vehicle | Vehicle use | Vehicle production | Fuel production | Infrastructure | Total |
|---|---|---|---|---|---|
| Euro/100 pkm | | | | | |
| Gasoline cars 1.2-2.0 l | 1.220 | 0.242 | 0.116 | 0.144 | 1.722 |
| Gasoline cars <1.2 l | 0.970 | 0.242 | 0.116 | 0.144 | 1.472 |
| Diesel cars < 2.0 l | 0.641 | 0.242 | 0.063 | 0.144 | 1.089 |
| Diesel coaches Early 90's | 0.170 | 0.034 | 0.010 | 0.008 | 0.223 |
| Diesel coaches EURO I | 0.126 | 0.034 | 0.010 | 0.008 | 0.179 |
| E95 coaches | 0.118 | 0.034 | 8.980 | 0.008 | 9.141 |
| Euro/100 tkm | | | | | |
| Diesel HDV 3.5-7 t | 0.543 | 0.186 | 0.072 | 0.137 | 0.938 |
| Diesel HDV 7-16 t | 0.411 | 0.094 | 0.037 | 0.069 | 0.610 |
| Diesel LDV | 2.966 | 0.891 | 0.280 | 0.353 | 4.490 |

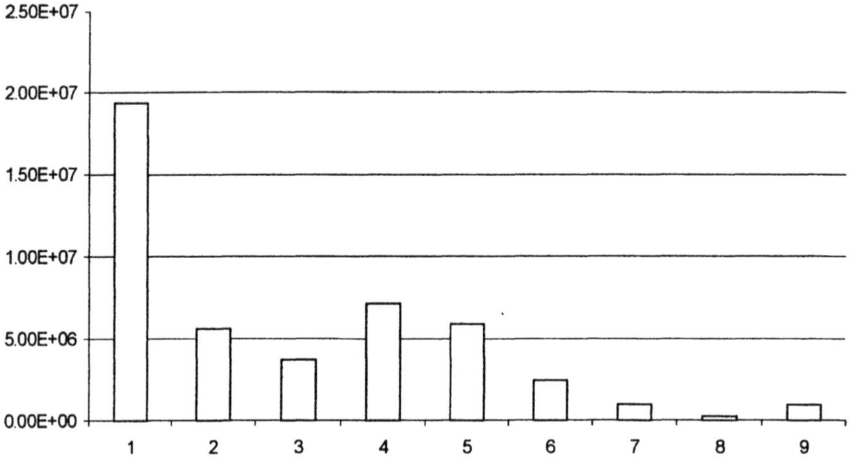


**Fig. 15.20.** Total damages in Euro per year

1. Gasoline cars 1.2-2.0 l
2. Gasoline cars <1.2 l
3. Diesel cars < 2.0 l
4. Diesel HDV 3.5-7 t
5. Diesel HDV 7-16 t
6. Diesel LDV
7. Diesel buses Early 90's
8. Diesel buses EUROI
9. E95 buses

The results show that diesel cars produce lower damages per passenger km than gasoline cars and that coaches produce lower damages than cars. Total damages for E95 buses are very high due to the high values obtained in the fuel production stage. Optimisation of this stage through a more appropriate selection of the crop and a better energetic balance of the process is essential in order to obtain environmental benefits from the introduction of biofuels.

### 15.8.3.3 Aggregation

Total damages per year have also been calculated, taking into account the number of vehicles driving along each road during all the days of the year. The numbers of vehicles driving along each road are shown in Table 15.16. Aggregated results are displayed in Euro in Fig. 15.20. The total value is estimated at 45.4 million Euro per year.

### 15.8.3.4 Comparison of External Environmental Costs of Both Scenarios

From the marginal damages obtained, the introduction of bioethanol obtained from wheat does not produce any environmental benefit. Furthermore, this introduction appears to have negative consequences in terms of total emissions and damages even considering the $CO_2$ sink effect. The reasons for these results have been explained in previous sections along with some ideas on how to improve the performance of the critical stages: agricultural production of biomass and biomass to ethanol transformation process.

**Table 15.16.** Number of Vehicles per Road

| | Type of vehicle | | | | | | | | |
|---|---|---|---|---|---|---|---|---|---|
| Road | 1 | 2 | 3 | 4 | 5 | 6 | 7 | 8 | 9 |
| A-7 | 14372 | 5295 | 5673 | 3782 | 2193 | 2527 | 228 | 80 | 288 |
| N-340 | 4814 | 1773 | 1900 | 1802 | 1048 | 1027 | 70 | 25 | 89 |
| N-III | 15236 | 5613 | 6014 | 123 | 72 | 2551 | 197 | 69 | 249 |
| N-234 | 2770 | 1021 | 1093 | 947 | 551 | 448 | 50 | 17 | 63 |
| N-221 | 17026 | 6273 | 6721 | 856 | 498 | 2786 | 194 | 68 | 244 |
| N-225 | 1709 | 630 | 675 | 461 | 268 | 341 | 15 | 5 | 19 |

1. Gasoline cars 1.2-2.0 l
2. Gasoline cars <1.2 l
3. Diesel cars < 2.0 l
4. Diesel HDV 3.5-7 t
5. Diesel HDV 7-16 t
6. Diesel LDV
7. Diesel buses Early 90's
8. Diesel buses EUROI
9. E95 buses

However, the results do show benefits in the vehicle use stage from the introduction of bioethanol. Separate estimation of these benefits is informative as they remain the same regardless of the raw material from which ethanol is produced. Once the upstream stages of the fuel production are optimised, the benefits obtained could be added to the ones produced in the combustion stage estimated here.

These benefits have been calculated considering than bioethanol buses will substitute the oldest buses in the fleet: the early 90's buses and some of the EURO I buses (75%). Benefits obtained from this substitution amount to 0.2 million Euro, around the 0.5 % of the total damages of the transport activities in the area. These small benefits can be improved considerably if upstream stages are optimised.

## 15.9 UK: Taxation of Goods Vehicles Compared to Environmental Damage

The UK policy case study looked at the current level of environmental costs from heavy goods vehicles in the UK using the ExternE methodology and compared these costs against the current levels of fuel duty and vehicle tax.

### 15.9.1 Assessment of External Costs

Heavy Duty Vehicles (HDVs) are goods vehicles with a laden weight of more than 3.5 tonnes. In 1996, HDVs accounted for 7% of road traffic in Great Britain, for around 95% of freight moved by road, and for around 65% of goods moved by all modes.

Within the UK, HDVs are categorised into seven broad categories including rigid vehicles (3.5 - 7.5 tonne, 7.5 - 17 t, 17 - 26 t, 26 - 32 t) and articulated vehicles (up to 33 tonne, 33 - 38 t, over 38 t). For each of these seven categories, data for 1997 are available for vehicle-kilometres, for year of first registration (used to estimate EURO standard) and for speed-related emissions for $NO_x$, $PM_{10}$, CO, NMVOC and $CO_2$ (Hickman, 1997).

This data was used as the starting point of the analysis. However, in order to look in detail at current taxation levels, a greater level of dis-aggregation was needed. The study therefore split the seven vehicle types into a total of 33 categories by Gross Vehicle Weight (GVW). The categories ranged from 3.5 tonnes up to 44 tonnes and included 13 rigid vehicle types and 20 articulated vehicle types. The base emission factors were interpolated to give factors for all 33 categories, dis-aggregated by EURO standard (pre- EURO, EURO I and EURO II vehicles) and by speed (so that speed related emissions for different technologies, on different road types, could be assessed).

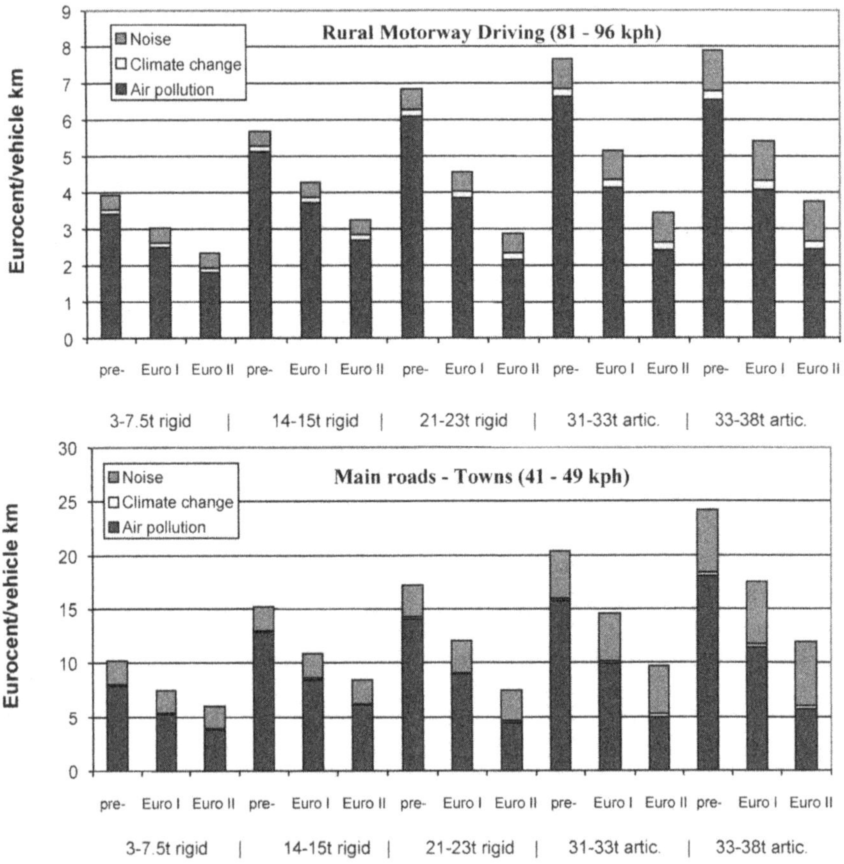


**Fig. 15.21.** Cost per Vehicle km for Rural and Urban Routes

Data on the split of goods vehicle kilometres by road type were used to assign vehicle km, by vehicle type, to urban, rural and motorway routes. Rigid and articulated vehicles were assigned separately across these routes (each using representative speeds). Average road speeds for goods vehicles on each road type were based on published UK road speed data. Vehicle emissions were converted to external costs using the marginal cost results for the UK.

The study also assessed noise damages from heavy goods vehicles. Data on noise emissions from different sized goods vehicles were assessed for urban, rural and motorway routes. In each case, incremental noise levels were calculated using noise dispersion models and differentiating by local population density and background noise levels (from typical traffic flows on each road type). The resulting populated weighted noise increases were valued using the approach recommended in ExternE (European Commission, 1995a) using a Noise Depreciation Sensitivity Index (NSDI) to assess changes to local property prices. A threshold of 55 dB(A) was used, as this is a common value set as a health goal for outdoor noise. As noise is a local effect, a similar variation by area type was seen, with much higher damages in urban areas.

Examples of air pollution and noise damages per km for a number of goods vehicles and technologies are shown below for urban and rural road/area types. The results show the increasing externalities with vehicle size and the advantages of newer vehicles (EURO I/II).

Annual costs were estimated by multiplying the costs per km on different road types by the mileage distribution for each vehicle class. The study assumed average mileage of 60,000 km/year for rigid vehicles (split across urban (32%), rural (44%) and motorway (24%)) and 100,000 km per year for articulated (split across urban (10%), rural (41%), motorway (49%)).

Data on total air pollution costs for rigid and articulated goods vehicles in the UK were presented in Chapt. 14 (see Fig. 14.2) and equate to 2046 million Euro for rigid vehicles and 906 million Euro for articulated vehicles. The additional damages from noise from goods vehicles in the UK were estimated to be 258 million Euro for rigid vehicles and 95 million Euro for articulated lorries. At first glance the noise damages may seem high, though it should be remembered that goods vehicles are particularly disruptive, especially on roads with intermittent traffic. They are also the major source of night-time road noise and have an additional impact through vibration. The case study also considered life-cycle costs from fuel production, vehicle manufacture and infrastructure for lorries. The average life-cycle costs added a further 745 million Euro for all goods vehicles in the UK. Total annual external costs, including air pollution, noise and life-cycle emissions, for a number of goods vehicles are shown in the Fig. 15.22.

Note, the urban component for the above values were derived assuming average air pollution costs for small to medium sized urban areas (i.e. towns). Because of the sensitivity of the results to urban size/population density, a sensitivity analysis was also included using pollution costs derived from a city. These give much higher unit pollution values, which in turn impact on the average weighted values in Fig. 15.22. As an example, the weighted costs (including urban, motorway and rural km) for a pre-EURO 3.5 – 7 t rigid vehicle rise to >8 000 and for a pre-EURO 33 - 38 t articulated vehicle rise to ~16 000 Euro.

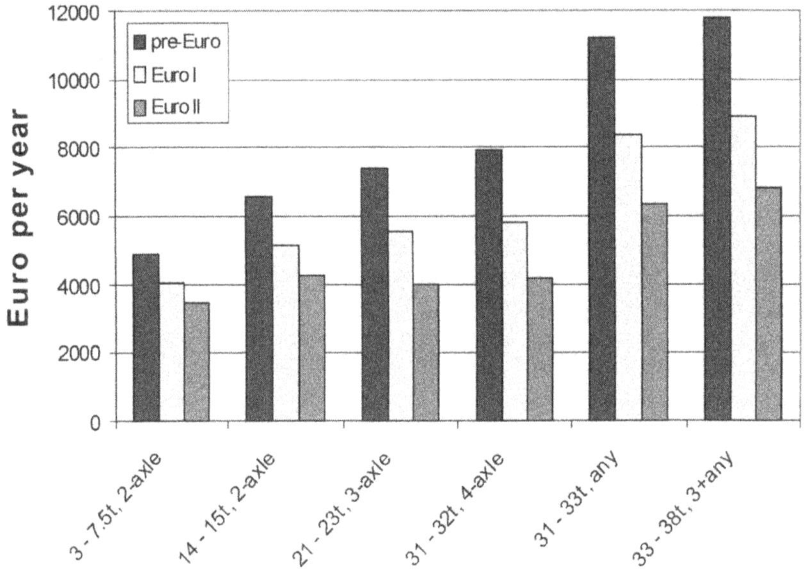


**Fig. 15.22.** Annual External Costs (air pollution, noise and life-cycle emissions) for HDVs

## 15.9.2 Assessment of Current Fiscal Instruments and Levels

These environmental costs were then compared to current duty levels on goods vehicles in the UK. The first duty assessed was fuel tax. The levels of duty on diesel fuel in the UK are the highest in Europe. Excise duty is ~ 0.77 Euro/litre (European Commission, 2000), and there is an additional tax of 17.5% Value Added Tax. Based on a typical end-use price (untaxed) for diesel of 0.17 Euro/litre, the total end use price is ~1.1 Euro/litre. (Note duty levels are based on current (year 2000) levels due to the large increases in fuel duty since 1997).

The current levels of fuel duty, per vehicle per year, were calculated by assessing fuel consumption at different speeds (i.e. on urban, rural and motorways). The mileage distribution across these roads was then taken to estimate the approximate annual fuel duty paid by different vehicle types. Average values for rigid and articulated vehicles are shown below.

The effects of EURO I, II and III technologies on fuel consumption are difficult to estimate as $CO_2$ is not a regulated pollutant. Indeed, most emission factors studies do not record any significant effects on fuel consumption from the introduction of these technologies. This assumption has been used in the following assessment.

In addition to fuel tax, a graduated system of vehicle excise duty (VED) exists for the annual registration of lorry ownership, determined primarily by gross vehicle weight, but also by numbers of axles. For example, the VED on a goods

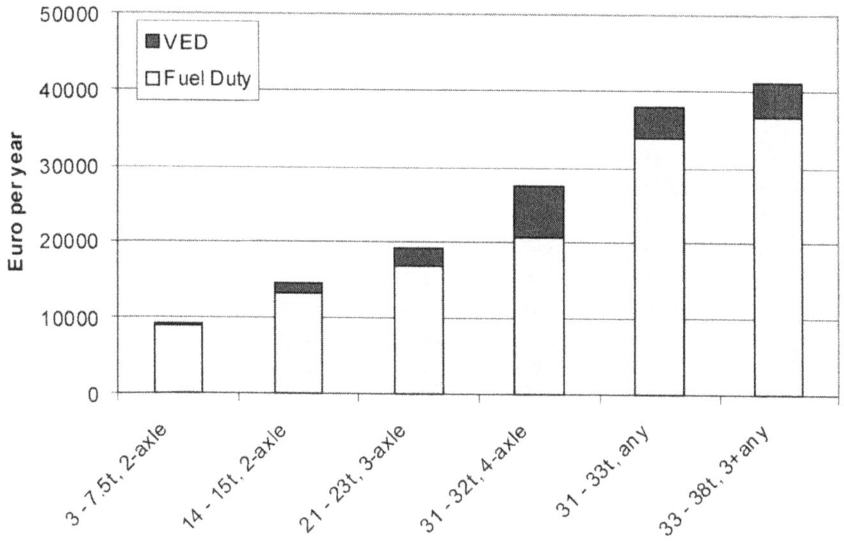


**Fig. 15.23.** Annual levels of Duty from Fuel Tax and Vehicle Excise Duty for UK Lorries

vehicle between 3.5 and 7.5 tonnes is £165 per year (Euro 260), whilst that on a rigid lorry between 31 and 32 tonnes (3 axle) is £4,400 per year (Euro 7,000). These rates have historically been based on road damage costs (track costs), hence vehicles with different numbers of axles are taxed differently in recognition of the different levels of damage they cause. In recent years, however, road track costs have not been used as the sole basis for setting VED, and greater emphasis has been placed on environmental and other factors. Indeed, proposals exist for a differentiated or graduated VED system that would more adequately reflect the environmental impacts of lorries and a VED incentive already exists of £1000 per year for retrofitting older vehicles. The total levels of duty (fuel tax and VED) for a number of goods vehicles are shown in Fig. 15.23.

**Table 15.17.** Annual Duty Paid on Fuel in the UK for Rigid and Articulated Vehicles

| | Fuel consumption (g/km) | | | Duty - Eurocent/km | | | Vehicle | Euro/yr |
|---|---|---|---|---|---|---|---|---|
| | Urban | Rural | M-way | Urban | Rural | M-way | km/yr | /vehicle |
| Rigid | 251 | 164 | 164 | 27.96 | 18.27 | 18.22 | 60 000 | 12,598 |
| Articulated | 502 | 358 | 297 | 55.80 | 39.3 | 33.08 | 100 000 | 37,563 |

### 15.9.3 Comparison of Duty Levels and Environmental Damage

From the comparison of the above results it would appear that current duty levels differ significantly to environmental costs. This is not surprising since current duty levels are not an environmental tax. What is perhaps more surprising is that fuel duty levels are significantly higher than environmental costs. However, it must be remembered that there are additional external costs to environmental pollution that are not considered in this analysis. These include road damage, congestion and accidents. Road track costs are particularly significant for goods vehicles (as these vehicles dominate road damages) and studies indicate annual damage costs from 1,500 Euro/year for a small rigid vehicle through to 15,000 Euro/year for a large articulated vehicle. When these values are added to environmental damages, the overall externalities are very close to current duty levels. If the additional costs from congestion and accidents were added, the results would show that the current levels of duty, although high, still do not fully capture the potential externalities from these goods vehicles.

A detailed analysis of the results showed that differences in the relative levels of duty and external costs varied with vehicle size. The relative levels also varied per tonne km for different vehicles - larger vehicles tend to have lower environmental costs per tonne km as they have a much higher relative carrying capacity. These results provide interesting numbers for looking at the most beneficial strategies for reducing overall freight haulage impacts.

One other area where a large difference existed between duty levels and the actual damage was for different vehicle technology standards – the environmental benefits (and lower external costs) of newer EURO I and EURO II vehicles are not captured in the current system. As a result, the current fiscal regime makes little incentive for vehicle owners or fleet managers to switch to less polluting vehicles.

Measures aimed at altering the absolute levels of fuel duty provide a strong fiscal incentive to reduce overall fuel use (i.e. increase energy efficiency) and therefore have an important role. There is also a role for duty differentials e.g. between ordinary diesel and ultra-low sulphur diesel fuel. However, although these have a direct bearing on the operating costs of vehicle use, such measures fail to capture the emissions benefits of improvements from EURO I and II legislation, as there is little or no benefit in fuel consumption rates between these vehicles. Alternative ways of capturing these environmental benefits are therefore needed.

The recent VED incentives for retrofitting vehicles show one such way forward. In this case it is the direct costs of vehicle ownership that is affected, rather than the costs of vehicle use. An environmentally based VED system (i.e. one that included a component that differentiated by emissions within a graduated scheme) should have an effect on the vehicle investment decisions undertaken by HDV operators. For example, a VED system which differentiated clearly in favour of vehicles meeting EURO II emission standards might encourage a more rapid uptake of purchase and more extensive operation of these vehicles.

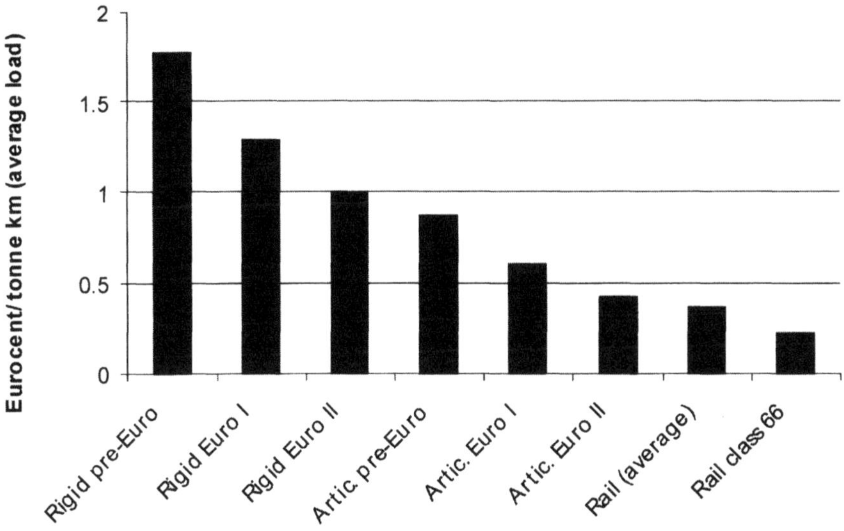


**Fig. 15.24.** Environmental Costs of Road and Rail Freight per tonne km in the UK

Other alternatives include using fiscal measures such as road pricing. Such an approach could achieve similar objectives and could also have the benefit of taxing urban trips at a higher level, to reflect the much higher externalities in these areas.

There is one final issue which was considered in the policy case study, albeit briefly. One of the key factors in internalisation strategies is to address externalities between different transport modes and so give the correct signals for modal choice. A comparison of the environmental costs from rail freight transport was therefore undertaken. These costs were based on the UK marginal cost analysis of diesel locomotives, i.e. for class 47, 56, 58 and 60 (i.e. 1997 freight mix). The numbers did not include the proportion of freight km by rail which are electric, though this was low for the base year. Life cycle emissions were included in the assessment.

Based on the trip km and external costs for each train type, the average external cost per tonne transported was 0.36 Eurocent. Note however that since 1998, the UK rail industry has been introducing newer cleaner locomotives (class 66) with the aim of replacing the existing freight fleet. These new locomotives have a significant environmental benefit, with the estimated external costs of 0.22 Eurocent per tonne km.

The comparison with road freight is shown below. For road, the average freight loads were calculated based on estimates that the UK freight sector averages load factors of 44% (including empty running). The rigid and articulated vehicles shown are specific values for 14-15 tonne rigid vehicle and a 33-38 tonne articulated vehicle.

Figure 15.24 shows that rail has significantly lower externalities per tonne km compared to road (adjusted average load). This suggests that greater fiscal incentives should be introduced to switch goods moved by mode, in addition to any measures considered for reducing those of road freight alone, for example through use of larger vehicles with lower environmental costs per tonne.

## 15.10 Summary

This chapter presents the results of the policy case studies. The studies have set out to demonstrate the potential application of the ExternE accounting framework for policy and decision making. They include a wide range of examples covering different types of issues, at different aggregation levels – showing the application of the numbers from local to national levels.

Local level examples have been undertaken in Finland, France and Greece. The studies here have assessed the potential for the introduction of different technologies or transport measures for improving urban air quality. The studies include an assessment of the benefits of electric vehicle passenger cars in Paris, both in terms of environmental costs and operating costs. Studies in Helsinki and Athens have looked at the urban network level, assessing the potential benefits of running CNG buses across the entire bus networks in both cities. Both studies are extremely interesting as they have used the ExternE values as part of larger cost-benefit studies (scheme appraisals) for transport. Both studies also illustrate the value of the results in comparing different options. In Helsinki the study also looked at how potential policies could be targeted – showing that the largest benefits would be achieved by introducing alternative fuelled buses on the central city routes (an example of how the numbers can be used to prioritise different options). In Athens a similar study was undertaken – indeed this analysis even showed the level of sophistication possible in local decision making by evaluating the optimal number of buses (in cost-benefit terms) to introduce onto the network.

Regional policy case studies were undertaken in Spain and in Germany. Both studies looked at quantifying the environmental benefits of regional policies for introducing alternative fuels to improve current transport related air pollution. The example in Spain looked at the introduction of biofuels into the *Castellón de la Plana region* and the study in Germany looked at the introduction of CNG into the federal state of Baden-Württemberg.

National level policy case studies were undertaken in Belgium and the UK. Both these studies looked at the road transport sector, and looked at which vehicles and fuels were responsible for most damages. The results provide interesting comparisons at the national level, showing how different national policies have led to different levels of externalities from certain vehicle types. Both studies also assessed the current levels of air pollution related external costs and compared these to current levels of fuel duty and road tax. For Belgium the study shows that the current duty levels did not reflect the relative levels of environmental damage from different fuels. For the UK, the study showed how current fiscal and regulatory policy instruments for freight transport could be altered to stimulate greater use of cleaner and more efficient goods vehicles. The final case study was a pan-European study from the Dutch team, undertaking a cost-benefit analysis of different policies for reducing environmental impacts by influencing tyre design.

In summary the policy case studies demonstrate the use of the ExternE methodology in decision making at local, regional and national level. At local level, the examples have included assessments of local transport options and scheme appraisal for improving air quality. At regional level the examples have included assessments of fuels and technologies for reducing the environmental impacts of the transport sector and providing justification for regional plans. Finally, at national level the examples include studies into how current policies can impact on road sector externalities and also show how the results could be used for internalisation strategies or for policies to encourage cleaner and more efficient vehicles. At all levels, from local to European, the case studies also successfully demonstrate the use of external costs in cost-benefits analyses of policy options.

# 16 Summary

Transport is an important source of emissions of many pollutants to the air, e.g. nitrogen oxides, fine particles, carbon monoxide, carbon dioxide, sulphur dioxide, various volatile organic compounds and others. The emissions of these substances cause considerable damage affecting a wide range of receptors including human health, flora and fauna and materials. These damages are in general not adequately reflected in the transport prices; as long as third parties and not the transport user himself is effected, they represent external effects. If such effects exist but are not adequately considered, when making decisions - for example about the number and destinations of trips, the choice of the transport mode, or investments in infrastructure - a non-optimal allocation of resources leading to welfare losses may occur. Hence, external effects should be taken into account when making decisions. This makes it necessary to quantify and monetise the external effects

In this book, a methodology that quantifies the external effects caused by air pollution from transport is presented. It is termed 'ExternE'-methodology, named after a series of EC funded research projects, where the methodology has been developed. It is a bottom-up approach – which means that the marginal damage caused by an additional transport activity is calculated by following the chain (impact pathway) from emission of pollutants, dispersion and chemical conversion in the atmosphere, impacts due to increased concentrations and deposition to transformation of the impacts into monetary units. This process is supported by a computer tool called EcoSense.

Computer tool and methodology were used to quantify external costs for all modes, for current and future technologies and for various locations in the EU (in Belgium, Germany, Spain, Finland, France, Greece, the Netherlands and the United Kingdom). Location-specific marginal external costs were aggregated for transport in and outside urban areas. Finally, the resulting methodology and data were applied in a number of policy case studies.

## 16.1 Emission Calculation

Comparability of case studies of externalities is an important issue in the context of European policy decision making. In this context, the use of incompatible approaches for emission calculation is often criticised. Therefore it was attempted to avoid the use of national emission factor methodologies as far as possible, and to present a list of reference technologies for road transport, where the same basis was used. A major source of information was the MEET project (Methodologies for Estimating Air Pollutant Emissions from Transport) funded by the European

Commission, DG TREN, where a set of methodologies for emission calculation had been developed, accepted by most of the European experts. However, in some cases additional sources had to be used in order to cover all priority pollutants and all modes.

## 16.2 Dispersion Modelling

For atmospheric dispersion modelling of non-reactive pollutants on the local scale and non-reactive pollutants and acid species on the regional scale two well established models were applied. The ROADPOL model is a Gaussian line source dispersion model predicting the annual or seasonal average concentrations of pollutants and the population exposure in a range of up to 20 km away from the emission source. Outside this local scale the Windrose Trajectory Model (WTM) is used to estimate the concentration and deposition of primary pollutants and acid species on the European ("regional") scale. The model is a receptor-orientated Lagrangian plume model employing an air parcel with a constant mixing height of 800 m moving with a representative wind speed.

Due to strong non-linearities associated with ozone formation the assessment of the effect of emission changes on ozone concentrations at the local scale is the most difficult problem within atmospheric transport modelling for ExternE. Current literature was reviewed, but no "simple" method is available to include the effects of emission control on local ozone changes into the ExternE framework. On the regional scale, a major problem of previous project phases, the lack of a regional model of ozone formation and transport applicable to the European situation, could be solved. This was possible by using the source-receptor ozone model (SROM) which allows to calculate ozone statistics. The SROM is based on source-receptor relationships from the EMEP MSC-W oxidant model for five years of meteorology.

## 16.3 Exposure-Response Relationships

### 16.3.1 Impacts on Human Health

A set of exposure-response (E-R) functions for acute and for chronic health effects of air pollution was provided, both for the 'classical' pollutants (particulate matter, PM; ozone, $SO_2$, $NO_2$, CO) and for carcinogenic substances. New epidemiological and other studies have illuminated not only exposure-response relationships between ambient air pollution and adverse health, but also possible mechanisms of action. The continuing information brings with it a need to keep under review not only what specific functions might be considered best, but also the strategic judgements which underlie the present quantification of impacts. This has guided the extent to which it was attempted to fine-tune the values provided. As a consequence, the E-R functions for chronic mortality (Pope et al., 1995) and chronic bronchitis (Abbey et al., 1995) were scaled down by 30% and 50% compared to earlier studies to account for the difference between recent and historical estimates of exposure and the higher particle effects in time series studies in the USA compared to Europe.

The recommended set of E-R functions includes functions for estimating increased mortality due to primary particles, nitrates, sulphates, $SO_2$, ozone, benzene and diesel particles. Morbidity effects considered include endpoints such as congestive heart failure due to particles and CO, respiratory hospital admissions due to particles, $SO_2$ and ozone, non-fatal cancer due to benzene and diesel particles, to mention only a few. A set of E-R functions which draws heavily on European studies is presented here; the use of life-table methods in estimating mortality impacts from cohort studies is consolidated.

### 16.3.2 Impacts on Building Materials

New dose-response functions are available from ICP Materials, an international exposure programme, describing the degradation of a wide range of materials including metals, stone materials, paint coatings, electric contact materials, glass materials and polymeric materials.

Selected functions from the programme, valid for unsheltered positions, have been adapted and exposure-response functions suitable for inclusion into the calculations have been derived for limestone, sandstone, zinc/galvanised steel, painted steel and painted galvanised steel. They include $SO_2$, temperature and relative humidity for quantification of dry deposition effects, and amount and acidity of precipitation for quantification of wet deposition effects.

A critical review of dose-response functions for soiling of materials was performed including additional analysis on strategies for selecting maintenance intervals. The functions quantify the loss of reflectance in terms of the particulate concentration times time of exposure, i.e. the particulate dose, which means that each exposure-response function can be summarised in the form of a critical dose. No other characteristics of the environment are included in the soiling functions. Critical soiling doses in sheltered conditions have been estimated. Compared to the degradation functions from ICP Materials the uncertainties of these critical doses are high. Among others this was the reason why soiling effects were not included in the monetary estimates.

### 16.3.3 Impacts on Terrestrial Ecosystems

Localised, impacts of transport on terrestrial ecosystems were considered. That is where the impacts are concentrated by the use of fixed routes such as roads, rail lines and airports.

Impacts of transport on (semi-) natural ecosystems were explored by a number of examples. Due to the site-specificity a general relationship between transport activity and impact cannot be derived. But results show that damages due to marginal changes can be expected to be only of minor importance. Even though, further work should be done to fill the existing gaps.

In the monetary cost estimates crop losses from $SO_2$ and ozone, costs due to acidification of agricultural soils, and fertilisational effects of nitrogen deposition were considered. The dose-response functions are given below; detailed information can be found in European Commission (1999).

#### 16.3.3.1 Effects from $SO_2$

The assessment of effects from $SO_2$ on crops an adapted function from the one suggested by Baker et al. (1986) is used. The function assumes that yield will increase with $SO_2$ from 0 to 6.8 ppb, and decline thereafter. Changes in crop yield are quantified for wheat, rye, barley, potato, sugar beet, and oats. The function is defined as

$$y = 0.74 \cdot [SO_2] - 0.55 \cdot [SO_2]^2 \qquad \text{for } 0 < [SO_2] < 13.6 \text{ ppb} \tag{16.1}$$

$$y = -0.69 \cdot [SO_2] + 9.35 \qquad \text{for } [SO_2] > 13.6 \text{ ppb} \tag{16.2}$$

with

$y$ = relative yield change
$[SO_2]$ = $SO_2$-concentration in ppb

#### 16.3.3.2 Effects from Ozone

For the assessment of ozone impacts, a linear relation between yield loss and the AOT40 value (Accumulated Ozone concentration above a Threshold of 40 ppb) is assumed. The relative yield loss was calculated with following equation; the sensitivity factors are given in Table 16.1:

$$y = 99.7 - \alpha \cdot AOT40_{crops} \tag{16.3}$$

with

$y$ = relative yield change
$\alpha$ = sensitivity factors

**Table 16.1.** Sensitivity factors for different crop species

| Sensitivity | $\alpha$ | Crop species |
|---|---|---|
| Slightly sensitive | 0.85 | rye, oats, rice |
| Sensitive | 1.7 | wheat, barley, potato, sunflower seed |
| Very sensitive | 3.4 | Tobacco |

#### 16.3.3.3 Acidification of Agricultural Soils

The amount of lime required to balance acid inputs on agricultural soils across Europe is assessed. The analysis of liming needs should be restricted to non-calcareous soils, but the percentage of the agricultural area on non-calcareous soils has not been available Europe-wide. Thus, the quantified additional lime required is an over-estimation giving an upper limit to the actual costs. The additional lime requirement was calculated as:

$$\Delta L = 50 \cdot A \cdot \Delta D_A \qquad (16.4)$$

with

| | | |
|---|---|---|
| $\Delta L$ | = | additional lime requirement in kg/year |
| $A$ | = | agricultural area in ha |
| $\Delta D_A$ | = | annual acid deposition in meq/m$^2$/year |

#### 16.3.3.4 *Fertiliser Effects of Nitrogen Deposition*

Nitrogen is an essential plant nutrient, applied by farmers in large quantities to their crops. The deposition of oxidised nitrogen to agricultural soils is thus beneficial (assuming that the doseage of any fertiliser applied by the farmer is not excessive). The reduction in fertiliser requirement is calculated as:

$$\Delta F = 14.0067 \cdot A \cdot \Delta D_N \qquad (16.5)$$

with

| | | |
|---|---|---|
| $\Delta F$ | = | reduction in fertiliser requirement in kg/year |
| $A$ | = | agricultural area in ha |
| $\Delta D_N$ | = | annual nitrogen deposition in meq/m$^2$/year |

## 16.4 Economic Valuation

### 16.4.1 Mortality Risks

A simple approach was recommended to estimate the costs of air pollution via mortality for all cases by using a *Value of a life year lost* (VLYL). The methodological advantages of using a VLYL measure are that it allows greater flexibility in valuation, and that it is one that clinicians are more comfortable in estimating in that it brings the WTP approach closer to the QALY approach. In addition, it is more amenable to valuing chronic mortality as it is easier to express chronic impacts in terms of life years lost.

Critics of the VLYL approach have argued that the VLYL approach is invalid because it assumes a constant value of a life year over the remaining lifetime, and the evidence suggests that life year values vary with age (among other factors). It is correct to say that VLYL values should not be constant and independent of age, but that does not make the method invalid. It is perfectly possible to develop values for life years that vary with age, and indeed the simple model presented gives a way of doing so. The other issue raised by critics of the VLYL approach is that it is invalid to add up life years, i.e. saving one life year for two people is not equal to saving two life years for one person. It is correct to say that the adding up of life years cannot be carried out in a simplistic fashion. The value of future years is less than that of present years on account of the discount factor and any application of the VLYL method has to take account of that. As far as adding up values across individuals, there are problems associated with such a procedure but they are no more or less than adding up VSL (value of a statistical life) values across individuals. Hence in our opinion, the use of VLYL as a measure of risk is both valid and useful as a means of valuing changes in mortality risk.

### 16.4.2 Morbidity

In the time that has passed since the last Externe Task Report on Economic Valuation there have been three new European projects using the Contingent Valuation (CV) method for valuing morbidity impacts from air pollution, one in Helsinki (Finland), one in Strasbourg (France) and Kehl (Germany), and a 5-country study (Amsterdam/NL, Lisbon/P, London and surroundings/UK, Oslo/NO and Vigo/ES). The results of each of these three studies were reviewed. Since the 5-country study is the most comprehensive study in terms of European countries covered and is a well designed CV survey, most weight was put on the results from this study when updating the list of economic values used in ExternE. However, for those morbidity symptoms that are not covered in this study, we had to rely on the results from the other two studies.

### 16.4.3 Evaluating Non-Marginal Changes

When a policy has a significant ("non-marginal") effect and prices change as a result, it must be resorted to the relevant supply and demand curves in order to attach an appropriate "social" valuation to the resulting changes in output or inputs. Therefore, depending on the questions to be answered by the economic analysis, an extended partial equilibrium analysis, or a full general equilibrium analysis may be required. It is recommended that the way in which the welfare effects of each policy option considered by ExternE are estimated, should be determined on an individual basis since there are no hard and fast guidelines on this issue.

### 16.4.4 Incorporating Non-Monetised Impacts into the Decision Making Process

In cases where an appropriate quantification of impacts in monetary terms is not possible, ways have to be found to include such effects in decision making. Therefore, ways were considered in which the ExternE results may be combined with data on other decision factors, which are not expressed in money terms, but are equally important determinants of policy selection. It should be emphasised that the intention was not to identify a single methodology to adopt in ExternE. Rather, the intention was to reflect on the main options available to decision makers whom might wish to use the ExternE results in a multi-attribute decision making context. Some general guidance on the use of avoidance costs in the absence of willingness to pay monetary values is given and the possibilities for multi-attribute decision-making are considered.

### 16.4.5 Overview of Monetary Values Used

**Table 16.2.** Monetary values used for economic valuation (Euro – base year 2000)

| Impact | Monetary value |
|---|---|
| *Health effects* | |
| Year of life lost (acute effects) | 165,700 |
| Year of life lost (chronic effects) | 96,500 |
| Chronic bronchitis | 169,330 |
| Cerebrovascular hospital admission | 16,730 |
| Respiratory hospital admission | 4,320 |
| Congestive heart failure | 3,260 |
| Chronic cough in children | 240 |
| Restricted activity day | 110 |
| Asthma attack | 75 |
| Cough | 45 |
| Minor restricted activity day | 45 |
| Symptom day | 45 |
| Bronchodilator usage | 40 |
| Lower respiratory symptom | 8 |
| *Crops* | |
| Barley – yield loss in decitonnes (dt) | 6.3 |
| Oats – yield loss in dt | 6.6 |
| Potato – yield loss in dt | 9.6 |
| Rice – yield loss in dt | 254.9 |
| Rye – yield loss in dt | 18.3 |
| Sugar beet – yield loss in dt | 6.6 |
| Sunflower seed – yield loss in dt | 25.8 |
| Tobacco – yield loss in dt | 3,414 |
| Wheat – yield loss in dt | 11.3 |
| Fertiliser – per kg | 0.53 |
| Lime – per kg | 0.018 |
| *Materials – maintenance costs per m2* | |
| Galvanised steel | country specific (17 – 55) |
| Limestone | 299.2 |
| Mortar | 33.0 |
| Natural stone | 299.2 |
| Paint | 13.4 |
| Rendering | 33.0 |
| Sandstone | 299.2 |
| Zinc | 26.9 |

## 16.5 Global Warming

Marginal costs of greenhouse gases, viz. carbon dioxide, methane, and nitrous oxide, were assessed and the marginal costs of sulphur emissions (via sulphate aerosols) and nitrogen emissions from aircraft (via ozone.) were estimated. The discounting and valuation procedures for marginal cost estimation were refined. The currently recommended values of 2.4 Euro/t $CO_2$, 750 Euro/t $N_2O$, and 45 Euro/t $CH_4$ are lower than those of Eyre et al. (1998), reflecting the more optimistic tone of recent impact literature.

It should be noted that with the inclusion of new insights into the impacts of climate change, it can no longer be excluded that marginal costs are negative, particularly for methane. The sign of the costs is model and region dependent. Despite their short life-time, the marginal costs of nitrogen and sulphur emissions are relatively large, primarily because they are not much discounted.

The results presented should not be taken as final estimates. The impacts covered by the models used are only a fraction (of unknown size) of all climate change impacts. Particularly, large scale disruptions, such as a breakdown of North Atlantic Deep Water formation or a collapse of the West-Antarctic Ice Sheet, are excluded from the analysis. The methodologies to estimate climate change impacts in a different future remain weak. Adaptation is not included in its full complexity. Valuation of impacts is still troublesome, particularly for nature and health. Our knowledge of atmospheric chemistry and climate has substantial gaps. The estimates reflect our current best knowledge, and indicate a stimulating research agenda.

One way of coping with these large uncertainties is to take into account avoidance costs to reach a certain $CO_2$ reduction target instead of damage costs. To illustrate this, the results presented in this summary are calculated by using avoidance costs of 19 Euro per tonne of $CO_2$ (whereas results in other chapters are calculated by using the damage costs mentioned above).

## 16.6 Treatment of Up- and Downstream Processes

In the spirit of life cycle assessment (LCA) one should account for all the environmental consequences, from cradle to grave, of a decision about a system, process or product. For example, the replacement of conventional cars by electric ones would eliminate air pollution from cars at the expense of increased pollution from power plants; in addition there would be increased pollution from the manufacture of batteries and their eventual disposal. Electricity production, and manufacture and disposal of batteries, are processes that occur upstream or downstream of the main process, i.e. driving the car.

Unfortunately a full site-specific impact pathway analysis is far too complex to be practical for LCA, quite apart from the fact that in most cases the sites of the upstream or downstream processes are not even known. Clearly a compromise is needed for the analysis of upstream or downstream processes, a compromise that involves typical damage estimates that are approximately right for typical processes.

Two possible approaches for obtaining typical damage estimates have been explored here. One approach is based on the software tool EcoSense for calculating external costs and the CORINAIR data base for emissions. The average damage cost per kg of a pollutant is calculated by calculating the total damage in Europe with and without emissions from the sector of interest (e.g. industry) and dividing by the total emission of that pollutant from that sector. The other approach is the "uniform world model" (UWM), a formula with five factors that follows from conservation of mass under certain simplifying assumptions. As the main processes involved in the considered up- and downstream process steps vehicle production, fuel production, and infrastructure provision are assumed to be located in Europe, unit damage factors per kg of pollutant, for each country calculated with EcoSense, were applied.

## 16.7 Uncertainty

Since the uncertainties are large, an analysis of the uncertainties is crucial for the credibility of the results. Uncertainties can be grouped into different categories, even though there may be some overlap:

- data uncertainty (e.g. slope of a dose-response function, cost of a day of restricted activity, and deposition velocity of a pollutant);
- model uncertainty (e.g. assumptions about causal links between a pollutant and a health impact, assumptions about form of a dose-response function (e.g. with or without threshold), and choice of models for atmospheric dispersion and chemistry);
- uncertainty about policy and ethical choices (e.g. discount rate for intergenerational costs, and value of statistical life);
- uncertainty about the future (e.g. the potential for reducing crop losses by the development of more resistant species);
- idiosyncrasies of the analyst (e.g. interpretation of ambiguous or incomplete information).

The first two categories (data and model uncertainties) are of a scientific nature. They are amenable to analysis by statistical methods, combining the component uncertainties over the steps of the impact pathway, to obtain formal confidence intervals around a mid estimate. For ethical choice and for uncertainty about the future a sensitivity analysis may be more appropriate, indicating how the results depend on these choices and on the scenarios for the future.

Quantifying the sources of uncertainty in this field is problematic because of a general lack of information. Usually one has to fall back on subjective judgment, preferably by the experts of the respective disciplines. That has been done, in an informal manner, for this project. In addition all the relevant information was used that could be found in the literature.

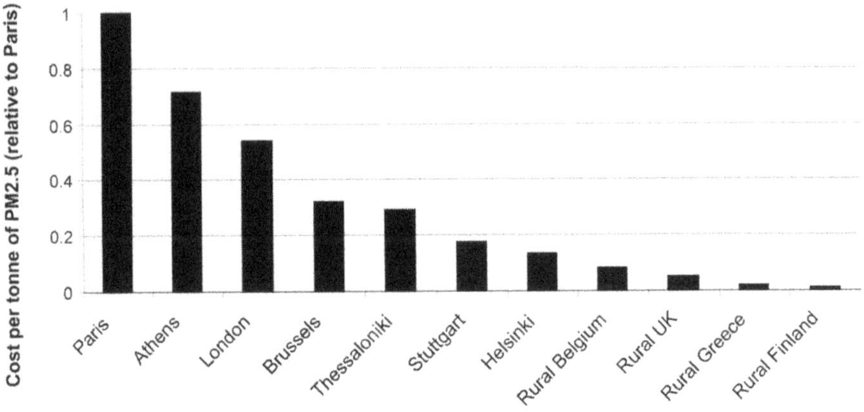


**Fig. 16.1.** Relative damage per tonne of $PM_{2.5}$ in different urban and rural locations

To find the resulting uncertainty of the damage cost from the uncertainties of the components of the analysis, an approach based on lognormal distributions and multiplicative confidence intervals was followed. However, not all uncertainties are amenable to such a quantitative analysis. For the presentation of the ExternE results it is desirable to provide some indication of the full uncertainty which is larger than the uncertainties that have been quantified explicitly. This point has been addressed by estimating comprehensive uncertainties, based on our subjective judgement about reasonable upper and lower limits of the damage costs.

## 16.8 Marginal Costs of Transport

The results show, that the quantified damages are dominated by health impacts, in particular mortality due to primary ($PM_{2.5}$) and secondary particles (nitrates, sulphates).

Figure 16.1 illustrates the location specificity of primary pollutant emissions. The damages are determined by the population density of the area where the emission takes place. For secondary pollutants (sulphates, nitrates, ozone), which were looked at on the regional scale, the geographic position determines the size of impact.

A comparison of damage costs for different car technologies shows that diesel cars cause much higher damages than petrol cars due to their high primary particle emissions. This is illustrated in Fig. 16.2 for the case of Stuttgart, but the picture is similar for other locations. The introduction of EURO I, EURO II, and EURO III emission standards reduces the damage costs per vehicle kilometre by 63%, 74% and 86% respectively compared to an uncontrolled diesel car. This is almost only due to the reduction of particle emissions. The reduction of damage costs caused by petrol cars by 77% is considerable from ECE 15/04 to EURO I.

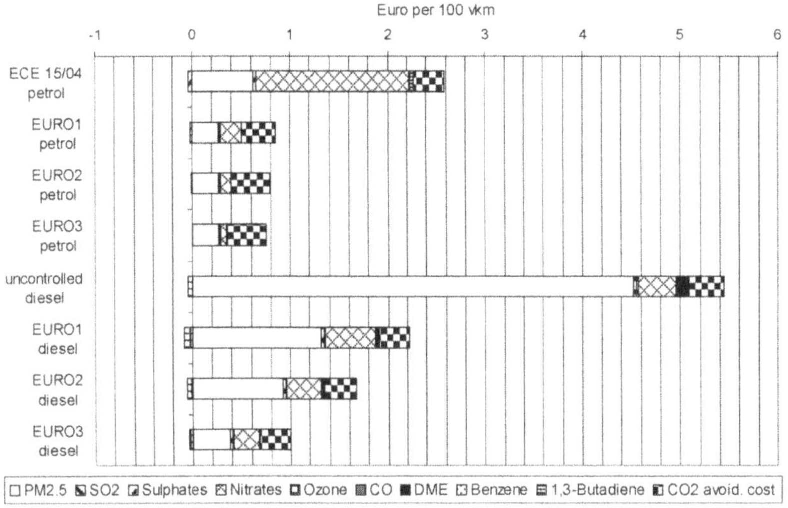


**Fig. 16.2.** Damage costs in Euro per 100 vkm for different car technologies in Stuttgart. $CO_2$ emissions are valued with avoidance costs of 19 Euro/t $CO_2$

The reductions due to the introduction of EURO II and EURO III are smaller, because particle emissions are only reduced marginally, but dominate the damages in the urban environment. Figure 16.2 shows negative costs (i.e. benefits) due to ozone impacts. As explained in Sect. 13.8.1, an additional unit of $NO_x$ emitted may lead to a reduction of ozone and thus a decrease in ozone damages But $NO_x$ emissions are far from being beneficial when taking into account the damage costs due to nitrates, which outweigh the ozone benefits by far.

For comparisons between vehicle types and modes it is important to include up- and downstream processes. This is illustrated in Fig. 16.3, where costs due to passenger transport in the city centre of Stuttgart are presented. In Stuttgart, rail vehicles for passenger transport are operated electrically only, that means there are no direct emissions from vehicle use. Of the remaining processes, fuel (electricity) production causes the highest share of costs. For vehicles with internal combustion engine, vehicle operation plays a very important role in the urban environment. The only exception is the petrol car complying with EURO II, where infrastructure use causes the highest cost per passenger kilometre (pkm). Costs per pkm due to vehicle production, fuel production and infrastructure use are much lower for busses and coaches than for cars. When looking at the total costs per pkm, the electric rail vehicles by far cause the smallest damage costs due to air pollution. They are followed by the coaches, this is mainly due to the high load factor of coaches. The load factor of buses for inner urban public transport is much lower. So the ranking in terms of increasing damage costs for modern road passenger vehicles is modern urban bus, EURO II petrol car and EURO II diesel car. For detailed results see Table 13.30 on p. 220.

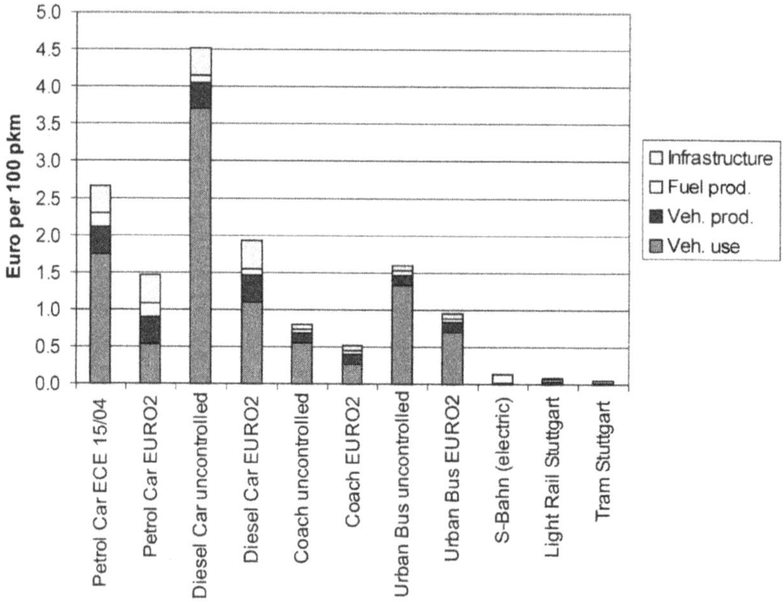


**Figure 16.3.** Comparison of air pollution costs due to urban passenger transport in Stuttgart. $CO_2$ emissions are valued with avoidance costs of 19 Euro/t $CO_2$ (pkm = passenger kilometre)

The goods transport vehicles presented in Fig. 16.4 are run by diesel fuel or electricity. The highest costs per tonne kilometre (tkm) are caused by light duty vehicles (LDV) due to their comparably small load. The heavy duty vehicles (HDV) considered have a total maximum weight of 32 to 40 tonnes and a maximum load of about 25 tonnes (compared to 1 - 2 tonnes for LDVs). As they are used for long-distance transport, the average capacity use is higher than for LDVs. For this reason the cost per tkm is comparably small. The same is true for the goods trains considered. The costs of the electric train are hardly visible on the given scale. For detailed results see Table 13.31 on p. 224.

A number of alternative fuels and technologies has been assessed for different locations. The use of alternative fuels or technologies offers high potential for reducing environmental damages, in particular in urban areas. Figure 16.5 illustrates the effect in London for cars and light duty vehicles. In the case of cars marginal damage costs can be reduced by up to 60% compared to a EURO II petrol car and up to 85% compared to a EURO II diesel car. The remaining damage costs are dominated by up- and downstream processes. The picture is similar for light duty vehicles with possible reductions of about 85% compared to a diesel engine complying with EURO II emission standards.

Figure 16.6 shows the quantifiable damages due to a landing and take-off (LTO) cycle at Berlin Tegel airport and at London Heathrow. The costs reflect the considerable variations in emission factors of the different engine types used. The costs for the Boeing 737-400 are lower in Berlin, mainly due to lower emissions in most components as a consequence of shorter times spent in different engine modes.

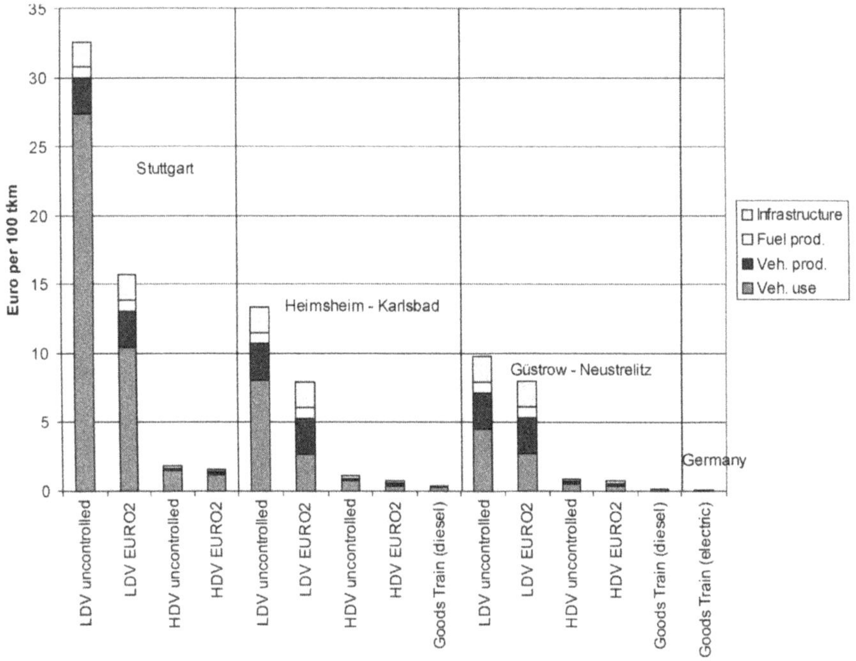


**Fig. 16.4.** Comparison of air pollution costs due to goods transport in Germany. $CO_2$ emissions are valued with avoidance costs of 19 Euro/t $CO_2$ (tkm = tonne kilometre)

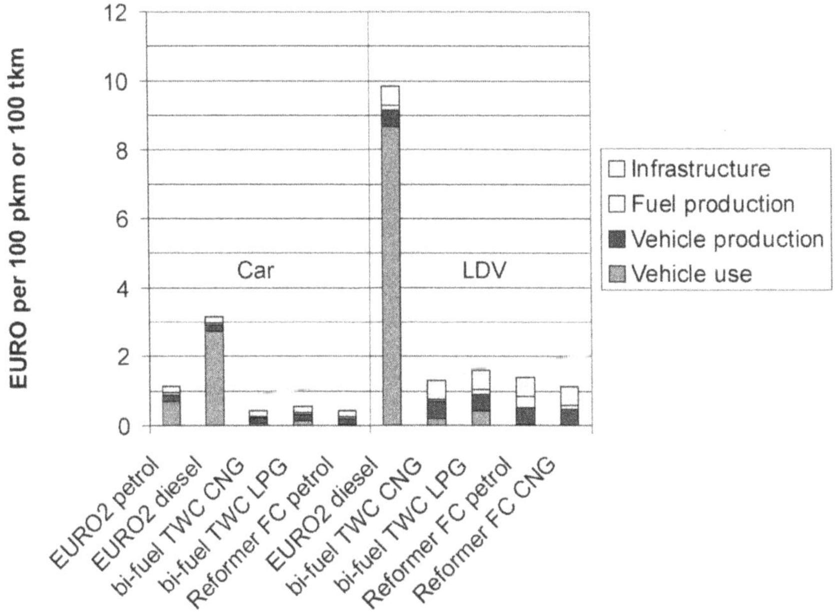


**Fig. 16.5.** Potential reduction of environmental costs due to use of alternative fuels/technologies in London

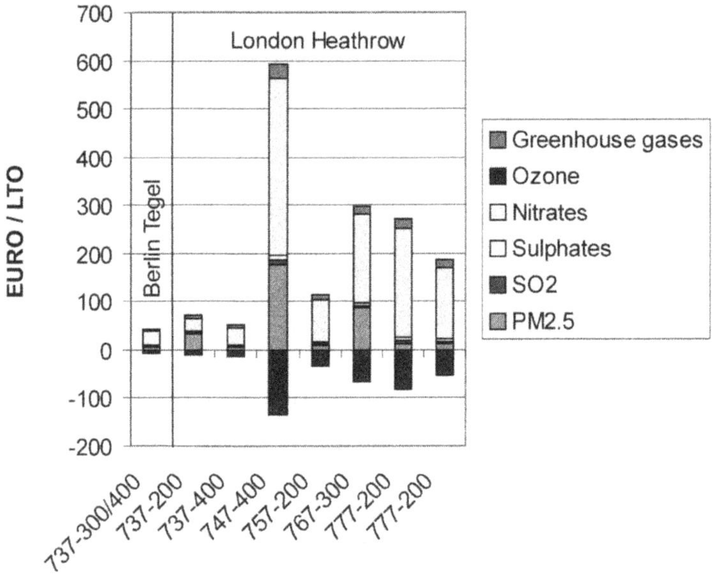


**Fig. 16.6.** Quantifiable damages due to a landing and take-off (LTO) cycle at Berlin Tegel and at London Heathrow in Euro per LTO cycle

## 16.9 Aggregated Costs

New issues arise when extending the process of estimating marginal costs to aggregating externality data to levels required by policy instruments or for the purpose of green accounting. The issue of geographical transferability becomes extremely important for transport, because of the importance of local impacts on damages. It is discussed how these issues can be addressed and the updated debate concerning the issue of economic valuation in aggregation including the discussion how these should be dealt with is presented. Finally, the issue of data compatibility (by location) between countries for national level aggregation is raised as an issue warranting further research. This is particularly important for producing consistent European wide results for future policy analysis.

The approaches proposed were applied in a number of aggregation studies at a national/regional level showing the potential usefulness of the numbers in policy and strategy studies.

The national level results are particularly interesting. The results show that the aggregated costs for the road transport sector are significant. The externalities of exhaust emissions from the road sector are estimated at 4.2 billion Euro (1.8% of GDP) for Belgium and 8.7 billion Euro (1% of GDP) for the UK. These values are interesting as they are higher than many estimates in the literature (which are of the order of 0.5% of GDP). This is to be expected, as these are the first national aggregation studies to take local impacts into account properly when estimating the external costs of the road transport sector (i.e. specifically taking into account the higher external costs in urban locations). All studies show that passenger cars are the dominant source of road transport sector damages, though heavy goods vehicles give rise to a disproportionately large share of damages relative to their share of vehicle kilometres.

The analysis of the entire transport sector in the Netherlands shows that road transport is also the dominant source of damages from all transport and mobile sources, responsible for around 82% of the total transport externalities of 3.9 billion (1.1% of GDP). Finally, the results show that life-cycle emissions are important: the aggregation studies in Belgium and the Netherlands show emissions from fuel production, vehicle manufacture and infrastructure are 23% (Belgium) and 40% (Netherlands) of total road sector externalities.

The regional analysis for Germany presents an elaborated approach to increase the resolution of local scale impact analysis. The central feature is the use of cost factors which were derived per exposed receptor. With these factors, aggregated costs can be calculated using a GIS (Geographic Information System) up to national or the European level. Thus, this approach allows to calculate damage costs on the national level for all European countries.

## 16.10 Policy Case Studies

The policy case studies undertaken have set out to demonstrate the practical application of the ExternE accounting framework for policy and decision making. They include a wide range of examples covering different types of issues, at different aggregation levels – showing the application of the numbers from local to national levels.

The case studies in Belgium were at a national level, looking at general transport sector trends and policies. The first study looked at the externalities of the Belgian road sector over time. The analysis showed that road sector externalities have shown modest benefits over recent years, despite the introduction of tightening legislation on emission standards for road vehicles. This contrasts with the same analysis over the same period in the UK (Chap. 14) which did show significant reductions in road sector externalities. Further analysis shows that there are specific reasons for this difference: both countries have seen rising annual mileage and numbers of vehicles but in Belgium there has been a greater uptake of diesel cars and a greater increase in goods vehicles. This shows how national level policies to influence vehicle technology uptake can have an important effect on transport related externalities and indicates current excise levels for certain vehicle types (e.g. diesel passenger cars, heavy duty vehicles) are sending the wrong signals with respect to the environment and transport.

The second study for Belgium looked at the external costs of air pollution and assessed them in relation to the current levels of duty (tax) on fuel. Interestingly the study shows that the externalities from air pollution are higher than level of duty on fuel tax for diesel in Belgium, but lower for gasoline – potentially justifying a raising of diesel duty and a lowering of gasoline duty. This issue is especially important as the shift from petrol to diesel cars in Belgium is one of the reasons why the total impacts from passenger cars have not dropped over recent years, even with the introduction of stricter environmental standards for new cars.

The Finnish policy case study looked at the introduction of alternative fuel buses in Helsinki, relative to the existing diesel fleet. The results show significant benefits from the use of natural gas in terms of the reduction in environmental costs. The study showed the change from diesel to gas buses would lead to slightly higher overall costs (operating and environmental), though it is stressed that the benefits gained with the gas buses might be greater than quantified in this study due to a number of additional benefits. The study also provided a good example on prioritising measures – an analysis of Helsinki showed that the natural gas buses had the greatest benefit on the central city routes – thus policies to introduce CNG buses into these areas should be a priority. The study also provides a good example of the use of the ExternE approach in cost-benefit assessments.

The French policy case study has looked at the potential for electric battery vehicles (EVs) in Paris. The study calculated the external cost of air pollution for electric vehicles and for conventional cars fuelled by gasoline or by diesel. The study showed that the air pollution related costs from cars in Paris are large, even for cars that meet the rather strict emission limits of the latest regulations. The damage cost amounts to approximately 4% of the private life cycle cost for new gasoline cars, and 70% for new diesel cars. If these costs were internalised, EV's would be competitive at current prices against diesel cars but not against gasoline. In terms of total social cost the EV is at least 50% more expensive than the gasoline cars, both <97 and new, but about 20 to 40% less expensive than diesel cars. Interestingly, the study indicates that if the cost of air pollution are not enough to give the EV a clear advantage against conventional cars in Paris (a very large metropolitan area where the external costs of electricity are small) then the potential for introduction may be low. It concludes that it may be more cost-effective to curb pollution by technologies other than the EV (e.g. hybrid or fuel cell vehicles) or by policies such as scrapping older highly polluting cars.

The German policy case study looked at alternative fuels for road transport in the federal state of Baden-Württemberg. The study looked at the introduction of CNG vehicles, over the time scale from 1995 to 2020 and assessed the levels of environmental benefits from potential levels of CNG introduction into different vehicle fleets. The study showed that in terms of the external costs, the introduction of CNG would have significant environmental benefits – with the potential reductions of more than 50 million Euro in the year 2010.

The policy case study in Greece looked at the use of the ExternE methodology in cost-benefit analysis for local transport decision making. The study assessed the externalities of the bus network of Athens, which is characterised by the existence of old buses with significant environmental burdens to the local environment. The assessment of the externalities resulting from the operation of the bus network, using different bus technologies, revealed a great potential for the substitution of the old buses with new technology or with alternative fuelled vehicles. A cost-benefit analysis was conducted for various scenarios of substitution of old vehicles with CNG buses. The analysis showed that such an investment would have positive results, if the resulting social benefits are included, and that the optimal number of CNG vehicles varies between 300 and 450 vehicles depending on the time preference rate.

The policy case study in the Netherlands looked at the private and external costs of different scenarios for tyres. These included a base case, the introduction of an eco-tyre (improving fuel consumption), correct tyre pressure inflation and increased tyre lifetime. The study showed that the best measure was to improve, monitor and manage tyre pressures, followed by the use of eco-tyres.

The Spanish Policy Case Study assessed the environmental benefits from the introduction of biofuels for transport in a region of Spain. The study showed that some benefits could be achieved in the vehicle use stage from the introduction of bioethanol, though the study highlighted that by optimising upstream stages of fuel production much greater overall environmental benefits could be achieved.

The UK policy case study looked at the current level of environmental costs from heavy goods vehicles in the UK using the ExternE methodology and compared these costs against the current levels of fuel duty and vehicle tax. The results suggest that greater fiscal incentives should be introduced to switch goods moved from road to rail, in addition to any measures considered for reducing those of road freight alone, for example through use of larger vehicles with lower environmental costs per tonne.

Overall the policy case studies provide examples of the use of the impact pathway methodology for decision making at a local, regional and national level. At the local level, the numbers have been demonstrated with examples as diverse as local scheme appraisal (the environmental benefits of different options and cost benefit analysis). At a regional level the examples show how the numbers can provide information on which fuels and technologies should be encouraged to reduce environmental impacts of the transport sector and also how they can be used to quantify the benefits of regional plans. At the national level the examples show how current policies can impact on road sector externalities and can indicate which vehicles or technologies should be targeted in order to reduce environmental impacts. They also show how the externality numbers could be used (through internalisation) with fiscal and regulatory policy instruments (such as fuel duty or vehicle taxation) to stimulate the greater use of cleaner and more efficient vehicles.

## 16.11 Outlook

In this book major remaining issues in the ExternE methodology were addressed. The transport accounting framework and the respective tools are available for broad dissemination and application. Marginal cost estimates were provided for all regions of Europe. A methodology for quantifying damage estimates on all aggregation levels was developed and applied. The practical application of the ExternE accounting framework for policy and decision making was demonstrated in a great variety of policy case studies, covering decision making on the local, regional and national level. Hence, the transport accounting framework and tools are ready to support activities such as

- assessing transport techniques in different areas (urban or regional transport),
- analysing single European transport tasks in order to compare different transport systems,
- application of environmental policy instruments in the transport sector,
- "green accounting" at the national and European levels.

# References


Aanen, P., Alberts, W., Bekker, G. J., van Bohemen, H. D., Melman, P. J. M. and al., e. (1991) Nature engineering and civil engineering works, PUDOC, Wageningen, Netherlands.

Abbey DE, Hwang BL, Burchette RJ, Vancuren T, Mills PK. (1995a). Estimated long-term ambient concentrations of $PM_{10}$ and development of respiratory symptoms in a nonsmoking population. Arch Env Health 50, 139-152.

Abbey DE, Lebowitz MD, Mills PK, Petersen FF, Lawrence Beeson W, Burchette RJ. (1995b). Long-term ambient concentrations of particulates and oxidants and development of chronic disease in a cohort of nonsmoking California residents. Inhalation Toxicology 7, 19-34.

Abbey DE, Mills PK, Petersen FF, Beeson WL. (1991). Long-term ambient concentrations of total suspended particulates and oxidants as related to incidence of chronic disease in California Seventh-Day Adventists. Env Health Perspectives 94, 43-50.

Abbey DE, Petersen F, Mills PK, Beeson WL. (1993). Long-term ambient concentrations of total suspended particulates, ozone and sulfur dioxide and respiratory symptoms in a non-smoking population. Arch Env Health 48, 33-46.

Anderson HR, Ponce de Leon A, Bland JM, Bower JS, Strachan DP. (1996). Air pollution and daily mortality in London: 1987-92. BMJ 312: 665-9.

Antonovics, J. (1971) The effects of a heterogeneous environment on the genetics of natural populations, American Scientist, 59, pp. 593-599.

ArcData Publishing Program (1995) ArcGemeinde (Population data on a community level for the Federal Republic of Germany), ESRI: Gesellschaft für Systemforschung und Umweltplanung mbH, Kranzberg.

Armstrong B and Tremblay C. (1994). Lung Cancer Mortality and Polyaromatic Hydrocarbons. American Journal of Epidemiology; 139; 3: 250-62.

Arrow, K. J., Cline, W. R., Maeler, K.-G., Munasinghe, M., Squitieri, R. and Stiglitz, J. E. (1996) Intertemporal Equity, Discounting, and Economic Efficiency. In Climate Change 1995: Economic and Social Dimensions -- Contribution of Working Group III to the Second Assessment Report of the Intergovernmental Panel on Climate Change (Eds, Bruce, J. P., Lee, H. and Haites, E. F.) Cambridge University Press, Cambridge, pp. 125-144.

Ashley, E. P. and Robinson, J. T. (1996) Road mortality of amphibians, reptiles and other wildlife on the Long Point causeway, Canadian Field Naturalist, 111, pp. 403-412.

Asman, W. A. H. and Janssen, A. J. (1987) A long-range transport for ammonia and ammonium for Europe, Atmospheric Environment, 21, pp. 2099-2119.

Baker, C.K., Colls, J.J., Fullwood, A.E. and Seaton, G.G.R. (1986) Depression of growth and yield in winter barley exposed to sulphur dioxide in the field. New Phytologist 104, 233-241.

Barrett, K., Seland, Ø., Foss, A., Mylona, S., Sandnes, H., Styve, H. and Tarrason, L. (1995) European Transboundary Acidifying Pollution, EMEP/MSC-W Report 1/95, Norwegian Meteorological Institute, Oslo.

Bates DV, Baker-Anderson M. Sizto R (1990). Asthma attack periodicity: A study of hospital emergency visits in Vancouver. Environ Res 51, 51-70.

Bates DV. (1992). Health indices of the adverse effects of air pollution: the question of coherence. Environ Res; 59: 336-349.

Bell, S. and Ashenden, T. W. (1997) Spatial and temporal variation in nitrogen dioxide pollution adjacent to rural roads, Water Air and Soil Pollution, 95, pp. 87-98.

Beloin, N. J. and Haynie, F. H. (1975) Soiling of building surfaces, Journal of Air Pollution Control Association, 25, pp. 393-403.

Bennet, A. F. (1991) Roads, roadsides and wildlife conservation: a review. In Nature Conservation 2: the role of corridors (Eds, Saunders, D. A. and Hobbs, R. J.) Surrey Beatty, Chipping Norton, Australia, pp. 99-117.

Benson, P. (1982) Modifications to the Gaussian Vertical Dispersion Parameter sZ Near Roadways, Atmospheric Environment, 16, pp. 1399-1405.

Beyschlag, W., Ryel, R. J., Ullmann, I. and Eckstein, J. (1996) Experimental studies on the competitive balance between two central European roadside grasses with different growth forms. 2. Controlled experiments on the influence of soil depth, salinity and allelopathy, Botanica Acta, 109, pp. 449-455.

Broome, J. (1992) Counting the Cost of Global Warming, White Horse Press, Cambridge.

Brown, P. G. (1997) Stewardship of Climate -- An Editorial Comment. In Climatic Change, Vol. 37, pp. 329-334.

Brühl, C. and Crutzen, P. J. (1993) MPIC two-dimensional model. In The atmospheric effects of stratospheric aircraft: Report of the 1992 models and measurement workshop, Vol. I (Eds, Prather, M. and Remberg, E.) NASA Reference Publication, Washington D.C., pp. 103-104.

Brunekreef B. (1997). Air pollution and life expectancy: is there a relation? Occupational and Environmental Medicine; 54: 781-784.

Buckley, G. P. (Ed.) (1989) Biological Habitat Reconstruction, Bellhaven, Saffron Waldon.

Burnett RT, Dales RE, Krewski D, Vincent R, Dann T, Brook J. (1995). Associations between ambient particulate sulfate and admissions to Ontario hospitals for cardiac and respiratory diseases. Am J Epidem; 142:15-22.

Burnett RT, Dales RE, Raizenne ME, Krewski D, Summers PW, Roberts GR, Raad-Young M, Dann T, Brook J. (1994). Effect of low ambient levels of ozone and sulfates on the frequency of respiratory admissions to Ontario hospitals. Environ Res; 65:172-194.

Buser, H. (1999) Internalizing the external costs of traffic to wildlife and the landscape, European Nature, 3, pp. 17-18.

Buser, H., Breitenstein, C., Buser, A., Loppacher, S., Masé, G. and Tester, R. (1999) Externe kosten des verkehrs im bereich natur und landschaft, Vorstudie II, Ökoskop, Gelterkinden.

Butlin, R., Yates, T., Murray, M., Paul, V., Medhurst, J. and Gameson, T. (1994) Effects of pollutants on buildings, DoE/HMIP/RR/94/030, Building Research Establishment.

Capros, P. and Mantzos, L. (2000) Kyoto and technology at the European Union: costs of emission reduction under flexibility mechanisms and technology progress, Int. J. Global Energy Issues, 14, pp. 169-183.

CBS (1995) in Dorland and Jansen (1997), ExternE Transport - The Netherlands, IVM final report, p.23.

Certu (1997) Pollutions atmosphériques et circulation routière: Données de base (Air pollution and road traffic: basic data), Dossier du CERTU no. 74, Centre d'études sur les réseaux, les transports, l'urbanisme et les constructions publiques Ministère de l'Equipement, des Transports, et du Logement, Paris, France.

CGS (1999) Jaarverslag `98, Commissie Geluidhinder Schipol, Haarlem.

Chandler, K., Malcosky, B., Motta, R., Norton, P., Kelly, K., Schumacher, L. and D. Lyons (1996) In International Spring Fuels and Lubricants MeetingSociety of Automotive Engineers.

Christensen, P., Beaumont, H., Dunkerley, C., Lindberg, G., Otterström, T., Gynther, L., Rothengatter, W. and Doll, C. (1998) Pricing European transport projects (PETS). Interalisation of externalities: Summary report, ITS, Leeds University for European Commission DG VII, Leeds.

Climate Research Unit, Norwich.

Cody RP, Weisel CL, Birnbaum G, Lioy PJ. (1992). The effect of ozone associated with summertime photo-chemical smog on the frequency of asthma visits to hospital emergency departments. Environ Res 58, 184-194.

Collins, J. and Glysson, E. (1980) Multi-attribute Utility Theory and Environmental Decisions, Journal of Environmental Engineering Division, 106, pp. 815-829.

COMEAP (Department of Health Committee on the Medical Effects of Air Pollutants) (1995). Non-biological particles and health. London: HMSO.

COMEAP (Department of Health Committee on the Medical Effects of Air Pollutants). (1998). Quantification of the effects of air pollution on health in the United Kingdom. London: HMSO.

Concawe (1998) A study of the number, size & mass of exhaust particles emitted from european diesel and gasoline vehicles under steady state and european driving cycle conditions, Report No. 98/51, CONCAWE (the oil companies' European organization for environment, health and safety), Brussels.

Corbo, P., Gambino, M. and S. Iannaccone (1999) Methane Fuelled Urban Bus Italian Experiences Overview, International Workshop, Naples.

Craps, R. and D. De Keukeleere (1999) Natural Gas Buses Improve the Quality of Life in Brussels, International Workshop, Naples.

Creighton, P. J., Lioy, P. J., Haynie, F. C., Lemons, T. J., Miller, J. L. and Gerhart, J. (1988) In 81st APCA Annual Meeting, Dallas, Texas, pp. 17 pp.

Criqui, P. and Viguier, L. (2000) Kyoto and technology at world level: costs of $CO_2$ reduction under flexibility mechanisms and technical progress, Int. J. Global Energy Issues, 14, pp. 155-168.

Cropper, M. (1994) Preferences for Life Saving Programs: How the Public Discounts Time and Age, Journal of Risk and Uncertainty, 8, pp. 243-265.

Crump K. (1994). Risk of benzene induced leukaemia: A sensitivity analysis of the pliofilm cohort with additional follow-up and new exposure estimates. Journal of Toxicology and Environmental Health; 42: 219-242.

CSERGE, IOS-NLH, IVM, CAS and DAE-UoV (1999) Benefits transfer and the Economic Valuation of Environmental Damage in the European Union.: With Special reference to Health, Centre for Social and Economic Research on the Global Environment (CSERGE), University College London and East Anglia ; Depart of Economics and Social Sciences, Agricultural University of Norway (IOS-NLH), Institute for Environmental Studies (IVM), Free University, Amsterdam; Centre for Applied Studies (CAS), Catholic University of Portugal; Department of Applied Economics, University of Vigo (DAE-UoV).

Curtiss, P. S. and Rabl, A. (1996) Impacts of air pollution : general relationships and site dependence, Atmospheric Environment, 30, pp. 3331-3347.

Dab W, Quenel SMP, Le Moullec Y, Le Tertre A, Thelot B, Monteil C, Lameloise P, Pirard P, Momas I, Ferry R, Festy B. (1996). Short term respiratory health effects of ambient air pollution: results of the APHEA project in Paris. J Epidem Comm Health 50 (suppl 1): S42-46.

De Vlieger, I. (1995) Wegverkeersemissies in België: evolutie 1990-1993, ENE.RA 9511.

De Vlieger, I. (1996) Wegverkeersemissies in Vlaanderen: 1990-2010.

De Vlieger, I. and Debruyn, J. (1992) Kosten/effectiviteitsstudie van de driewegkatalysator, ENE.RA 9220.

Derwent, R. G. and Nodop, K. (1986) Long-range transport and deposition of acidic nitrogen species in north-west Europe, Nature, 324, pp. 356-358.

Derwent, R. G., Dollard, G. J. and Metcalfe, S. E. (1988) On the nitrogen budget for the United Kingdom and north-west Europe, Q.J.R. Meteorol. Soc., 114, pp. 1127-1152.

Derwent, R. G., Hov, Ø., Asman, W. A. H., Jaarsfeld, J. A. v. and Leuw, F. d. (1989) An intercomparison of long-term atmospheric transport models; the budgets of acidifying species for the Netherlands, Atmospheric Environment, 23, pp. 1893-1909.

Deutscher Wetterdienst (1999) Ausbreitungsklassenstatistik für den Zeitraum 1981 bis 1990, Offenbach.

Dijkstra (1998) personal communication, Centre for Energy Conservation and Envi-ronmental Technology (CE), Delft.

Dijkstra, W. J. and Dings, J. M. W. (1996) ZEMIS, emission model for marine transport, (in Dutch), Centre for Energy Conservation and Environmental Technology (CE), Delft.

Dings, J. M. W. (1996) Kosten en milieu-effecten van technische maatregelen in het verkeer:., Centrum voor energiebesparing en schone technologie, Delft.

Dings, J. M. W., Dijkstra, W. J., Moorman, S. A. H. and Janse, P. (1997) Clean ship in the Dutch inland shipping sector (in Dutch), Centre for Energy Conservation and Environ-mental Technology (CE), Delft.

Dockery DW, Pope CA III, Xiping X. Spengler JD, Ware JH, Fay MA, Ferries BG Jr, Speizer FE. (1993). An association between air pollution and mortality in six US cities. N. Engl J Med; 329(24): 1753-1759.

Dockery DW, Pope CA III. (1994). Acute respiratory effects of particulate air pollution. Ann Rev Public Health; 15:107-132.

Dockery DW, Pope CA III. (1994). Acute respiratory effects of particulate air pollution. Ann Rev Public Health; 15:107-132.

Dockery DW, Speizer FE, Stram DO, Ware JH, Spengler JD, Ferries BG. (1989). Effects of inhalable particles on respiratory health of children. Am Rev Respir Dis 139, 587-594.

Dockery, D. W. and Spengler, J. D. (1981) Personal exposure to respirable particulates and sulfates, J. Air Pollution Control Association, 31, pp. 153-159.

Donnan P and Hurley JF. (1997). ExternE CORE Project, Maintenance Note: Exposure-response (E-R) functions for the acute and chronic public health effects of air pollutants. ExternE Project, DGXII, European Commission.

Dop, H. V., Addis, R., Fraser, G., Girardi, F., Graziani, G., Inoue, Y., Kelly, N., Klug, W., Kulmala, A., Nodop, K. and Pretel, J. (1998) ETEX: a European Tracer Experiment; observations, dispersion modelling and emergency response, Atmospheric Environment, 32, pp. 4089 - 4094.

Dorland, C. and Olsthoorn, A. A. (2000) Inland shipping emission factors - Working paper for ExternE Core/Transport project Tast 5.3, Institut for Environmental Studies, Vrije Universiteit Amsterdam.

Downing, T. E., Eyre, N., Greener, R. and Blackwell, D. (1996) Projected Costs of Climate Change for Two Reference Scenarios and Fossil Fuel Cycles, Environmental Change Unit, Oxford.

Downing, T. E., Greener, R. A. and Eyre, N. (1995) The Economic Impacts of Climate Change: Assessment of Fossil Fuel Cycles for the ExternE Project, Environmental Change Unit

Droste-Franke, B., Krewitt, W., Friedrich, R. and Trukenmüller, A. (1998) in A. Markandya: Green Accounting Research Project (GARP II), Contract No. ENV4-CT96-0285, DG-XII-D5, University of Bath, Bath.

Duerinck, J. (2000) Personal communication based on: Duerinck, J. et al. (1999) Prospective study of emissions in Belgium until 2008/2012 of the greenhouse gases included in the Kyoto Protocol. Costs and potential measures and policy instruments to reduce GHG emissions, Vito & KU-Leuven.

Dusseldorp A, Kruize H, Brunekreef B, Hofschreuder P, de Meer G, van Oudvorst AB. (1995). Associations of $PM_{10}$ and airborne iron with respiratory health of adults near a steel factory. Am J Respir Crit Care Med 152, 1932-9.

ECE, U. (1996) Manual on methodologies and criteria for mapping critical levels/loads and geographical areas where they are exceeded, Umwel Bundes Amt, Berlin.

Egmond, N. D. v., Jaarsveld, J. A. v. and Onderdelinden, D. (1986) The Dutch aerosol study: general overview and preliminary results. In In Aerosols: Research, Risk Assessment and Control Strategies (Eds, Lee, S. D., Schneider, T., Grant, L. D. and Verkerk, P. J.) Lewis Publishers, Chelsea, pp. 269-282.

Eliassen, A. and Saltbones, J. (1983) Modeling of long-range transport of sulfur over Europe: a two-year model run and some model experiments, Atmospheric Environment, 17, pp. 1457-1473.

EMEP (1999) Transboundary Photo-oxidants in Europe, EMEP Summary report 2/99, Norwegian Meteorological Institute, Oslo

EMEP/CORINAIR (1996-1999) Atmospheric emission inventory guidebook, UNECE/EMEP Task Force on Emissions Inventories and Projections.

EPAQS: Expert Panel on Air Quality Standards. (1998). Lead. Department of Transport, Environment and the Regions/ HMSO; London

Euler GL, Abbey DE, Hodgkin JE, Magie AR. (1988). Chronic obstructive pulmonary disease symptoms effects of long term cumulative exposure to ambient levels of total oxidants and nitrogen dioxide in California Seventh Day Adventist Residents. Arch Environ Health; 43(4):279-285.

European Commission (1995a) ExternE Externalities of Energy. Vol 2 - Methodology, A Report produced for the EC - DG XII, Luxembourg, Office of Publications for the European Communities, Brussel, Luxembourg.

European Commission (1995b) Towards Fair and Efficient Pricing in Transport, COM(95)691.

European Commission (1996) Methodologies for Impact Assessment, EC - DG VII, Transport Research APAS Strategic Transport, Office for Publications of the European Communities., Luxembourg.

European Commission (1998a) ESTEEM (European Scenarios on Transport Energy Environment in Metropolitan Areas), DG XII, Science, Research and Development.

European Commission (1998b) MEET (Methodologies for Estimating Air Pollutant Emissions from Transport), EC/DG VII, Transport.

European Commission (1999) ExternE Externalities of Energy. Vol 7 – Methodology 1998 Update, A Report produced for the EC - DG XII, Luxembourg, Office of Publications for the European Communities, Brussel, Luxembourg.

European Commission (2000) Excise Duty Tables. Excise Duties and Transport, Environment and Energy Taxes, European Commission Directorate General Taxation and Customs Union Tax Policy.

Eyre, N., Downing, T. E., Hoekstra, R. and Rennings, K. (1997) Global Warming Damages, ExternE, Brussels.

Eyre, N., Downing, T., Hoekstra, R., Rennings, K. and Tol, R. S. J. (1998) Global Warming Damages, Contract JOS3-CT95-0002, ExternE Program of European Commission.

Fahl, U., Läge, E., Remme, U., Schaumann, P. (1999) $E^3$Net. In: Forum für Energiemodelle und Energiewirtschaftliche Systemanalysen in Deutschland (Hrsg.) (1999) Energiemodelle zum Klimaschutz in Deutschland. Physica-Verlag, Heidelberg.

Fahrig, L., Pedlar, J. H., Pope, S. E., Taylor, P. D. and Wegner, J. F. (1995) Effect of road traffic on amphibian density, Biological Conservation, 73, pp. 177-182.

Fankhauser, S., Tol, R. S. J. and Pearce, D. W. (1997) The Aggregation of Climate Change Damages: A Welfare Theoretic Approach, Environmental and Resource Economics, 10, pp. 249-266.

Fankhauser, S., Tol, R. S. J. and Pearce, D. W. (1998) Extensions and Alternatives to Climate Change Impact Valuation: On the Critique of IPCC Working Group III's Impact Estimates, Environment and Development Economics, 3, pp. 59-81.

Federal Statistial Office (1999) Statistical yearbook 1999 of the Federal Republic of Germany, Federal Statistical Office, Stuttgart.

Flückiger, W., Oertli, J. J. and Baltensweiler, W. (1978) Observations of an aphid infestation on hawthorn in the vicinity of a motorway, Naturwissenschaften, 65, pp. 654-655.

Forman, R. T. T. and Alexander, L. E. (1998) Roads and their major ecological effects, Annual Review of Ecology and Systematics, 29, pp. 207-231.

Forman, R. T. T. and Herpsberger, A. M. (1996) Road ecology and road density in different landscapes, with international planning and mitigation solutions. In Trends in addressing transportation related wildlife mortality, Vol. No. FL-ER-58-96 (Eds, Evink, G. L., Garret, P., Zeigler, D. and Berry, J.) Florida Department Transportion, Talahassee, Florida, pp. 1-22.

Forman, R. T. T., Friedman, D. S., Fitzhenry, D., Martin, J. D., Chen, A. S. and Alexander, L. E. (1997) Ecological effects of roads: toward three summary indices and an overview for North America. In Habitat fragmentation and infrastructure (Ed, Canters, K.) Ministry of Transport, Public Works and Water Management, Delft, Netherlands, pp. 40-54.

Foster, H. D. (1993) Road-Deicing Salt and Cancer - the Need For Further Study, Journal of the National Cancer Institute, 85, pp. 1603-1605.

Fried, G. (1999) Soot emission of jet engines (personal communication) (Personal communication).

Friedrich, R. and Reis, S. (Eds.) (2000) Tropospheric ozone abatement, Springer Verlag, Berlin Germany.

Friedrich, R., Bickel, P. and Krewitt, W. (Eds.) (1998) External costs of transport, IER, Stuttgart.

Fuller, R. M., Groom, G. B. and Jones, A. R. (1994) The Land-Cover Map of Great-Britain - an Automated Classification of Landsat Thematic Mapper Data, Photogrammetric Engineering and Remote Sensing, 60, pp. 553-562.

Funk, K. and Rabl, A. (1999) Electric versus conventional vehicles: Social Costs and Benefits in France, Transport and Environment, 4, pp. 397-411.

Gamble J. (1998). $PM_{2.5}$ and mortality in long-term prospective cohort studies: cause-effect or statistical associations? Environ Hlth Persp; 106: 535-549.

Gibbons, D. W., Reid, J. B. and Chapman, R. A. (1993) The new atlas of breeding birds in Britain and Ireland 1988-1991, T & D Poyser, London.

Giovannoni, J. M., Clappier, A. and Russell, A. (1995) Ozone control strategy modeling and evaluation for Athens, Greece: ROG vs. $NO_x$ effectiveness and the impact of using different different wind field preparation techniques, Meteorol. Atmos.Phys., 57, pp. 3-20.

Griffiths (1994) Errors in the use of the Briggs parameterisation for atmospheric dispersion coefficients, Atmospheric Environment, 28, pp. 2861-2865.

Haan, M. D. (1996) An Input-output Calculation of Avoidance Costs, Boston, Mass.

Haes, H. A. U. D., Jolliet, O., Finnveden, G., Hauschild, M., Krewitt, W. and Mueller-Wenk, R. (1999) SETAC-Europe: Second Working Group on LCIA (WIA-2) - Best Available Practice Regarding Impact Categories and Category Indicators in Life Cycle Impact Assessment: Background Document for the Second Working Group on Life Cycle Impact Assessment of SETAC-Europe (WIA-2), International J. of Life Cycle Assessment, 4, pp. 66-74.

Hagen, D. E., Trueblood, M. B. and Whitefield, P. D. (1992) A field sampling of jet exhaust aerosols, Particle Sci, & Techn., 10, pp. 53-63.

Hajnal, I. and Miermans, W. (1996) Onderzoek verplaatsingsgedrag Vlaanderen, analyse optracht., Eindverslag.

Hamilton, R. S. and Mansfield, T. A. (1993) The soiling of materials in the ambient atmosphere, Atmospheric Environment, 27A, pp. 1369-1374.

Hammitt, J. K., Lempert, R. J. and Schlesinger, M. E. (1992) A Sequential-Decision Strategy for Abating Climate Change, Nature, 357, pp. 315-318.

Hanson, P. J. and Lindberg, S. E. (1991) Dry deposition of reactive nitrogen compounds: a review of leaf, canopy and non-foliar measurements, Atmospheric Environment, 25A, pp. 1615-1634.

Hassel, D., Weber, F.-J. and Schmitz, T. (1998) Zusammensetzung der Kohlenwasserstoffe für unterschiedliche Fahrzeugkonzepte, Bundesministerium für Bildung, Wissenschaft, Forschung und Technologie; Förderschwerpunkt Troposhärenforschung, Leitthema 2, Projekt A.2, Cologne.

Haynie, F. H. (1986) In 2nd U.S.-Dutch Int. Symp.(Eds, Lee, S. D., Schneider, T., Grant, L. D. and Verkerk, P. J.) Lewis, Cheisea, MI, In Aerosol, pp. 951-959.

Hickman, J. (1997) Emission Functions for Heavy Duty Vehicles, SE/289/97, TRL Project Report.

Hitchcock, G., Collings, S. and Lewis, A. (1999) Alternative Fuels for Public Transport - The Costs and Benefits, Harwell, UK.

Hodson, N. L. (1962) Some notes on the causes of bird road casualities, Bird Study, 9, pp. 168-173.

Hodson, N. L. (1966) A survey of road mortality in mammals (including data for the grass snake and common frog), Journal of Zoology, London, 148, pp. 576-579.

Hofstra, G. and Hall, R. (1971) Injury on roadside trees: leaf injury on pine and white cedar in relation to foliar levels of sodium chloride, Canadian Journal of Botany, 49, pp. 613-622.

Holguin AH, Buffler PA, Contant CF jr, Stock TH, Kotchmar D, Hsi BP, Jenkins DE, Gwhan BM, Noel LM, Mei M. (1984). The effects of ozone on asthmatics in the Houston area. In: Evaluation of the scientific basis for ozone/oxidants standards (Lee SD, ed). Pittsburgh, PA; Air Pollution Control Association: 262-280.

Hollingsworth, A. (1987) Objective analysis for numerical weather prediction, Collection of papers presented at The WMO/IUGG NWP Sympsium, Tokyo.
Hunter D, Milton R, Perry KMA. (1945). Asthma Caused by Complex Salts of platinum. British Journal of Industrial Medicine; 2: 92-8.
Hurley JF and Donnan P. (1997). "An Update of Exposure-Response (E-R) Functions for the Acute and Chronic Public Health Effects of Air Pollution," Institute of Occupational Medicine (IOM), Edinburgh, UK. Internal paper for ExternE Project.
Hurley JF, Archibald RMcL, Colling PL, Fanning DM, Jacobsen M, Steel RC. (1983). The mortality of coke workers in Britain. Ann. J. Ind. Med. 4, 691-704.
Huygen, C. (1990) Plan van aanpak Schipol en Omgeving: Uitbreiding van de luchthaven Schipol, luchtverontreinigingaspecten, VROM, s'Gravenhage.
IARC. (1992). IARC Monographs on the evaluation of Carcinogenic Risk to Humans. Vol. 54. Occupational exposures to mists and vapours from strong inorganic acis, and other industrial chemicals. Lyon.
IARC. (1995). IARC Monographs on the Evaluation of the Carcinogenic Risk of Chemicals to Humans, Vol. 62. Formaldehyde. In: Wood dust and formaldelyde. Lyon.
ICAO/CAEP (1999) Database of aircraft jet engine emissions certification data, DERA (Defence Evaluation and Research Agency) / UK Ministry of Defence, on behalf of International Civil Aviation Organization (ICAO), Committee on Aviation Environmental Protection (CAEP), http://www.dera.gov.uk/aviation-emissions-databank.htm.
INFRAS/IWW (2000) External Costs of Transport, Study for the International Union of Railways, Paris.
Int Panis, L. (1999) Particle emission factors (Personal communication).
IPCC (1995) Economic and Social Dimensions of Climate Chage, Cambridge University Press, Cambridge.
Johannesson, M. and Johannesson, P. O. (1996) To Be or Not To Be, That Is The Question: An Empirical Study of the WTP for an Increased Life Expectancy at an Advanced Age, Journal of Risk and Uncertainty, 13, pp. 163-174.
Jones-Lee, M. W. (1989) The Economics of Safety and Physical Risk, Basil Blackwell., Oxford.
Jones-Lee, M. W. and al, e. (1985) The value of Transport Safety: Results of a sample Survey, The Economic Journal, 95, pp. 49-72.
Jørgensen, M. W. and Sorensson, S. C. (1997) Estimating Emissions from Railway Traffic, MEET Deliverable No. 17, Lyngby, Denmark.
Joumard, R. (1999) COST 319, Estimation of pollutant emissions from transport, European Commission, Directorate General Transport, Brussels.
Joumard, R., Vidon, R., Paturel, L., Pruvost, C., Tassel, P., Soete, G. D. and Saber, A. I. (1995) Évolution des Émissions de Polluants des Voitures Particulières lors du Départ Moteur Froid (Evolution of pollutant emissions from cars, beginning with cold start), 197, Bron France.
Katsouyanni K, Touloumi G, Spix C, Schwartz J, Balducci F, Medina S, Rossi G, Wotjtyniak B, Sunyer J, Bacharova L, Schouten JP, Ponka A, Anderson HR. (1997). Short term effects of ambient sulphur dioxide and particulate matter on mortality in 12 European cities: results from time series data from the APHEA project. BMJ 314, 1658-1663.
Katsouyanni K, Zmirou D, Spix et al. (1995). Short-term effects of air pollution on health: a European appraoch using epidemiologic time series data. Eur Respir J; 8: 1030-1038.
Kattenberg, A., Giorgi, F., Grassl, H., Meehl, G. A., Mitchell, J. F. B., Stouffer, R. J., Tokioka, T., Weaver, A. J. and Wigley, T. M. L. (1996) Climate Models - Projections of Future Climate. In Climate Change 1995: The Science of Climate Change -- Contribution of Working Group I to the Second Assessment Report of the Intergovernmental Panel on Climate Change (Eds, Houghton, J. T., Filho, L. G. M., Callander, B. A., Harris, N., Kattenberg, A. and Maskell, K.) Cambridge University Press, Cambridge, pp. 285-357.
Keeney, R. and Raiffa, H. (1976) Decisions with Multiple Objectives: Preferences and Value Trade-offs, John Wiley and Sons, Inc., New York.
Klein, J. A. P. (1993) Air pollution emissions by road traffic - Methodology to establish emission factors, Report M45, Centraal Bureau voor Statistiek (CBS), (in Dutch).

Knudsen, T., Svalheim, S. M. and Navrud, S. (1998) Miljøtiltak til havs er lønnsomt. (En sammenligning av livssykelanalysemetoder og nytte-kostnadsanalyse). (Marine environmental measures are profitable. A comparison of life cycle analysis mehods and cost-benefit analysis), Teknisk Ukeblad, 11, pp. 30-32.

Koopmans, T. C. (1967) Objectives, Constraints, and Outcomes in Optimal Growth Models, Econometrica, 35, pp. 1-15.

Krewitt, W., Trukenmueller, A., Mayerhofer, P. and Friedrich, R. (1995) EcoSense - an Integrated Tool for Environmental Impact Analysis. In Space and Time in Environmental Information Systems. Umwelt-Informatik aktuell, Vol. Band 7 (Eds, Kremers, H. and Pillmann, W.) Metropolis-Verlag, Marburg Germany.

Krüger, R. and Krewitt, W. (1999) STEEDS-LCM/EIM - Reference Guide -, IER, Stuttgart.

Krüger, R., Fahl, U., Obermeier, A. and Voß, A. (1998) Strategien zur Senkung der Feinstaub- und anderer Emissionen des Straßenverkehrs in Baden-Württemberg durch einen verstärkten Einsatz von Erdgas- und Brennstoffzellenfahrzeugen, Institut für Energiewirtschaft und rationelle Energieanwendung (IER), Stuttgart, (unpublished).

Krupnick AJ, Harrington W, Ostro B. (1990). Ambient ozone and acute health effects: Evidence from daily data. J. Environ Econ Manage 18, 1-18.

Krupnick, A., Alberini, A., Cropper, M., Simon, N., O'Brien, B., Goeree, R. and Heintzelman, M. (2000) What are Older People Willing to Pay to Reduce Their Risk of Dying?, Resources for the Future, Boston.

Krzyzanowski M, Quackenboss JJ, Lebowitz MD. (1990). Chronic respiratory effects of indoor formaldehyde exposure. Environmental research; 52: 117-125.

Kyriakis, N. A., Samaras, Z. S. and Andrias, A. E. (1997) Methodologies for estimating air pollutant emissions from transport: Road traffic composition, LAT 9718, Aristoteles University of Thessaloniki, Laboratory of Applied Thermodynamics, Thessaloniki.

Lanting, R. (1986) In 2nd U.S.-Dutch Int. Symp.(Eds, Lee, S. D., Schneider, T., Grant, L. D. and Verkerk, P. J.) Lewis, Cheisea, MI., In Aerosol, pp. 951-959.

Lee, D. and Watkiss, P. (1998) Atmospheric Modelling Note on Secondary Particulates, ETSU, Harwell, UK.

Leemans, R. and Cramer, W. P. (1991) The IIASA Database for Mean Monthly Values of Temperature, RR-91-18, International Institute for Applied Systems Analysis, Laxenburg.

Levene GM. (1971). Platinum Sensitivity. British Journal of Dermatology; 85: 590-93.

Lewis, C. A. and Gover, M. P. (1995) Life-Cycle Analysis of Motor Fuel Emissions, European Commission Directorate General for Transport COST 319, 27-28, Université Libre de Bruxelles, Belgium.

Liang KY, Zeger SL. (1986). Longitudinal data analysis using generalised linear models. Biometrika 1986; 73:13-22.

Lipfert FW. (1994). Air pollution and community health. New York: Van Nostrand Reinhold.

Lloyd's Register (1990a) Marine Exhaust Emissions Research Programme - Steady State Operation, Lloyd's Register of Shipping, London

Lloyd's Register (1990b) Marine Exhaust Emissions Research Programme - Steady State Operation - Slow Speed Addendum, Lloyd's Register of Shipping, London. U.K.

Lloyd's Register (1993) Marine Exhaust Emissions Research Programme - Phase II Summary Report, Lloyd's Register House, Engineering Services Group, Croydon, U.K.

Lloyd's Register (1995) Marine Exhaust Emissions Research Programme, Lloyd's Register House, Engineering Services Group, Croydon, U.K.

MacArthur, R. H. and Wilson, E. O. (1967) The theory of island biogeography, Princeton University Press, Princeton.

Mader, H. J. (1984) Animal habitat isolation by roads and agricultural fields, Biological Conservation, 29, pp. 81-96.

McKone, T. E. and Ryan, P. B. (1989) Human exposures to chemicals through food chains: an uncertainty analysis, Environmental Science and Technology, 23, pp. 1154-1163.

McNeill, S. and Southwood, T. R. E. (1978) The role of nitrogen in the development of insect/plant relationships. In Biochemical aspects of plant and animal coevolution (Ed, Harborne, J. B.) Academic Press, London.

Mech, L. D. (1989) Wolf population survival in an area of high road density, American Midland Naturalist, 121, pp. 389-389.
MEET (1999) Methodology for calculating transport emissions and energy consumption, Office for Official Publications of the European Communities, Luxembourg.
Mendelsohn, R. O. and Neumann, J. E. (Eds.) (1999) The Impact of Climate Change on the United States Economy, Cambridge University Press, Cambridge.
Milberg, P. and Persson, T. S. (1994) Soil Seed Bank and Species Recruitment in Road Verge Grassland Vegetation, Annales Botanici Fennici, 31, pp. 155-162.
MIRA-2 (1996) Leren om te keren pp 147-166.
Molhave L, Jensen JG, Larsen S. (1991). Subjective reactions to volatile organic compounds as air pollutants. Atmospheric Environment; 25A(7); 1283-93.
Moore, M. J. and Viscusi, K. (1988) The Quantity Adjusted Value of Life, Economic Inquiry, 26, pp. 368-388.
Morris RD, Naumova EN, Munasinghe RL. (1995). Ambient air pollution and hospitalization for congestive heart failure among the elderly in seven large American cities. Am J Public Health. Am J Public Health 85, 1361-1365.
Most, v. d. P. F. J. and Heijen, C. F. (1988) Emission factors for inland shipping (in Dutch), Institute of Environmental Science, Energy Research and Process Innovation (TNO-MT), Apeldoorn.
Moussiopoulos, N., Tourlou, P. M., Sahm, P. and Karatzas, K. (1996) Three dimensional local scale modelling of ozone in Europe: calculations for the EU Auto Oil Programme, Aristotle University of Thessaloniki, Greece.
Nauss KM and the HEI Diesel Working Group. (1995). Critical issues in assessing the carcinogenicity of diesel exhaust: A synthesis of current knowledge. In: Diesel Exhaust: A critical analysis of emissions, exposure and health effects. Health Effects Institute, Cambridge.
Navrud, S. (1997) External Costs of Transport in ExternE. Task 4.4: Economic Valuation., ENCO Environmental Consultants.
Navrud, S. (1998) Valuing Health Impacts From Air Pollution in Europe. New Empirical Evidence On Morbidity. Presented at the World Congress for Environmental and Resource Economists, Environmental and Resource Economics.
Navrud, S. (1999) Linking Weak and Economic Indicators of Environmental Damages: Acidic Deposition in Norway. Chap. 6 in C.L. Splash and S. McNally (eds.) 2000: Case Studies in Ecological and Environmental Economics. John Wiley & Sons Ltd.
NEA (1997) POINT, energy use and emission in freight transport (in Dutch), RIVM report nr 773002 010, National Institute of Public Health and Environmental Protection (RIVM), Bilthoven.
NERA and CASPAR (1997) Valuation of Deaths from Air Pollution: A Report for the UK Department of Environment, Transport and the Regions, London.
Neumann, J. V. and Morgenstern, O. (1947) Theory of Games and Economic Behaviour, Princeton University Press, Princeton, N. J.
Newby, P. T., Mansfield, T. A. and Hamilton, R. S. (1991) Sources and economic implications of building soiling in urban areas, The Science of the Total Environment, 100, pp. 347-365.
Nicholson, K. W. (1988) The dry deposition of small particles: a review of experimental measurements, Atmospheric Environment, 22, pp. 2653-2666.
Nordeng, T. E. (1986) Parameterization of physical processes in a three-dimensional numerical weather prediction model, 65, Technical Report, Oslo.
Nordhaus, W. D. (1994) Managing the Global Commons: The Economics of Climate Change, The MIT Press, Cambridge.
NRC, N. R. C. (1997) Highway de-icing: comparing salt and calcium magnesium acetate, 235, Transport Research Board, Washington DC, USA.
Oak Ridge National Laboratory (ORNL) and Resources for the Future (RFF). (1994). Estimating Fuel Cycle Externalities: Analytical methods and issues. Report No. 2 on the External costs of fuel cycles: A study by US DOE and the Commission of the European Communities. Oak Ridge, Tennessee, USA, ORNL.

Obermeier, A. (1999) VOC-profiles for gasoline conventional and three-way-catalyst, and diesel passenger cars (Personal communication).

Obika, B., Freerhewish, R. J. and Fookes, P. G. (1989) Soluble Salt Damage to Thin Bituminous Road and Runway Surfaces, Quarterly Journal of Engineering Geology, 22, pp. 59-73.

OECD (1997) Environmental Policies and Employment, OECD, Paris.

Ostro BD, Rothschild S. (1989). Air pollution and acute respiratory morbidity: An observational study of multiple pollutants. Environ Res 50, 238-247.

Ostro BD. (1983). The effects of air pollution on work loss and morbidity. J. Env Econom Management; 10:371-382.

Ostro BD. (1987). Air pollution and morbidity revisited: A specification test. J Environ Econ Manage 14, 87-98.

Otterström, T., Gynther, L. and Vesa, P. (1998) The willingness to pay for better air quality, Ekono Energy Ltd.

Pearce, D. W., Cline, W. R., Achanta, A. N., Fankhauser, S., Pachauri, R. K., Tol, R. S. J. and Vellinga, P. (1996) The Social Costs of Climate Change: Greenhouse Damage and the Benefits of Control. In Climate Change 1995: Economic and Social Dimensions -- Contribution of Working Group III to the Second Assessment Report of the Intergovernmental Panel on Climate Change (Eds, Bruce, J. P., Lee, H. and Haites, E. F.) Cambridge University Press, Cambridge, pp. 179-224.

Pilkington A, Hurley F. (1997). Cancer risk estimate. Institute of Occupational Medicine (IOM) Edinburgh, UK.

Pilkington A, Hurley JF, Donnan P. (1997). Health effects in ExternE Transport: assesment and exposure-response functions. Working Paper, ExternE Project, DGXII, Euiropean Commission.

Pio, C. A., Ramos, M. M. and Duarte, A. C. (1998) Atmospheric aerosol and soiling of external surfaces in an urban environment, Atmospheric Environment, 32, pp. 1979-1989.

Ponce de Leon A, Anderson HR, Bland JM, Strachan DP, Bower J. (1996). Effects of air pollution on daily hospital admissions for respiratory disease in London between 1987-88 and 1991-92. J Epidem Comm Health 50 (suppl 1): S63-70.

Pope CA III, Thun MJ, Namboodiri MM, Dockery DW, Evans JS, Speizer FE, Heath CW Jr. (1995). Particulate air pollution as predictor of mortality in a prospective study of US adults. Am J Resp Crit Care Med 151: 669-674.

Pope CA, Dockery DW. (1992). Acute health effects of $PM_{10}$ pollution on symptomatic and asymptomatic children. Am Rev Respir Dis 145, 1123-1126.

Port, G. R. and Thompson, J. R. (1980) Outbreaks of insect herbivores along motorways in the United Kingdom, Journal of Applied Ecology, 17, pp. 649-656.

Portney PR, Mullahy J. (1986). Urban air quality and acute respiratory illness. J Urban Econ; 20:21-38.

Proops, J. and al, e. (1996) The Lifetime Pollution Implications of Various Types of Electricity Generation: An Input-output Analysis, Energy Policy, 24, pp. 229-237.

QUARG (1993) Urban air quality in the United Kingdom, Department of the Environment, London.

Rabl, A., Curtiss, P. S., Spadaro, J. V., Hernandez, B., Pons, A., Dreicer, M., Tort, V., Margerie, H., Landrieu, G., Desaigues, B. and Proult, D. (1996) Environmental Impacts and Costs: the Nuclear and the Fossil Fuel Cycles, Report to EC, DG XII, Version 3.0 June 1996., Centre d'Energétique, Ecole des Mines, Paris, France.

Race, M. S. and Fonesca, M. S. (1995) Fixing compensatory mitigation: what will it take?, Ecological Applications, 6, pp. 94-101.

Ramsey, F. (1928) A Mathematical Theory of Saving, Economic Journal, 38, pp. 543-549.

Ray, A. (1984) Cost-benefit analysis: issues and methodologies, John Hopkins University Press for the World Bank, Baltimore, London.

Reh, W. and Seitz, A. (1990) The influence of land use on the genetic structure of populations of the common frog Rana temporaria, Biological Conservation, 54, pp. 239-249.

Reis, S., Simpson, D., Friedrich, R., Jonson, J. E., Unger, S. and Obermeier, A. (1999) Road traffic emissions - predictions of future contributions to regional ozone levels in Europe, 8th International Symposium Transport and Air Pollution, IER, Graz Austria.

Roberts AD. (1951). Platinosis. Archives of Industrial Hygiene and Occupational Medicine; 4: 549-59.

Roemer W, Hoek G, Brunekreef B. (1993). Effect of ambient winter air pollution on respiratory health of children with chronic respiratory symptoms. Am Rev Respir Dis 147, 118-124.

Ross, S. M. (1986) Vegetation change on highway verges in south-east Scotland, Journal of Biogeography, 13, pp. 109-113.

Rowe DR, Bernav SS, Bird LA, Callaway JM, Chestnut LG. Eldridge MM, Lang CM, Latimer DA, Murdoch JC, Ostro BD, Patterson AK, Rae DA, White DE. (1994). ESEERCO Project Report 2: Methodology, RCG Hagler, Bailly, Colorado.

Rozan, A. and Willinger, M. (1999) Willingness to Pay and knowledge of the health damage origin., Strasbourg.

Sahm, P. and Moussiopoulos, N. (1999) The OFIS model: an efficient tool for assessing ozone exposure and evaluating air pollution abatement strategies, Int.J. Env. and Poll.

Samaras, Z., Kyriakis, N., Joumard, R., André, M., Serié, E., Hassel, D., Weber, F.-J., Hickman, A.-J., Rijkeboer, R., Sturm, P., Sorenson, S., Lewis, C. A., Beckman, E., Tozzi, C., Vaccaro, R. and Kalivida, M. (1998) Methodologies for Estimating Air Pollutant Emissions from Transport: Emission Factors and Traffic Characteristics Data Set, LAT 9802, Aristotle University of Thessaloniki: Laboratory of Applied Thermodynamics, Thessaloniki.

Sargent, C. (1984) Britain's railway vegetation, Institute of Terrestrial Ecology, Abbots ripton.

Schimel, D., Alves, D., Enting, I., Heimann, M., Joos, F., Raynaud, M., Derwent, R., Ehhalt, D., Fraser, P., Sanhueza, E., Zhou, X., Jonas, P., Charlson, R., Rodhe, H., Sadasivan, S., Shine, K. P., Fouquart, Y., Ramaswamy, V., Solomon, S., Srinivasan, J., Albritton, D. L., Isaksen, I., Lal, M. and Wuebbles, D. J. (1996) Radiative Forcing of Climate Change. In Climate Change 1995: The Science of Climate Change -- Contribution of Working Group I to the Second Assessment Report of the Intergovernmental Panel on Climate Change (Eds, Houghton, J. T., Filho, L. G. M., Callander, B. A., Harris, N., Kattenberg, A. and Maskell, K.) Cambridge University Press, Cambridge, pp. 65-131.

Schmid, S. (Ed.) (1998) Task report emission factors. ExternE Core/Transport., IER, University of Stuttgart, Stuttgart.

Schouten JP, Vonk JM, de Graaf A. (1996). Short term effects of air pollution on emergency hospital admissions for respiratory disease: results of the APHEA project in two major cities in the Netherlands, 1977-89. J Epidem Comm Health 50 (suppl 1), S22-S29.

Schwartz J and Morris R. (1995). Air pollution and hospital admissions for cardiovascular disease in Detroit, Michigan. Am J Epidem 142, 23-35. Am J Epidem 137, 701-705.

Schwartz J, Slater D, Larson TV, Pierson WE, Koenig JQ. (1993). Particulate air pollution and hospital emergency room visits for asthma in Seattle. Am Rev Respir Dis; 147:826-831.

Schwartz J, Spix C, Wichmann HE, Malin E. (1991). Air pollution and acute respiratory illness in fiver German communities. Environ Res 56, 1-14.

Schwartz J. (1993). Particulate air pollution and chronic respiratory disease. Environ Res 62, 7-13.

Schwartz J. (1994). Low-level exposure to lead and children's IQ: a meta-analysis and search for a threshold. Environ Res; 65: 42-55.

Schwartz J. (1997). Health effects of air pollution from traffic: ozone and particulate matter. In: Fletcher T, McMichael AJ, eds. (1997). Health at the crossroads: transport policy and urban health. Chichester: John Wiley and Sons: 61-82.

Scott, N. E. and Davison, A. W. (1982) De-icing salt and the invasion of maritime plants, Watsonia, 14, pp. 41-52.

Sehmel, G. (1980) Particle and gas dry depostion: a review, Atmospheric Environment, 14, pp. 983.

Seinfeld, J. H. and Pandis, S. N. (1998) Atmospheric chemistry and physics: from air pollution to climate change, John Wiley and Sons, New York.

Sen, A. (1982) Choice, welfare and measurement, Harvard Press.

SETAC (1993) A Conceptual Framework for Life-cycle Impact Assessment, Society of Environmental Toxicology and Chemistry and SETAC Foundation for Environmental Education, Inc., Pensacola, Florida.

Shareef, G. S., Butler, W. A., Bravo, L. A. and Stockton, M. B. (1988) Air emissions species manual. Vol I: Volatile organic compound (VOC) species profile, Radian Corp.

Shine, K. P., Derwent, R. G., Wuebbles, D. J. and Morcrette, J.-J. (1990) Radiative Forcing of Climate. In Climate Change - The IPCC Scientific Assessment (Eds, Houghton, J. T., Jenkins, G. J. and Ephraums, J. J.) Cambridge University Press, Cambridge, pp. 41-68.

Simpson, D. (1992) Long period modelling of photochemical oxidants in Europe. Calculations for July 1985, Atmos. Environ., 26A, pp. 1609-1634.

Simpson, D. (1993) Photochemical model calculations over Europe for two extended summer periods: 1985 and 1989. Model results and comparisons with observations, Atmos. Environ., 27A, pp. 921-943.

Simpson, D. (1995) Biogenic emissions in Europe 2: Implications for ozone control strategies, J. Geophys. Res., 100, pp. 22891-22906.

Simpson, D. and Andersson-Sköld, Y. (1997) Regional and local scale modelling of ozone in Europe: calculations for the EU Auto/Oil Programme, Research Report No.47, Norwegian Meteorological Institute, Oslo Norway.

Spadaro, J. V. and Rabl, A. (1998) External Costs of Energy: Application of the ExternE Methodology to France, JOS3-CT95-0010, ARMINES, Paris.

Spadaro, J. V. and Rabl, A. (1999a) Estimates of real damage from air pollution: site dependence and simple impact indices for LCA, International J. of Life Cycle Assessment, 14, pp. 229-243.

Spadaro, J. V. and Rabl, A. (1999b) Estimation of health damage due to emission of air pollutants by cars: the canyon effect, California, USA.

Spadaro, J. V., Rabl, A., Jourdain, E. and Coussy, P. (1998) External Costs of Air Pollution: Case Study and Results for Transport Between Paris and Lyon, International J. of Vehicle Design, 20, pp. 274 - 282.

Spellerberg, I. F. (1998) Ecological effects of roads and traffic: a literature review, Global Ecology and Biogeography Letters, 7, pp. 317-333.

Spix C, Heinrich J, Dockery D, Schwartz J, Volksch G, Schwinkowski K (1993). Air pollution and daily mortality in Erfurt, East Germany, 1980-1989. Env Health Pers 101,518-519.

Spix C, Wichmann HE. (1996). Daily mortality and air pollutants: findings from Köln, Germany. J Epidem Comm Health 50 (suppl 1): S52-S58.

Staessen JA et al. (1993). Hypertension caused by low-level lead exposure: myth or fact? Journal of cardiovascular risk; 1: 87-97.

Stationary Office, T. (1999) Economic Appraisal of Health Effects., UK Department of Health.

Steil, D., Dameris, M., Brühl, C., Crutzen, P. J., Grewe, V., Ponater, M. and Sausen, R. (1998) Development of a chemistry module for GCMs: first results of a multiannual integration, Ann. Geophys., pp. 205-228.

Sugden, R. and Williams, A. (1978) The Principles of Practical Cost-benefit Analysis, Oxford University Press, Oxford.

Sunyer J, Antó JM, Murillo C, Saez M. (1991). Effects of urban air pollution on emergency room admissions for chronic obstructive pulmonary disease. Am J Epidemiol; 134:277-286.

Sunyer J, Castellsague J, Saez M, Tobias A, Anto JM. (1996). Air pollution and mortality in Barcelona. J Epidem Comm Health 50 (suppl 1): S76-S80.

Sunyer J, Saez M, Murillo C, Castellsague J, Martinez F, Antó JM (1993). Air pollution and emergency room admissions for chronic obstructive pulmonary disease: A 5-year study. Am J Epid 137, 701-705.

Svensson, G. (1998) Model simulations of the air quality in Athens, Greece, during the MEDECAPHOT-TRACE campaign., Atmospheric Environment, 32, pp. 2239-2268.

Terborgh, J. (1974) Preservation of natural diversity: the problem of extinction prone species, BioScience, 24, pp. 715-722.

Tidblad, J. and Kucera, V. (1998) Materials damage, in Urban Air Pollution - European Aspects. In (Eds, Fenger, J., Hertel, O. and Palmgren, F.) Kluwer academic publishers.

Tidblad, J., Kucera, V. and Mikhailov, A. A. (1998) Statistical analysis of 8 year materials exposure and acceptable deterioration and pollution levels, Report No 30, Swedish Corrosion Institute, Stockholm, Sweden.
Tol, R. S. J. (1995) The Damage Costs of Climate Change Toward More Comprehensive Calculations, Environmental and Resource Economics, 5, pp. 353-374.
Tol, R. S. J. (1996) The Damage Costs of Climate Change Towards a Dynamic Representation, Ecological Economics, 19, pp. 67-90.
Tol, R. S. J. (1999a) The Marginal Costs of Greenhouse Gas Emissions, Energy Journal, 20, pp. 61-81.
Tol, R. S. J. (1999b) New Estimates of the Damage Costs of Climate Change, D99/01, Institute for Environmental Studies, Vrije Universiteit, Amsterdam.
Tol, R. S. J. (1999c) New Estimates of the Damage Costs of Climate Change, D99/02, Institute for Environmental Studies, Vrije Universiteit, Amsterdam.
Tol, R. S. J. and Domning, T. E. (2000) The Marginal Costs of Climate Changing Emissions, Institut for Environmental Studies, Vrije Universiteit, Amsterdam, in press.
Tol, R. S. J. and Grewe, V. (1999) Climate-Related External Costs of Nitrogen and Sulfur Emissions.
Torkkeli, S. (1999) Speed-dependend particle emission function for gasoline vehicles. Taken from: Helsinki Metropolitan Area Council (Personal communication).
Touloumi G, Pocock SJ, Katsouyanni K, Trichopoulos D. (1994). Short-term effects of air pollution on daily mortality in Athens: A time-series analysis. Int J Epidem 23, 957-967.
Touloumi G, Samoli E, Katsouyanni K. (1996). Daily mortality and 'winter type' air pollution in Athens, Greece - a time series analysis within the APHEA project. J Epidem Comm Health 50 (suppl 1): S47-S51.
U.S. Environmental Protection Agency (1995) User's Guide for the Industrial Source Complex (ISC3) Dispersion Model, Office of Air Quality Planning and Standards Emissions, Monitoring, and Analysis Division Research, North Carolina.
Umweltbundesamt and INFRAS (1997) Handbuch Emissionsfaktoren des Straßenverkehrs. Grundlagenbericht, Berlin.
UN ECE Workshop (1997) In Proceedings of the UN ECE Workshop on Economic Evaluation of Air Pollution Abatement and Damage to Buildings including Cultural Heritage(Eds, Kucera, V., Pearce, D. and Brodin, Y. W.) Swedish Environmental Protection Agency, Report 4761.
US Environmental Protection Agency (US EPA). (1996). Air Quality criteria for particulate matter, Vol III. Report No. EPA/600/P-95/001cF. Washington DC:US EPA.
US Environmental Protection Agency (USEPA). (1990). Department of Health data on lead.
van den Brink, R. M. M. and van Wee, G. P. (1997) Energy use and emissions by means of transport (in Dutch), Report nr. 773002007, National Institute of Public Health and Environmental Protection (RIVM), Bilthoven.
van der Zande, A. N., ter Keurs, J. and van der Weijden, W. J. (1980) The impacts of roads on the densities of four bird species in an open field habitat - evidence of a long distance effect, Biological Conservation, 18, pp. 299-321.
van Wee, G. P. (1998) personal communication, National Institute of Public Health and Environmental Protection (RIVM), Bilthoven.
Vehovar, L., Kuhar, V. and Vehovar, A. (1998) Hydrogen-assisted stress-corrosion of prestressing wires in a motorway viaduct, Engineering Failure Analysis, 5, pp. 21-27.
Verhoeff AP, Hoek G, Schwartz J, van Wijnen JH. (1996). Air pollution and daily mortality in Amsterdam. Epidemiology 7, 225-230.
Viner, D. and Hulme, M. (1994) The Climate Impacts LINK Project -- Providing Climate Change Scenarios for Impacts Assessment in the UK, University of East Anglia
Viscusi, W. K. and Moore, M. L. (1989) Rates of Time Preference and Valuations of the Duration of Life, Journal of Public Economics, 38, pp. 297-317.
Volta, M. and Finzi, G. (1999) In EUROTRAC Newsletter 21/99.
Watson, R. T., Zinyowera, M. C. and Moss, R. H. (Eds.) (1996) Climate Change 1995: Impacts, Adaptation, and Mitigation of Climate Change -- Scientific-Technical Analysis --

Contribution of Working Group II to the Second Assessment Report of the Intergovernmental Panel on Climate Change, Cambridge University Press, Cambridge.

Way, J. M. (1977) Roadside verges and conservation: a review, Biological Conservation, 12, pp. 65-74.

Whittemore AS, Korn EL. (1980). Asthma and air pollution in the Los Angeles area. Am J Public Health 70, 687-696.

Wietlisbach V, Pope CA, Ackermann-Liebrich V (1996). Air pollution and daily mortality in three Swiss urban areas. Soc Prev Med 41, 107-115.

Wigley, T. M. L., Hulme, M. and Raper, S. C. B. (1993) MAGICC: Model for the Assessment of Greenhouse Induced Climate Change, University Consortium for Atmospheric Research, Boulder.

Wilcox, D. A. (1989) Migration and control of purple loosetrife (Lythrum salicaria L.) along highway corridors, Environmental Management, 13, pp. 365-370.

Williams, D. D., Williams, N. E. and Cao, Y. (2000) Road salt contamination of groundwater in a major metropolitan area and development of a biological index to monitor its impact, Water Research, 34, pp. 127-138.

Winkler, R. (1972) An Introduction to Bayesian Inference and Decision, Holt, Rinehart, and Winston, Inc., New York.

Wittwer, D., Masé, G. and Buser, H. (1998) Externe kosten des verkehrs im bereich natur und landschaft, 322, Ökoskop, Gelterkinden.

Wordley J, Walters S, Ayres JG. (1997). Short term variations in hospital admissions and mortality and particulate air pollution. Occup Environ Med; 54: 108-116.

World Health Organisation. (1987). Air Quality Guidelines for Europe. Copenhagen: WHO.

Zannetti, P. (1990) Air Pollution Modeling: Theories, computational methods and available software, Van Nostrand Reinhold, NY.

Zerbe, J. R. and Dively, D. (1994) Benefit-cost Analysis: In Theory and Practice, Harper Collins College Publishers, New York.

Ziomass, I. C., Tzoumaka, P., Balis, D., Melas, D., Zerefos, C. and Klemm, O. (1998) Ozone episodes in Athens, Greece. A modelling approach using date from The MEDCAPHOT-TRACE., Atmospheric Environment, 32, pp. 2313-2321.

www.ingramcontent.com/pod-product-compliance
Ingram Content Group UK Ltd.
Pitfield, Milton Keynes, MK11 3LW, UK
UKHW021607110926
15112UKWH00017B/107

* 9 7 8 3 6 6 2 0 4 3 3 0 1 *